Library of
Davidson College

ALIENATION, PRAXIS, AND TECHNĒ

IN THE THOUGHT OF KARL MARX

Kostas Axelos

Alienation, Praxis, and Technē in the Thought of Karl Marx

Translated by Ronald Bruzina

University of Texas Press, Austin & London

The publication of this book was assisted
by a grant from the Matchette Foundation.

Originally published as *Marx, penseur de la technique:
De l'aliénation de l'homme à la conquête du monde*,
Arguments 2, 3d ed. (Paris: Les Editions de Minuit, 1969).
Library of Congress Cataloging in Publication Data
Axelos, Kostas.
    Alienation, praxis, and technē in the thought of Karl
Marx.
    Translation of Marx, penseur de la technique.
    Includes index.
    1. Marx, Karl, 1818–1883. 2. Alienation (Social
psychology). 3. Alienation (Philosophy). 4. Technology–
Philosophy. I. Title.
B3305.M74A9313   301.6'2   76-5429
ISBN 0-292-78013-3

Translation copyright © 1976 by the University of Texas Press

All rights reserved

Printed in the United States of America

Modern philosophy has only continued the work begun by Heraclitus and Aristotle.
> —Karl Marx in the *Rheinische Zeitung*,
> July 14, 1842

For the world to become philosophic amounts to philosophy's becoming world-order reality; and it means that philosophy, at the same time that it is realized, disappears.
> —Karl Marx in his doctoral dissertation,
> 1839–1841

CONTENTS

Translator's Introduction ix

Preface 3

Introduction. On Understanding Marx's Thought 7

Part I. From Hegel to Marx
1. From Absolute Knowledge to Total Praxis 29

Part II. Economic and Social Alienation
2. Labor, the Division of Labor, and the Workers 53
3. Private Property, Capital, and Money 67
4. The Machine, Industry, and Technicist Civilization 77

Part III. Political Alienation
5. Civil Society and the State 89

Part IV. Human Alienation
6. The Relationship of the Sexes and the Family 113
7. Being, Making, and Having 123

Part V. Ideological Alienation
8. Thought and Consciousness: True and Real? 145
9. Religion and Ideas 159
10. Art and Poetry 175
11. Philosophy (Metaphysics) and the Sciences 195

Part VI. The Prospect of Reconciliation as Conquest
12. The Premises for Transcending Alienation 217
13. Communism: Naturalism, Humanism, and Socialism 229
14. The Abolishing of Philosophy by Its Realization 271

Conclusion. Open Questions
The Economic Problematic 293
The Political Problematic 311
The Anthropological Problematic 314
The Ideological Problematic 319
The Problematic of Reconciliation as Conquest 324

Afterword 333

Notes 337

Index 377

TRANSLATOR'S INTRODUCTION

Kostas Axelos's study of Karl Marx, here presented in translation, has the singular merit of working simultaneously in two of the more intense fields of recent research, writing, and publishing, namely, that of technology, particularly as an all-pervasive modern phenomenon, and that of the Marxian analysis of the human condition, especially on the issue of alienation. But though this is the book's merit, it at the same time poses its greatest difficulty; for those whose interest is technology may not find themselves led to get involved in Marx's theoretical studies and, conversely, those who move deeply into Marx's thinking may not find technology something that has to be questioned in any focal way, being an issue subordinate to much larger questions. In other words, Axelos here focuses on a double topic whose two constituents represent points of weakness in studies of technology and of Marxism, respectively. For the book's choice of subject, then, to prove itself meritorious in the eyes of its readers, the author has to show reason why technology is a key issue in Marx's thinking and, conversely, why Marx's thinking is a vitally important contribution to understanding technology—all of which means he has to locate the proper terrain for explaining his double focus and making his position persuasive. Axelos recognizes that the only adequate approach to understanding simultaneously the meaning of technology and of Marxism is an explicitly *philosophic* one. Moreover, it has to be a philosophic perspective that not only is appreciative of the historical character of human being as such—for technical achievements show dramatic change and development, while Marx's thought is precisely a study of historical movement in human society—but also is equally aware of the historical condition of the human enterprise which philosophy itself is. The reader of Axelos's book, then, must be prepared to take the analysis and argument here as together a historically sensitive *philosophic* study of technology and a historically sensitive *philosophic* study of Marx's principle. For herein lie the specific virtue of the work and the basis of its claim to a hearing on the subject of Marxism and technology, amid the plethora of excellent materials available in either area.

Axelos himself takes the time to make clear his own method in the book with respect to Marx's own philosophic undertaking,[1] and a glance at the table of contents shows that he is going to treat point by point the principal concepts and conclusions of Marx's position. It quickly becomes clear, then, that Axelos will be probing deeply into Marx's writings, particularly as they compose a culminating moment and turning point in the history of Western thought. What may not be so quickly apparent, however, is the very thing Axelos must establish clearly and solidly, namely, as just mentioned, the fundamental significance (*a*) of technology for Marx's own program for mankind and (*b*) of Marx's overall analysis for understanding technology. This double lesson is worked out much more gradually; it is a cumulative argument rather than a thematic discussion all in one place. It would be useful, then, to draw together some of the main points Axelos will be making about technology in the course of his study of Karl Marx, thus helping to make clear the rationale for the book's existence.

Axelos on Marx's Treatment of Technology

For Marx, as Axelos presents him, technology can be explained only if one first develops an adequate definition of human being. Human being, now, is to be defined not as a set of fixed descriptive characteristics in a taxonomic ordering of observable features but in terms of a *relationship to a setting* which is at the same time an *activity*. Human being begins and continues as a project of seeking to fill basic life needs[2] by engaging actively with an environment to produce the things which meet those needs; and this active engagement, presupposing as it does for Marx that the basic relationship to the setting is one of *opposition*, is seen as an enterprise of achieving mastery over that setting, that is, over the world of nature. To be more precise, the engagement is an action of *making*, an action of *production*, an action by which portions of the setting are made into objects suitable to the satisfaction of human needs.

Here, now, one would have to add a number of essential points that would more fully spell out all that is involved in this concept of human being as a project that is accomplished through the production of objects outside itself but conformed to it and in service to it; and of these points the most important would be that the project which constitutes human being is necessarily *social*.[3] That is, the acts of making take the form they do as acts of a coherent group. For the present purpose of outlining the place of technology in Axelos's study of Marx, however, what has to be emphasized is (1) that with the filling of needs by acts of making *new needs* are created, (2) that filling needs by acts of making in the first place is itself *the elemental technical performance*, and (3) that this action of filling needs by making things is *the action that begins human being as such*. For, given that "life involves before everything else eating and drinking, a habitation, clothing and many other things," as Marx says, "the first historical [i.e., properly *human*] act is thus the production of the means to satisfy these needs, the production of material life itself."[4] That is, for the human species the operation that converts portions of the setting in which it finds itself into objects suitable for satisfying its needs is an operation consisting of the use of other objects, the use of instrumental means; and the production and use of these instrumental objects are the first historical act, the act that *institutes human being* properly speaking.[5]

Now, in this fact that filling needs, thereby creating new needs, through an action of making is the performance that institutes the human species *as human*, that is, as having its own proper sphere of activity with its own proper characteristics, among them that of being a *history*,[6] three important points have to be emphasized. First of all, technology has its roots in the very constitution of mankind as such. Second, that same constitution contains a relationship of opposition/mastery with respect to the world of nature; so that to establish and maintain itself as human being the human species has constantly to impose species-suitable characteristics upon items in and from its environment—that is, in order to achieve its own humanness, human being must transform nature into *nature-for-man*. But one must not fail to include a third point here, namely, that inasmuch as making things to fill needs *creates new needs*, which in being filled by further making lead to *still more new needs*, the project of human self-achievement through engagement with nature to transform it into a human world is *a progressively expanding process* whose dimensions are limited only by limitations that might lie within the capabilities of human activity. Now, if human capabilities as a whole are such as to encompass the whole of things in all its aspects, then the human transformation of nature, fundamentally necessary to the very existence of the human species to begin with, is a process intrinsically ordained to include the whole of the world of nature and to involve the whole of mankind. In other words, given the relationship of active, transformative opposition to the initial setting for that relationship itself, that is, to nature, as the central structure of human being; given, second, that the productive process by which needs arising from this relationship are filled leads expansively to new needs, thus to new production, and so on; and given, last, the capability in this curious species, human being, to be aware of the whole both of things around it and of its own kind as human and to set goals for itself in terms of all-inclusiveness, in terms of totality, then we are brought before the fact of *a thrust toward the universal* which is potential in man's very constitution. For Marx, as Axelos reads him, the move to actualize that potentiality of universality is essential to the central dynamism of history, particularly in the realm of the transformative creation of a world for man. And that is what is at the heart of technology.

The summary just sketched, however, still leaves out whole dimensions of the interpretation of human being in Marx; for it has yet to mention the conceptual means he uses to account for *evils* in the process being examined, that is, in historical development as the building of the human world. In Marx's view, as Axelos follows him, the very activity that wins man his humanness, the very action of making that gives him the means to fill his needs, that gives him the instrumental objects whereby to impart a form to things that makes them appropriate to him, this very activity which should be his self-achievement is disrupted from the very first and leads instead to self-negation. Here we have the Marxian concept of alienation, so central in much of the present-day study of Marx's thought and of primary importance in Axelos's treatment. It is not my intention to try to explain that concept here, especially since Axelos himself discusses it numerous times.[7] Here I wish merely to indicate in general terms how the concept of alienation enters into the understanding of technology. In the simplest terms, it amounts to this: the character of the product of an activity of making and of that making action itself, as that character is interpreted in Marx's historical analysis, will be the same character that must be attributed to technology; for technology is a form of human making (see the next section). If, then, *alienation* is the character of human making as it has gone on historically from the beginning until the present, then technology in its whole history is itself endemically characterized by alienation. Likewise, the significance for the historical development of mankind that alienation holds will correlatively qualify the role technology plays for mankind; and this will be particularly true for the present, modern world, which for Marx represents the culminating stage of alienation as well as the culminating stage of technical performance. This is precisely the knot of topics that Axelos wishes eventually to trace out for us, and he does so particularly in his later chapters.[8] There he shows clearly how technology is the concrete power that effects simultaneously the complete technical transformation of nature out of a merely natural character into a specifically human one—such is the work of technology as human making—and the uttermost alienation of human being through the total capture of human capabilities and meanings in the form of objects and objective processes which are identified, organized, and put to use in terms of a system of abstract, monetary value—such is the effect of technology as sustaining and sustained by capitalistic enterprise.

All this would be stark indeed were Marx, and Axelos's discussion of him, to end simply with the unmitigated prospect of this kind of simultaneous culmination of technological power and human alienation. But Marx's purpose in his brilliant analysis of human history is to show how that culmination, terrible as it may appear, is a necessary phase of the very process by which human society can and will achieve genuine humanness. For once the historical process contains some factor of contravention as an element of its very developmental movement—and Marx's analysis shows alienation to be just that—then that factor can be overcome only when it has exhausted its developmental possibilities, that is, when it has reached its maximum growth and effectiveness in human history, its culmination. Furthermore, since human being is human in its activities precisely in that it possesses a potential for universality, a potential for extending to all things in the mobilization of all human powers and all human individuals, culmination has to be utterly all-inclusive, right in its negativeness, in its wrongness, before it can be made to transmute positively into the actual achievement of full and genuine humanness. This understanding of things, of course, is nothing less than the Hegelian concept of dialectical process, but Marx adapts it to the realm of practical action, and this adaptation is what makes Marx's theory a theory of *technology*. For understanding history in terms of the culmination of a process means, for Marx, understanding it in terms of the culmination of the process of the development of *the concrete action of human making as at the same time contrahumanly alienated.* This development at its maximum attainment in full universality, that is, technology and industry as a world-wide phenomenon, as the concrete underpinning for all-inclusive relations binding human institutions together around the world, is the phase of culmination through which humanness will at last be won—if in fact mankind can move beyond it by negating its negative contrahuman character and transforming it into positive human realization. This, at least, is what Axelos finds Marx to be saying, even though in the end he sees the lineaments of Marx's vision here to be deeply problematic—thus his massive closing chapter, "Conclusion. Open Questions."

In briefest outline, such is Axelos's philosophic treatment of the question of technology in a reading of Marx on alienation so as to demonstrate the deep mutual importance of the two issues. An outline, of course, is no substitute for the book itself. Too much is left out, and the clarification and nuance that give a position persuasiveness are missing. One must now turn to the book itself. Yet there are certain matters that it would be well to probe further here, not as summarization of the text itself but as discussion beyond what it explicitly says. There are only two topics I wish to go into somewhat: the first is the issue of modern technology as somehow different from performances of making in the past and the second concerns the way Axelos finds the culmination phase Marx spells out for us to represent an opening to something other than what Marx is generally taken to envision.

Technique . . . Technology

Let me begin the first topic with a question that, on first sight, may seem to be merely a matter of translation. In French, Axelos uses *la technique* predominantly in talking about the operation of human making. Why have I used the corresponding English word *technique* instead of simply *technology*?[9] Does not the latter do as well in referring to the ensemble of technical performances that surely is what the French word represents?

In fact this is not an issue merely of imitating French lexical form. The very nature of the phenomenon in question is involved in Axelos's use of *la technique* as distinguished from *la technologie*. In the first place, Axelos himself reserves one strict usage of the word *technology* (*la technologie*) for a quite specific meaning, which the common usage in English would not represent, namely, a *logos* of *technē*. Here, in a manner reminiscent of Martin Heidegger, from whom he gains certain fundamental insights, Axelos relies upon an order of primordial meaning in the Greek roots of a modern term in order to delineate a concept: *logos*—word, meaning, reason, study; *technē*—a making that produces something, art, skill, method. *Techno-logy* in a strict sense, then, means an articulate thinking (*logos*) turned toward production and making (*technē*) so as to probe their meaning in an appropriate, radical, and comprehensive way. As Axelos has explained elsewhere, technology is the thinking of technique,[10] while technique is the productive transformation of being.[11] And, to Axelos's own mind, adequate reflective understanding of that productive transformation is still to be achieved: "Techno-*logy* is still wanting; the *logos* of technique is still lacking."[12] Here, then, we have the sense of the term *technology* as Axelos uses it in the final paragraphs of the conclusion when characterizing Marx's thought as a reflection on human making: "Perhaps it is time to begin understanding Marx's thought as *Technology*, provided we take the term in its full amplitude and true depth."[13]

There is another way, however, in which Axelos uses *technology* (*la technologie*) so as to require us to distinguish it clearly from *technique* (*la technique*). Technique, that is, any operation of making, has been with mankind from its beginnings. But, as Axelos details it, technique *in its modern form alone* has come to play its full role as the concrete driving power for historical development.[14] Technique has come *actually* to possess an almost absolute creative power,[15] to undertake the transformation of all being,[16] to be inherently unrestrained by limits of any sort from any source,[17] and to promote a universally integrated and similar plan, purpose, and set of methods and means in a single whole of historical movement.[18] In short, what we today call *technology* is simple *technique* now seen finally in full possession of the role it plays within the perspective of historical development Axelos credits to Marx, a role that while placing it in basic continuity with the past also makes it *unlike* technique in the past.[19] For now the performance of the transformative work which technique is to begin with is *actually consummate and all-inclusive*. In a word, technology is

xvi *Translator's Introduction*

technique *as total and universal.*

The question now must be, *What has made the difference in the concrete*, empowering technique to be total and universal in the present historical period, to be technology, in distinction from its previous condition? Here we might look for some help from Jacques Ellul's widely read book, *La technique ou l'enjeu du siècle* [Technique: The stakes of the century],[20] for Ellul's concept of "technique" seems to resemble Axelos's. For Ellul technique is a generic phenomenon that characterizes human history from the beginning. But as a generic phenomenon it must not be defined exclusively in terms of precontemporary stages. Technique is and has been "nothing more than *means* and the *ensemble of means*,"[21] but technique at an earlier stage was different from technique now, in at least two ways. In the first place, the determination of means now is in terms of explicit, rational concepts formulated in the most rationally effective way, that is, quantitatively; and, second, this way of determining means is now operative for *all* areas of human existence. Accordingly, Ellul proposes a definition for technique specifically in the contemporary age: "The term technique . . . does not mean machines, technology, or this or that procedure for attaining an end. In our technological society, *technique* is the *totality of methods rationally arrived at and having absolute efficiency* (for a given stage of development) in *every* field of human activity."[22] Technique, then, in the present world is more than simply industrial productive apparatus and methods, whether extremely simple or extremely complex, which is what the term *technology* generally suggests as its meaning, particularly if such productive means are taken as an isolable or subordinate fact in society;[23] this is Ellul's distinction between the two concepts. Technique is more than just technology, that is, technological apparatus in industry. It includes as well and more importantly the similarly operative procedures for determining means and operations in economic planning, in the organization of groups of all kinds, and in the handling of all aspects of the human constitution, physiological as well as psychological; and this total phenomenon, *Technique*, is what Ellul tries to describe in his book.

The totality character of the phenomenon of technique today is, of course, deeply important. But Axelos in studying Marx considers technique from the point of view of the base-level determining function that, precisely, *making things* exercises in social development. Consequently, Ellul's reflections do not offer us the conceptual clarification needed to distinguish between operations of *making* in the past and such operations in the present, as this distinction might be indicated in the difference between the very words *technique* and *technology*. Here we gain considerably more help from some remarks made by Maurice Daumas, general editor of the recent comprehensive work in French on the history of technology, *Histoire générale des techniques*.[24] For Daumas, one can meaningfully talk about *technology* as a "new form of activity . . . as distinct from both simple applied technique and from science as discovery. Technology lies between science and technique, and is characterized by their interpenetration."[25] Daumas goes on to explain:

> The term "technology" can lead to confusion particularly because of the equivalent term in English. We use the term to designate, somewhat arbitrarily, a kind of higher technique, a kind of science-minded technique (*technique savante*), or better, the science of technique. Without going any further in the search for a precise definition, the reader will understand that we wish to call attention to that area of activity that is common to the sciences and to techniques but which at the same time differs from each of them, the area within which their contacts and reciprocal collaboration [are] established for their respective greater profit. We shall later see this area take on an ever-increasing importance in both quality and extent.[26]

So, while cautioning against exaggerated notions regarding the role of science in early modern invention, specifically, before the nineteenth century,[27] Daumas points to the emergence of the specifically new phenomenon of *technology* as precisely the product of an intimate wedding of science (as theoretical knowing) and technique (as efficacious making and doing): technology as *a third area* in which science and technique fuse into something neither exclusively science (pure science) nor simply technique (practical, methodic application).[28]

Now, without claiming that Axelos and Daumas agree entirely on the precise interplay of factors in the birth of modern industry and of the historically distinct, somewhat later modern phenomenon of technique-as-technology, I do wish to point out how the character of Daumas's *technology* coincides to a great extent with Axelos's second sense of *technology*, namely, technique as the unification into a single system of both the practical and the theoretical means for productive activity on a limitless and universal scale.

Furthermore, inasmuch as Daumas suggests the term *technology* to designate this new third thing precisely as a *"science* of technique," he indicates the element that makes technique *technological*, namely, that (in Axelos's terms), on the one hand, it is a practice conditioned and promoted by a *"logos" of a specific sort*, that is, by explicitly theoretical insights in the form of rational science, and that, on the other hand, it is a practice, a *"technē,"* that conditions and promotes that same science. In other words, technique becomes technology because, in distinction from the kind of technique that is either not at all or only to a minor degree involved with explicit science, it fuses with science in an intimate mutual engendering and conditioning.[29] At the same time, for Axelos this is not a *proper logos*, because it is a *logos* of abstraction- and power-minded scientific reason (Marx would say of an alienating and alienated reason)[30] which corresponds exactly to and fuses with the performances of making that, as themselves dehumanized power-minded operations, compose that industrial system we call in more offhand fashion "modern technology." Technology, then, for Axelos is the consummately "errant"[31] phenomenon of the modern world, that is, an agency that takes itself to be supremely true and efficacious, whereas it mistakes its own character and significance in the scheme of the whole, even while it is the very power that achieves dominance on the scale of the total and the universal. Technology, being a practical achiever precisely on the scale of the concepts in theoretical science, because the two now form a single thing, is *the* concrete power not only capable of but actually on the way to achieving a transformation of the world and of human beings everywhere and in all respects.

Now, as we can easily see, very much at the center of Marx's, and Axelos's, way of reading history so as to understand technology (in the sense of the fusion of science and technique) is the insistence that a *wrong* has to be redressed, that somewhere along the line (or even at its point of origin) a *break* occurred that has yet to be mended. It is important to realize, now, where *Marx* sees the "break" so as to understand the kind of "mending" *his* program is meant to achieve, in distinction from the way others in today's proliferating literature on technology may theorize about a wrong that needs correction and, particularly, in distinction from Axelos's own thought on the matter as it gradually emerges in his study of Marx. One should recall first of all that the very concept of human being operative in Marx's thinking is predicated upon a kind of break, namely, a relation of opposition to the setting which is the natural world.³² But this break is the very thing that promotes man's being as a project of self-achievement, as the activity of transforming the merely natural into the human, the "for-itself" into the "for-man." In Marx's view, only with the existence of something-in-opposition, something *negative* with respect to some order, can there be a drive to establish and accomplish that order; and human being is nothing if it is not the drive to actions of making which thereby establish it as the order it is to be. (This, of course, once again, is Marx's adaptation of the Hegelian concept of dialectical process.) When nature, then, is finally transformed into the utterly "for-man," the initial break with nature will have been mended, but the productive operation of making, the process of self-realization through the making of objects, will go on, only now the materials to be made human are themselves already "humanized" rather than merely "natural." Yet, because human needs have grown vaster and more complex, these already humanized materials are no longer suitable and adequate, and they stand once more in opposition to the order of human self-achievement.³³

The difficulty in human development does not lie for Marx, then, in some "opposition to nature," for that is the very thing that starts off the process of human self-achievement. Rather, the basic wrong lies in a condition of the process *with respect to itself as a whole*, namely, *alienation*. Alienation is a break *of the entire process* away from *what it is supposed to be*, namely, the actualization of humanity through the making and use of the means for filling needs.³⁴ Making and using objects are to go on (to put it technically: dialectical mediation through objectivity is to go on), which is to say that human history and development are to go on, but with alienation basically overcome.³⁵

Now, this fundamental distinction between the process of human self-achievement through making as it ought to be and that same process as it has historically in fact gone on—thus the *Marxian* concept of a qualitative break—does not imply that man's activity has ever *been* as it should be, that is, not alienated. In Axelos's reading of Marx, human being has been alienated *from the very beginning*.³⁶ From the very first human act the very process by which human being was to realize itself (viz., through making things distinct from itself yet as fulfillment of itself, that is, *through technique*) resulted instead in its being set in opposition to its own activities, its own products, its whole social community, and its whole worked-out environment as all external to it *in a hostile and alien way*,³⁷ and it has done this in an ever-increasing measure down through history. As we saw earlier, the road to overcoming alienation must lead precisely through the historical development of this situation to its fullest.[38] But the crucial fact remains that Marx sees a *radical break* in the historically developed situation of technical (and therefore especially of *technological*) activity that both *is fundamentally wrong* and *can be overcome*—indeed that *is to be overcome* by historical necessity—a break, once again, between the *actual* condition of human being in the world and the kind of condition the very structure of human being calls for in order for it to be fulfilled.[39] At the same time, this kind of original radical break to be overcome calls correspondingly for *an equally radical break as its overcoming*. István Mészáros puts this point in the clearest terms: " 'Alienation' is an eminently historical concept. If man is 'alienated,' he must be alienated *from* something, as a result of certain *causes*—the interplay of events and circumstances in relation to man as subject of this alienation—which manifest themselves in a historical framework. Similarly, the 'transcendence of alienation' is an inherently historical concept which envisages the successful accomplishment of a process leading to a *qualitatively different state of affairs.*"[40] And, though in Marx it is not some fixed human essence, some fixed anthropological entity of permanent properties, that is to be finally restored, there is a basic condition—that of historical self-accomplishing action through making other things (through self-mediation, as it would be put in dialectical terms)—to be put *properly* in motion, that is, without any form of alienation such as has been inherent in historical forms of development up to the present consummation phase of capitalism. Thus, only if this basic condition of self-realization is fulfilled—that is, "insofar as there is an effective *break* in the objective ontological continuity of capital in its broadest Marxian sense—can we speak of a *qualitatively* new phase of development: the beginning of the 'true history of mankind.' "[41]

In sum, then, for Marx there is a symmetrical relationship between (1) the radical, original break dividing, on the one side, the historically actual, that is, alienated, condition of human self-realization, and, on the other, the condition of human becoming that *ought to* and *will* take place; and (2) the break between the consummation phase of that historically actual development of the human world of alienation, that is, technological capitalism, and the beginning of true human history. But the problem, obviously, is that one must find grounds in the actual, presently historical state of affairs for asserting this radical break of alienation to be historically original, and at the same time one must discern those aspects and factors of alienation in present, actual technical development that are to be made no longer determinative, that is, that are to be overcome and transcended in favor of other, finally authentic factors. In other words, one must be able to identify those aspects and factors in contemporary technical activity—technology—which, as constituting the alienation dimension, the "break" character, are to be overcome and transcended by *the qualitatively new* in a human, historical development that continues from that point.

Now, on this score of trying to see what a mending of the break (or, put more accurately, a transcending of the consummation phase of human alienation) might consist of, one matter deserves a little more discussion, particularly in view of the need finally to move to some brief discussion of Axelos's own thoughts on this movement beyond the culmination of technology. For, in his treatment of Marx, Axelos regularly shows openings to an alternative view *of his own* on technology and human activity; and it would leave matters unfinished not to make those ideas a little more explicit.

We have seen that technique in the modern world, technology, finally has the actual capability of being all-inclusive, of involving the totality of human beings and of transforming the totality of the natural world. As totalizing, technology in Axelos's reading of Marx offers the prospect of an activity that has the following total effects upon the man-nature relationship which, in Marx's own telling, is the point of departure for human self-achievement. On the side of nature, the material setting (nature) for scientific/technical activity becomes a source for the utterly usable, for that which is utterly open to human manipulation and use and can be totally transformed into objects-for-man.[42] On the side of man, that is, of man as agent of this total transformation of nature into the human (i.e., into the usable in technological production), activity in general becomes an activity of gaining and exercising utter mastery, utter conquest on the part of human will under the guidance of a knowledge of the inherent structure and properties of all things, that is, through science. Now, for Marx, both these effects have so far been produced *within a condition of alienation*, that is, within a condition in which operations are performed in abstraction from and contrary to authentic human development. But the serious question is, What will the character of science be, and correlatively the character of the objects of human technique, when the alienation factor in projects of rational knowing and technical transformation is overcome? Will human knowing of reality *still* be basically that of science as we know it, will objects that we make *still* be utter "manipulables," and, consequently, will our basic activity in the world *still* be *technological* production? These are the sorts of questions Axelos begins asking of Marx quite early in the book but with culminating intensity in the later chapters, particularly the last one, "Conclusions. Open Questions."

Once again, on these questions one has to try to reach some clear conception of the specific character of technique today, that is, technology. Some thinkers who have turned to this issue seem, in effect, to find the principal factor to be the effect *scientific rationality as such* has upon technical operations. For example, in the eyes of Jacques Ellul, who was mentioned earlier, "the technique of the present has no common measure with that of the past."[43] The "specifically modern problem" lies in the *scope* and *depth* of determinancy exercised by technique (in his sense).[44] "Herein lies the inversion we are witnessing. Without exception in the course of history, *technique belonged to a civilization*. Today *technique has taken over the whole of civilization*."[45] As to what *produces* this change, Ellul's "characterology" of *modern* technique,[46] compared to that of previous ages and civilizations, seems to suggest the principal factor to be the emergence of a totally autonomous rationality, a rationality that, on the one hand, excludes any factors of individuality, especially those of nonrational spontaneity and feeling, and, on the other, forms a total, self-sustaining and self-augmenting system that is devoted—by both its methods and its aims—to the universal. "Geographically and qualitatively, technique is universal in its manifestations. It is devoted, by nature and necessity, to the universal. It could not be otherwise. It depends upon a science itself devoted to the universal, and it is becoming the universal language understood by all men."[47] Another author, Hans Jonas, is more precise and much clearer. "The scientific revolution changed man's ways of thinking, *by* thinking, before it materially changed, even affected, his ways of living. It was a change in theory, in world-view, in metaphysical outlook, in conception and method of knowledge.... Technology, historically speaking, is the delayed effect of the scientific and metaphysical revolution with which the modern age begins."[48] And the reason the scientific revolution of the seventeenth century could lead in the nineteenth to the emergence of what Daumas, as we have seen, would properly call *technology* is simply that, as Jonas explains, "to modern theory in general, practical use is no accident but is integral to it, or that 'science' is technological by its nature."[49] For modern science, application can always eventually be made "only because of the manipulative aspect inherent in the theoretic constitution of modern science as such."[50] Jonas develops his argument in some detail, but for our purposes the point to be retained is this: *the very form of rationality* which is modern science, the empirical, experimental investigation of the natural world, is a suitable mate for technique in the wedlock of technology—as we might put it— because in both, modern science and technique, at least as Marx here interprets the latter, human being poses itself *in opposition to* the world of natural phenomena and because in both human being takes the stance of *exercising power over* its setting.[51]

Now, clearly, the change in the character of the stance implicit, or explicit, in the pursuit of knowledge about the world of nature is quite important. Yet the fact remains that, as Daumas and other historians of human technique have taken pains to point out, were technical industry not ready in incredibly manifold ways to adopt the immense control and development possibilities offered by modern science, there would have been no birth of technology in the nineteenth century. The history of human technique shows a long, gradual growth in technical conditions that were prerequisites for the adoption of scientific guidance, but that came about largely without that guidance. But, in granting that, what we find significant in the evolution of technical industry in the West independently of theoretical, experimental science is the fact that, in great part and in a fundamental way, it was an evolution in the mechanical utilization of sources of power. And this aspect of the matter, namely, the exercise of power as primary in a relationship to the surrounding world, is what Axelos finds himself led to question most pointedly, whether on the practical side of things or on the theoretical.

On the theoretical side, Axelos repeatedly underlines the extent to which Marx's understanding of human being implies this power/ opposition relationship. It is intrinsic and even constitutional to Marx's thinking. But Axelos wishes to question *the very concept of dualistic relationship* which underlies Marx's way of analyzing human being and human activity—not only this *oppositional* duality between human power-exercising productive activity and the natural world but also dualities of all sorts as constituting a system of implied metaphysical distinctions.[52] In other words, Axelos wants to criticize not only the modern notion of science but also *any conceptual undertaking that implies certain metaphysical distinction/oppositions*. Axelos's reading of Marx finds Marx in the end implying a radical overcoming of all theoretical systems of the past; but *radically* overcoming them must, as Axelos finds Marx's dialectics to demand, result in a *unity* where before there were duality and opposition. Axelos, in other words, finds Marx to imply a movement not simply through and beyond the one oppositional condition of *alienation* as it can be revealed and analyzed within a certain system of concepts but, more than that, *through and beyond any system of concepts itself that is built up on a base of certain dualistic distinctions and oppositions*. Similarly, it is not just some *one* form of rationality, such as empirical science, that merits being superseded[53] but *the very form of rational, theoretical understanding as a whole*, for it begins with dualistic distinctions in order to articulate an explanation that has to be transcended. For Axelos, then, Marx's position implies the eventual transcendence of the very approach to reality embodied in the concepts that compose Marx's own thinking. The ultimate lesson of Marx's thinking is that it is itself to be radically transcended.

This, however, is only half the matter. Not only is a *thought-*approach to reality that is founded in a set of concepts of dualistic opposition to be superseded, but also, for Axelos, who has read his Marx, a *practical action* approach that is based in a stance of opposition is *to be transcended*. In other words, what Axelos in reading Marx ultimately finds that is to be transcended and superseded more than anything else is *the whole stance before the world* that begins with dualistic, oppositional distinction, however that oppositional distinction may manifest itself, whether in thought or in action. And in the modern world whose primary, characterizing activity and thinking are *technology*—an approach to the world that, both theoretically and practically, is one of power-asserting opposition which is alienation, particularly in capitalism—what to Axelos's mind is needed are a thinking of the world and a concrete, active involvement with it that go beyond that division to find a base in a stance of unity. This is the perspective within which Axelos critically reviews not only Marx[54] but also the whole history of Western philosophy, which he sees culminating in Marx; and it constitutes that alternative to which he keeps suggesting openings in the course of his study of Marx. Some remarks on that alternative, then, are in order before this introductory essay can be closed.

Axelos: Play as the Transcendence of All Duality

This direction in Axelos's own thinking, really only announced in his book on Marx, is more decidedly apparent in the larger context of his writings. Moreover, it lies deeply operative in the history of his own thinking. In an interview held in 1961, Axelos was asked how he came to choose the subject matter of the present book on Marx as his main doctoral thesis in philosophy. His reply was: "Because, from the time I was sixteen, Marx has been my major preoccupation. He is at the same time the culmination of the Western philosophic tradition since Heraclitus, the most significant critical questioning of the contemporary world, and the denunciation of all alienations, and he himself prepares the way for being himself superseded. Concrete historical events and meditation had their part in it as well."[55] As can be seen, this is virtually a résumé of the path of Axelos's own thinking, but it needs some elaboration. In the first place, there is his involvement in "concrete events." On Axelos's own account, his earliest intellectual efforts in his native Greece were both theoretical and militant—as "organizer, journalist, and communist theoretician" for the Greek resistance during World War II and in the early stages of the subsequent Greek civil war.[56] As things developed, however, he ended up being both excluded from the Greek Communist party and condemned to death by the rightist government that came into power in 1945; and so, at the age of twenty, he fled to Paris, where he has lived ever since. He first studied at the Sorbonne, then did independent research under the auspices of the Centre National de Recherche Scientifique (1950-1957), and finally took the positions of teacher of philosophy at the University of Paris and editor of various publication projects (1957 to the present).[57] All this, of course, is reflected in the way Marx figures as an important figure in Axelos's thinking, for in the end Marx represents anything but ideological orthodoxy demanding conformity. Accordingly, in coming to terms with Marx in the form of the present book, Axelos takes a standpoint that emphasizes critical, open questioning, as he himself explains in the same interview quoted a little earlier:

> [In this book] the problem of alienation is taken up in a threefold way: first, the description of economic, social, political, existential, and ideological alienation, that is, of everything that makes man a stranger (alien) to himself and

to the world. Second, there is the prospect of transcending alienation, in what Marx called fully achieved naturalism, humanism, and socialism. Marx, in fact, and with a great deal of optimism, thought that man could wholly coincide with nature, with his own being, and with the society of men. In the third place, the problems presented by the project of conquering the world by and for man are dealt with. In the domains of economy and politics—of the deepest parts of human life, of art, of poetry, and of philosophy—are hidden gaping problems that offer no satisfactory solution. It is perhaps our task, at this threshold point in a history beginning to become world-wide, to maintain a serenity in the face of the demands of technique and practice and to learn how to continue asking embarrassing questions that remain open.[58]

The Marx book, however, does not stand altogether by itself. Axelos has conceived it within a trilogy of philosophic writing that he entitles *Le déploiement de l'errance* [The deployment of errance] [59] and whose three component pieces are arranged thus: (1) *Héraclite et la philosophie: La première saisie de l'être en devenir de la totalité* [Heraclitus and philosophy: The first grasp of the being-in-becoming of totality];[60] (2) *Marx, penseur de la technique: De l'aliénation de l'homme à la conquête du monde* [Marx, the man who thinks out technique: From the alienation of man to the conquest of the world];[61] (3) *Vers la pensée planétaire: Le devenir-pensée du monde et le devenir-monde de la pensée* [Toward planetary thinking: Thought becoming world, world becoming thought].[62] Now, what this complex of writings attempts to work out is an approach to reality that, through a meditative investigation concentrating on the three critical stages of the history of Western thought—(1) the beginnings (Heraclitus); (2) the culmination (Marx); and (3) the transcending passage to a new way (. . . ?)—would fulfill the following exigencies: (*a*) that it be based on *unity* rather than division and opposition, even while celebrating multiplicity, variety, and change; (*b*) that it be *concrete and actual*, rather than merely abstract and speculative; and (*c*) that it not impose some final, fixed character or structure but remain an *open* process. For Axelos, these exigencies are met in the overall stance-involvement that can be best characterized as *play*. The last matter to consider briefly then is how Axelos understands play, and this is made a little easier when one realizes the extent to which, beyond Marx, the most dominant figure influencing Axelos's thinking in this direction is Martin Heidegger.[63]

Heidegger is scarcely named in Axelos's book on Marx, appearing only in a few footnote references. Yet Heidegger's way of pursuing a radical reflection on the meaning of a movement that would transcend the culmination phase of Western history permeates the book,[64] providing Axelos with the perspective and the vocabulary for articulating his own attempted solution to the "riddle of history." In particular, the very title Axelos gives to the trilogy in which he examines that riddle, *Le déploiement de l'errance*, makes use of a Heideggerian term, *die Irre—l'errance—errance*,[65] and Axelos uses it frequently in trying to describe play. So, while not at all attempting to represent the meaning network that Heidegger himself would find in the usages of this term (or others), I shall offer at least a summary of Axelos's sense of the word in the context of *his* movement beyond the culmination phase of Western history into play.

For Axelos, transcending the phase of completion of Western history, the history of both metaphysics and technology, ultimately means moving beyond one foundational, comprehensive requirement: that one understand things *by finding an ultimate reason* or *basis* for them, and that one take a position and make assertions about things in terms of reasons and bases such that that position and those assertions lay claim to being unequivocally true.[66] What is needed instead of this requirement is an attitude that recognizes the peculiar, irreducible multivalence of the world and consequently sees that an asserted, definable *position* regarding the world is necessarily as much error as truth, and in the very same respect. Truth and error are not mutually exclusive; "they are the two faces of the same coin."[67] In other words, reality is not characterizable in terms of unequivocal basic elements or principles which one could accordingly represent in unequivocal, reason-stating assertions. The world is rather the perpetual *play* of aspects, elements, and powers,[68] and the play-character of it is simply not made accessible by efforts to find unequivocal reasons or principles. If one has to make use of such terms as *truth* and *error*—and indeed one does if one is to link up at all with Western intellectual tradition—then one can best characterize the world as *errance*,[69] the copenetration of truth and error. A "new thinking" is needed here in order to match the world's own character and engage in its processes for what they are. Since there is no determinate ultimate structure to seek by one's thinking, play itself being the ultimate process, thinking must itself become play, precisely to the degree it wishes to be a thinking of what is ultimate in things. And this play mode of thinking, done in an altogether comprehensive way, Axelos calls "planetary thinking."[70] "What a new thinking, one that is without a point of view, without one-sided direction (neither spiritualistic nor idealistic nor materialistic nor realistic), must take under consideration is the play of the *planetary* world, for the being-in-becoming of the whole of the World is play."[71] And, in this way, the "new thinking" answers to the "errance" inherent in the play scheme of things.

> Planetary thought answers to the errance of the being-in-becoming of the totality of the world. What, then, happens to truth? Errance does not mean error and aberration, falsity, vagrancy, and lying. There is no longer any reference to an absolute—something absolved from what?—but just the play of "It." Everything does not become relative—relative to what?—but constitutes an approach to "It," to the ungraspable, which is neither an idea, nor a person, nor a thing. It "is" the play of time which does not allow stabilizing being, hypostasizing becoming, positing totality, and taking the world as a supposition. . . . Truth . . . becomes question, problem, interrogation, in the play of errance. . . . Reality becomes itself question, problem, interrogation, and even play. . . . What is to be done? Play the play. Let ourselves be carried by the play of time which is—at the same time?—movement and rest, concentration and dispersal, gathering and shattering.[72]

In a word, "beyond significance and absurdity, truth and error, our planet will perhaps have to think and experience the world as play."[73]

A change in the way one *thinks*, however, is insufficient for achieving the unity-based stance in the world that Axelos feels alone can constitute a genuine transcendence over the culmination phase of Western, division-bound history. The change in thinking has to involve just as totally a change in *acting*, and this is where *technique* enters Axelos's project of transcendence. Technique (i.e., making things) is preeminently concrete involvement; and technique today is involvement with things through the application of the search for reasons in an all-inclusive empirical, manipulative way—that is, technique is technology, as was explained earlier. But, in addition, "technique appears more and more as the dominant configuration taken by the play of man and the play of the world."[74] Here, indeed, we meet the principal points of our whole discussion of the meaning of technology and of Marx's importance for understanding it. Technique, "while not being the Whole, is the constellation which encompasses and forges everything. It is under this constellation that man's fate is being played out."[75] It is crucially important, then, in Axelos's eyes, that "a new opening in regard to technique" be achieved, "another kind of play with and within the total compass of technique in its encirclement of all that was and was made and is and is made, in its enrolling of us into its play."[76] Technique is not to be abolished, for technique is "the instrument and weapon in the dialogue-duel between man-and-world."[77] Rather, it is to be opened up, that is, thought and experienced, and then performed, *as play*,[78] for play is the master scheme and matrix of all phenomena of power and energy.[79] In brief, technique has become the consummating, all-enfolding agency in and through which man and world are engaged in the play of reality; it is technique as all-encompassing which has to be confronted, understood, and brought to its inner authentic fulfillment. The final question, then, for Axelos, is just how this is actually to be done—and he is ultimately simply unable to answer. He can only pose the question and offer his vision of the necessity of moving questioningly into the planetary play he finds himself led to promote.[80] Indeed, in the end, paradoxically, he must even bring into question this very same pivotal notion of his, "Play."[81]

Such is the overall prospect of a movement beyond Marx which Axelos opens but does not actually pursue in his study of Marx's thinking on technology. There remains, of course, the serious question of how justifiable it is to ground that opening in the Marxian project, not to mention other sorts of problems, such as the inherent difficulties of the paradoxical, transrational, self-undermining nonposition that Axelos ultimately seems to take, the same difficulties that adhere to the Heideggerian project. But those are not matters for this introduction to pursue. My purpose in this section has been simply to show what lies beyond the recurrent indications Axelos gives in the present book regarding his own thinking on transcendence beyond the division-fraught realities of our history up to the present day.

The Translation: Some Explanations

One more task remains, a practical one, namely, to give some explanation of difficulties involved in the passage of thought from one language to another. A translator's position with regard to a writer's ideas is something like that of a valet to a gentleman's personal traits. Like the valet, the translator becomes acquainted with both serious defects and simple idiosyncrasies in his charge, while having to maintain a professional posture that reveals, preferably, only such forgivable faults as serve to render the man more human and understandable. My remarks in this last section deal with things that, as translator, I should bring to the attention of a reader to enable him or her to get over peculiarities or difficulties arising from Axelos's text. But it may be that in doing this I shall have to point out some things that a gentleman's translator perhaps ought not mention.

The first difficulty involved in the Englishing of Axelos's French arises from his predilection for certain kinds of philosophic expressions. Here most notably we find the endless use of *devenir*—"becoming"—used as a noun. *Être*—"being"—is also frequently used, but it is not as unusual in English. When the two are joined together into a single phrase, keeping close to the French results in an awkward English phrase: *être en devenir*—"being-in-becoming." Yet I have retained this strange turn of phrase in order to keep the kind of metaphysical flavor that links Axelos, especially in his own mind, to a tradition identifiable at its extremes by the figures of Hegel and Heraclitus. For a similar reason, I have retained the structure of *ce qui est* in "that which is," instead of rendering it by "entity" or something similar; for these are expressions that by their form betray an awareness both of Greek origins and of the possibility that the question of Being is not reducible to the question of entitative determination, all of which is very Heideggerian by explicit purpose on Axelos's part. (In the same vein, I have kept closely to Axelos's usage of *la technique* and *l'errance*, but this has been explained in the preceding sections of this introduction.)

On another score, a particularly thorny term is the French *sensible* and the noun form, *sensibilité*. Here the difficulty is that three languages are involved, French, English, and German, because Axelos often enough uses the words to render Marx's *sinnlich* and *Sinnlichkeit*. Consistency in translation in this matter is impossible because of the various philosophic meanings that the words may have and that are variously prominent in the usage in Axelos, in Marx, and in the different translations of Marx quoted in this English translation of Axelos. In reading Axelos's book, then, it is suggested that, when the words *sensuous, sensory, perceptual* (or, in some standard translations of Marx's texts, *sense-perceptible*), and the like occur, all being attempts to cope with *sensible* and *sinnlich* (and their nominal forms), the reader should keep in mind the following critical remarks made on the issue by Nicholas Lobkowicz:

> Marx, following Feuerbach, argues that "sensibility (*Sinnlichkeit*) must be the basis of all *Wissenschaft* [knowledge]." It would be very misleading to translate, as some have done, the expression '*Sinnlichkeit*' by 'sense perception.' For even though Feuerbach had argued that philosophy has to proceed from sense data, not from abstract ideas, and even though Marx basically agrees with Feuerbach on this point, he is far from supporting the empiricist-positivist claim that all scientific knowledge has to proceed from or even be reduced to sense observation. Rather, what Marx seems to have in mind is Kant's notion of *Sinnlichkeit* as receptivity, that is, man's essential dependence upon external objects. Thus the point which interests him is an anthropological rather than an epistemological one. What he means is simply that all *Wissenschaft* has to proceed from the one basic fact that man depends upon, has a need of, desires, and therefore also acts upon objects independent of himself.[82]

Perhaps here also should be mentioned one set of terms whose distinction could be more clearly explained by Axelos, especially since they are repeatedly used as key concepts in Marx's theory of alienation. Here, again, there are three languages involved and varying interpretations possible, interpretations that may be different not only in the different languages but also in different translations of the same passage in the same language. Under these circumstances it is again impossible to achieve completely unambiguous consistency, unless one undertook to reformulate and reexpress the entire work being translated, quoted passages included. For want, then, of clear, explicit definition on Axelos's part and for want of consistency among the various languages, the best one can offer is a set of possible distinctions for use here, which conform more or less to Axelos's reading of Marx but which themselves would be debated by students and exponents of Marx in their various persuasions.

1. Axelos: *objectification* (Marx: *Vergegenständlichung*)—translated "objectification": the process by which human being actualizes itself through the action of producing a product identifiable as an object for discernment by the senses.

2. Axelos: *extériorisation* (Marx: *Entäusserung*; but also sometimes equally well *Veräusserung*)—translated "externalization": objectification seen as a matter of constituting an object as externally distinguishable from the human agent itself.
3. Axelos: *aliénation* (Marx: *Entfremdung*; though sometimes also *Entäusserung* or *Veräusserung* in certain contexts)—translated "alienation": objectification (or externalization) as converted into an alien, hostile, antagonistic phenomenon.
4. Axelos: *réification* (Marx: *Verdinglichung*)—translated "reification": alienation inasmuch as it is a process of conversion into a fixed thing, especially a thing as an item of property defined by the value established for it in terms of money, that is, in capitalistic selling and buying. The important factor here is the fact that the order created by this conversion, the capitalistic system, possesses immense power of its own and works according to necessary laws to compel human activities and relationships in all-inclusive ways. It is a kind of second-order "nature" which, while resulting from human activity, systematically contravenes the intrinsic thrust of human being toward self-realization.[83]

In the English versions of the texts from Marx that I provide, when the terminology departs from the above usage on Axelos's part, I modify it to conform to that usage. This is always indicated by putting angle brackets (⟨ ⟩) around the changed words. Modifications made for other reasons in the wording of quoted English versions, also indicated by angle brackets, are generally explained in the notes, as are other difficulties regarding texts and translations.

Under quite a different heading are a few things I should point out as peculiarities of Axelos's style that may annoy the English-speaking reader. Axelos uses italics and parenthetical phrases and remarks in ways a stylist in English would not, but they have been retained here. It is his book, after all, employing his way of qualifying or explaining things. The same holds for the way he capitalizes some words (such as *Play* and *World*); and, because here I have not been entirely able to discern his purposes for doing so one time and not another, I have simply followed his practice. Obviously too he likes to use Greek words as charged with seminal meaning, in a way directly reminiscent of Heidegger. These must of course be strictly kept as Greek in the English. Lastly, Axelos is fond of the recapitulatory generalization, particularly in those recurring passages in which he summarizes the history of philosophy.

Here too one finds a liberal employment of "-ism" terms, sometimes in long chains. In both these matters, instead of using some kind of paraphrastic equivalent, I have opted to let Axelos's way of putting things prevail.

Two final notices about the translation remain. The first concerns the biblical texts that occasionally occur (see Axelos's introduction). For these I have used the English of the Revised Standard Version (New York: Thomas Nelson, 1953). A second, more important matter is the fact that it has been deemed unnecessary to include in the English translation the long bibliography Axelos draws up at the end of the French edition under the title "Itinéraire Bibliographique" [A bibliographical itinerary]. The catalog given in it of source writings by both the primary and the secondary figures in Marxism (Hegel, Marx, Engels; Lenin, Stalin, Trotsky, etc.) is well enough or better served by other works in English, especially as there will also be indications of available English translations. The secondary works listed reflect mainly French scholarship and hence have an interest too restricted for a general English-speaking audience. If this information is desired, one can easily turn to the original French edition of Axelos's book. Given these considerations, the length of the bibliography does not justify its inclusion.

A final word in this introduction has to be one of acknowledgment. First of all, Joseph Bien deserves the credit for deciding this text ought to appear in English, and it was at his urging that I have undertaken the translation. It was originally to have appeared in a series in contemporary political and social philosophy which he had planned but that has for various unfortunate reasons come to naught. I am grateful to the assistance he offered me wherever he could. Second, the University of Kentucky Research Foundation has been most generous in providing funds for obtaining materials needed to complete the translation and for preparing the typescript. It was in great part owing to their support that I was able to meet with Axelos in the summer of 1973 and discuss his book with him. Lastly, I have had considerable help from my colleague, Dan Breazeale, in solving the inevitable difficulties that result from trying to handle concepts from German Idealism in English.

PREFACE

The aim of this study is to bring before us the philosophic thinking of Karl Marx, for this is the thinking that, subsequently become Marxism, has played such an active role in the movement of history in Europe, in Asia, and now in a world on its way to becoming a single planetary phenomenon.

Marx's philosophic thought is not very well known despite all the studies that bear upon it; and this is simply because these studies are usually either too narrowly historical or too dogmatic and apologetic. The center of his thought, the foundation from which it rises, and the principle of the movement that springs from it remain veiled. They are unrecognized, underestimated, opposed, admired, but they are not thought through. It is time perhaps to try to engage in dialog with that philosophy and dialectic of the development of Technique, which is the riddle element of universal history. But what we have to do before beginning this dialog, in the interest of that very purpose, is listen to and understand Marx's words.

Someone who wants to read what the founder of Marxism himself has written finds himself facing the imposing bulk of the *Complete Works*. It is not easy to find the way into it. There is a doctoral thesis on Democritus and Epicurus. There are critiques and commentaries on the thought and writings of Hegel. There are philosophic and even "metaphysical" studies violent in what they disclose (*The Economic and Philosophic Manuscripts of 1844*; *The German Ideology*); polemics against the Left Hegelians (*The Holy Family*) and against Proudhon's socialism (*The Poverty of Philosophy*); monumental works in economics (*A Contribution to the Critique of Political Economy*; *Capital*) and programs of political action (*The Communist Manifesto*; *Critique of the Gotha Program*); sociological and historical analyses of class struggles in France

(1848-50; 1871); journalistic articles on European and oriental questions, lyric poems, numerous letters, notes, and outlines of all kinds. From one end to the other a single thought inspires and dominates these texts, even if their unity is hidden by the working out of many different themes; and this thought, philosophic in its origin, aims for the *overthrow* of traditional Western metaphysics, it aims to complete, abolish, and supersede philosophy itself precisely by bringing it to realization in practice and technique. Beginning from the analysis and critique of the *alienation* of human being, of the alienation of work, of economy, of politics, of human existence, and of ideas, this central core of thinking culminates in the technique-structured prospect of *universal reconciliation*. The vision is one of man's reconciliation with Nature and with himself brought about through the historical and social community of men, a reconciliation that finally makes possible the full satisfaction of basic needs, the reign of abundance, and a world in which everything which is or comes to be is fully transparent. It is a (re)conciliation that means the *conquest of the world* by and for man in the unlimited deployment of technical forces.

We shall try to make Marx's thinking heard as it speaks in all its coherence, all its depth, and all its limitations. Any important thinking always contains problematic dimensions and multivalency. Bringing it really before us means that we have to submit it to serious questioning. But it is not a matter of simply opposing it, which would mean nothing. Rather here the idea is that entering into the rhythm of a thought and reawakening all the radical intensity that is too often forgotten or dulled are far more fruitful than reading a philosophic text as if it were a sacred book, a scholarly dissertation, or a newspaper.

One cannot put on one side a man's systematic thought, his "opinions," and on the other the history of the genesis and development of his thought as located within the general history of philosophy. Neither can one separate a man's systematic thought from the method he follows or from the whole of the history of thought. There is not first Thought and then History alongside it, or History and then Thought. What we have to do here is try to grasp a thinking that implies history, belongs to history, opens out into history. Our aim, then, is neither "systematic" nor "historical": it is simply to prepare dialog with a thinking whose full import still escapes us. For this thinking, turning on the pivot of Technique, forms a single thing with the being-in-becoming of the history of the world at the same time that it raises the serious problem of the essence of thought on the one hand and on the other the meaning of man's practical activity in the setting of historical and political reality.

In short, the design behind this study is that it serve as an introduction to meditation on the *work* of Marx. But is not this *work* inserted in the very heart of the totality of being, even if by way of illuminating but one aspect of the World and bringing it to realization? It is a work rooted in time, in the becoming of being, and it expresses a certain movement. But perhaps it is still too early to tell if this movement, under the name *dialectic*, is principally a dialectic of what is called *thought* ("subjective" dialectic) or a dialectic of what is called *reality* ("objective" dialectic). The distinction between reality and thought, as well as the Marxian claim of their unification, will remain problematic for us as the study proceeds, just as will the terms *subjective* and *objective*. As long as we do not get beyond the subjectivity of the subject and the objectivity of the object in a movement toward their unitary foundation and common horizon, we fail to say anything essential about them.

INTRODUCTION.

ON UNDERSTANDING MARX'S THOUGHT

The thought of Karl Marx has, since Marx, become Marxism, a method and doctrine, a theory and practice turned into an official system. But Marx's thought is also a continuation and extension of the whole of Western metaphysics, even while it wants to abolish philosophy in a radical way so that it can realize itself in real, material action. Like links in the chain of its history, each new opening that thought makes as it ever moves along established tracks—by taking new steps, of course—is brought to readiness by Philosophy and History. Thought in its history is the central fiber of what is generally called the history of ideas. In the form of philosophic reflection it becomes indeed the light of history.

Greek thought forms the first stage in the course of movement that leads to Marx. "Modern philosophy has only continued the work begun by Heraclitus and Aristotle," the founder of Marxism will say.[1] The Pre-Socratics are the first in the history of the world to grasp the being-becoming of the Whole (of what is) as the Λόγος ("word, reason") and the Ἀλήθειδ ("truth") of divine and indestructible *Physis* ("nature"). For Heraclitus, to love wisdom (τὸ σοψόν) means in friendship, through war and harmony, to grasp the One-All (ἓν πάντα εἶναι) which is *Physis*, the unity of all that conceals and reveals itself, the being-in-becoming of the whole of the world, and to express in human language the rhythm of the *Logos*. Plato and Aristotle, the two classical philosophers of Hellas (and not only of Hellas), followed Socrates in the fight against the Sophists and found philosophy as philosophy, that is, as metaphysics; and the way is thus opened to dualism. Greek thought thinks—contemplates—the phenomena of *Physis* and, upon Plato's sundering of ideas from things, tries to grasp their metaphysical, indeed their theological, significance. Yet in Aristotle's eyes the whole of first philosophy still culminates in the question, "What is being (i.e., the being of the totality of the world)?"

The Skeptics turned the weapons of thought against thought, and, original unity being shattered, the Stoics and Epicureans tried to find support in self-consciousness. Then Greek thought met the Christian faith, which was itself preceded by Jewish revelation; and Plotinus tried again to reestablish the sense of (lost) unity. Rome, victorious over Hellas, developed on an essentially social and practical level, turning everything that is into a *res* ("thing"). But Rome did not thus escape ruin. What has always been true for every form of existence was true as well for the Roman republic and the Roman empire: every victory is the prelude to defeat.

The ancient world died and the *Judeo-Christian tradition* unfolded, opening a new horizon. Here the whole of what is, the world, was now seen as having been created *ex nihilo* by God, the being par excellence. The world was no longer the becoming of *physis* but Creation destined to experience the Apocalypse. Mankind now lived under the sign of sin, awaiting a final Redemption, made possible by the sacrifice of God made Man and dying for men. The human person thus carried the divine message and became the theater for the drama. Man, created in the image of God, himself created works in a constant fight against nature and the flesh to bring about the triumph of the Spirit. Though he moved upon the earth, creating and procreating, man was to direct his eyes to heaven alone, knowing that the true world was beyond the present one. The Jewish prophets, the Old and New Testaments, and St. Augustine and the whole tradition of mysticism and scholasticism maintained this idea, and it continued to exercise its power right up to the Reformation. It is very difficult to talk of Christian philosophy, for something that depends upon faith in a Revelation and supports itself on the authority of the Church is no longer philosophy. Nevertheless, there was a kind of thought still working in the Christian world.

The third stage takes up where the two preceding ones stop, to continue them, of course, but showing a new face, that is, showing itself precisely as thought. The unity of the whole in the Greek *physis* was already put in question by Christianity, but *modern European thought* simply dissolves it, objecting in turn to the Christian's order of creation. The *ego* of the human subject is posed as *res cogitans* standing opposed to the (objective) world of *res extensa*. Representation and *ratio*, consciousness and science set the basis for the possibility of grasping the infinite world, and thought here enters upon the course of development which will see consciousness, science, and technique unfurl the will to power. Mankind enthusiastically sets off to conquer the whole of reality in terms of objective works. Man's self-awareness and will no longer encounter any limits. History becomes the theater of the struggle against nature and Technique proposes to transform everything there is. Rationalism and humanism, the natural sciences as well as the historical and human sciences, both practical and theoretical activity are alike caught up in a course destined to envelop the entire planet. Not even *crisis* seems to halt the advance, even when the threat is that of nothingness, of nihilism; for the first and ultimate foundation of the whole enterprise is nowhere to be seen, and to the questions, What is it all for? *Why* is there everything there is and that happens? no answer is given.

The founding thinkers of the Renaissance, Descartes, the French encyclopedists, the philosophers of the Enlightenment, and the philosophers of "German idealism" (from Leibniz to Kant to Hegel)—here are the heroes of the victorious progress of the *ego cogito*, the transcendental Ego, absolute Subject, absolute Spirit. Yet in the midst of this same progress we hear other voices that we can hardly understand, the voices of tragedy.

In *Greek thought*, the world, the totality of everything that is in becoming, remains One: *Physis* illuminated by *Logos*, *Logos* animating the language and thought of mortals. The being of the eternal *Cosmos* is manifest in its truth through beings, finds expression in the myths and acts of the immortals, and reveals itself by and in the city-nation, art, and poetry. Man is a being of *physis*. (The word *physis* hardly meant for the men of Hellas what the word Nature means for us, for *physis* to them was totality.) He is bound by "physical" ties and obeys a cosmic rhythm. He may explore reality, bring it to language, and raise it to the level of knowledge, he may put into practice a *technē* ("art, technique") coessential with *physis*, but he does not set himself up as master of the Cosmos. His works do not try to move beyond his own order. What he grasps as a being—in all its natural and sacred fullness—remains both the tribunal of final appeal and the first foundation. The men of Hellas know how to face the threat of nonbeing, of a nothingness which as threat does exist. Now and then they even dare find words for it, without for all that inquiring into the power of nothingness itself and questioning the foundation of being.

In the eyes of *Judeo-Christian Revelation*, which after a point becomes more Christian than Jewish, totality as world is drawn from Nothingness by the creative act of God. All creation and all creatures stand subordinated to the Creator. This world is only a temporary stage leading to the other, true world. Nature is that against which men struggle in creating in turn new beings. Nature is to be mastered, for above it hovers Spirit, and over mortals reigns the Man-God, God become man, son to the absolute master, who brings their labors and their works to completion for the glory of God. All that is lies subject to the plan of Divine Providence pursuing its aims.

For *modern thought*, the world is at once Nature and Spirit, reality and idea; and it comes to be grasped in representational thought by man, whose will is to know it, enrich it, explain it, and transform it. Man is this quasi-absolute subject, who, in concert with his fellow men, in Society, works and toils at building objects, setting in motion the terrible powers of technique; and it is technique that, once engaged in the struggle against nature, is destined to become the lever that will move the entire planet.

Modern European thought, the third phase of Western thought, bent on world-wide expansion, is essentially and perpetually in *crisis*. It searches for the ultimate foundation upon which its theoretical and practical activity could base itself, it raises the problem of *why?* with regard to the Whole of things and with regard to everything in particular, and yet it reaches no radical and total answer. Even while it is given over to the search for the being of the unified whole of things (*l'être de l'unité de la totalité*) as well as of that which appears within it, it shows nonetheless a constant tendency to think the World not in its plenitude but in terms of special regions viewed within certain perspectives. The totality of being is thus dissolved, and thought institutes and then explores independent groupings as making up individually special dimensions.

Modern thought is in its very essence set up as *logical, physical, historical, sociological, psychological*, and *aesthetic*. These are the perspectives within which and through which it moves as thinking, each being a relatively autonomous dimension and a special region of reality as a whole. One by one in turn, or through the selection of some one direction in particular, each of these perspectives ends up in a position of privilege in thought as *the* view of the Whole. And so we find the established positions of logicism, physicalism, historicism, sociologism, psychologism, aestheticism, and all the minglings, alliances, and combinations thereof, real and possible.

Each for itself, these partial and segmentary outlooks put into practice a technique of reflection to explore those regions of reality called thought, nature, history, man, and art as so many special spheres, as so many groupings having their own proper laws and structures. Unquestionably this is all necessary in order for investigation and research and for technical theory and practice as the agency of transformation to advance, yielding positive and useful results. Yet where is the central thought that grasps all that exists in unity and plenitude? *Is there no longer any plenitude?* Have individual totalities and the sum of them replaced the circle of totality? For inevitably, even when thought tries to be multidimensional (or simply tries to encompass the lot of one-sided lines of thought), as long as it does no more than link up individual dimensions, it can hardly result in a global comprehension.

Obviously modern thought is still looking in its own way for the unified Whole, shattered as it is; but this task really belongs to one single discipline that would have generality as its theme. We have already mentioned metaphysics (or, more positively, general philosophy), but it degenerates too often into a general history of metaphysics or even into a simple history of (great) ideas. First philosophy, ontology, never succeeds in being truly fundamental. It never escapes the dangerous palliative of scholarly systematization, whether theological, idealist, or positivist. Just to intend to give a discourse on being does not mean at all that one does in fact search the meaning of the *logos* of being in its totality and in all the richness of its manifestations.

Sprung from philosophy, the sciences have developed by pushing specialization as far as possible. Restructuring themselves as technique wherever possible, they pursue their work of discovery and invention within the limits they themselves set, yet without for all that giving up the wish to submit everything to their point of view and their method. Nevertheless, science takes in only one aspect of the world and not the world as a whole itself. Even if it be granted that one aspect and truth of reality is indeed taken hold of here, still it will not be universal Reality that is thus grasped. *Language, thought, nature, history, man, poetry,* and *art* come more and more under attack from two directions that nonetheless share the common denominator of technique-structured science. One is the historical (or historicist) point of view, the other is the systematic (i.e., the point of view that wishes to make each of these things the subject matter of a treatise). According to these two insistences each of the domains in question has its "history" and to each belongs a scientific treatment claimed to be perfectly adequate. But does the problem of history and of man, the problem of the meaning of historicity and of human being, gain clarification by these means? Do the historical and human sciences resolve the problem they are thought to be concerned with or are they simply exploring it technically?

The modern world and modern thought have made *History* and *Man* privileged centers. Their study holds a special and almost dramatic importance. Everything in existence is approached as revealing itself in the space and time of *Human History*. Human History thus undertakes a scientific appropriation of the whole of mankind's historical past, while transforming *Nature* through technique in preparation for a better future beyond the endlessly somber and desolate present. The *Man of History* becomes in this way the chief hero of the tragedy of totality. His thought aims to grasp all of being, and his works of art and technique express his essence, all the while showing that man realizes himself and objectifies himself, in universal history, through alienation and suffering. Notwithstanding, men do not give up the will to be victorious, whether individually or collectively, over what escapes them. They believe themselves the protagonists of becoming, while forgetting too easily that, if they are the subject of this becoming, they are ruled by it; and becoming is quite another thing from a flow of things carrying "subjects" and "objects" toward a destination that can be decided and organized according to a plan. But they do not wish to delay; they have no inclination to take *errance*[2] seriously.

Thought in its modern stage, and for the first time in the history of the world, assigns man a role that is unqualifiedly privileged. It posits the *ego* endowed with *ratio* and self-consciousness in the form of the *res cogitans*, assigning as its field of activity the whole realm of *res extensa*. Human subjectivity becomes thus the (objective) foundation of what is. The World comprises the world of the subject and the world of objects, with man as the (objective) Subject, an active knowing thing dealing with objects and grasping them in representation. *Cogitatio* goes hand in hand with action and even *agitation*. History is henceforward the earth of men, the terrain of all action, the true place; and thought sets out to examine the riddle of man's being and the meaning of his history. Yet the relationships between the concrete and finite, on the one hand, and the universal and absolute, on the other, become agonizing questions. Efforts to answer the great questions in a systematic way fail to justify either what is or what comes to be—a state of affairs that hardly keeps action already under way from following its path.

Realism and idealism, rationalism and empiricism, individualism and universalism, and the theory of knowledge and the technique of action make their paired appearance in this era of humanism and moralism that is ready to spread over the whole surface of the globe. In the eyes of the great modern thinkers (and *only* in the eyes of the great), man and history do not permit being treated in isolation and autonomy in relation to everything else, in relation, that is, to the totality that contains them; they remain within a circle far more encompassing.

Essentially it is in the driving will to scientific mastery and practical transforming action that man and history become compartmentalized and cut off from any ties that may bind them to something grounding them or going beyond them. This new direction occurred about the middle of the nineteenth century, just after the completion of the last great philosophy of this stage in the process by which "philosophy becomes world" and "world becomes philosophy." With Hegel as the point of departure, historicism (as well as sociologism) and practical humanism (whether individualist or collectivist) make their appearance. We see here the rise of the general tendency to study man and history in an "anthropological" and "historical" autonomy, despite the efforts of certain thinkers who refuse to divide totality into sections, who refuse to pursue the search for the essence and roots of reality in all its manifestations by way of ontological dismemberment and methodological segmentation. Toward the end of the nineteenth century and the beginning of our own, an "ontological" schema for ordering reality and knowledge achieved victory. According to it, the links in the chain of becoming consist of *matter, minerals, plants, animals, men,* and *historical societies,* and a corresponding ordered set of sciences goes from *mathematics* and *physics* to *biology* and *sociology;* this linked sequence of becoming culminates in historical man, men in society, who through work, technique, and science aim to master and transform both themselves and all else besides.

Man, consequently, ends up being a problem, an open question one tries to clear up any way one can in one's haste to find an answer. But the human subject is essentially a historical being. History as well, then, offers itself to exploration, and the effort is made to penetrate the secret of its meaning. Anthropological science tries to take in human reality, and historical and social science, political economy, and politics endeavor to comprehend the historical development of human societies. These studies all address themselves to that which is principally human and historical, namely, mankind's past historical, social, and political structures in their making and unmaking; and their aim is to perform careful analyses for the sake of a better way of doing things both now and in preparation for a future that, in their projections, will operate on a planetary scale.

The ontic realities called *man* and *history* are thus dealt with on both a theoretical and a practical level by a whole network of anthropological, historical, and sociological sciences; and this is something that has come about since the end of a great period of speculative philosophy, of ontological metaphysics. Has philosophy no longer anything to say? Has it become real science? Unless we set the thought of Marx in the world-historical horizon of thought, we shall hardly come to understand it; for it ties in at a precise moment of history. This is why we have prefaced a presentation of Marx's thought by a survey of the intellectual movements that have led to it. Marx himself thinks philosophy as philosophy is a theoretic activity that had to be surpassed through its realization, even though "its realization is its destruction." Marx emphasized, and gave shape to, what was happening around him. In his own way, he took issue not with the being-becoming of the world as totality but with the being of man and the becoming of history and society.

Human being and historical becoming constitute precisely the pivot of the investigations whose lineaments we shall try to sketch. Here we find this being and this becoming studied in order to be transformed and bettered by the active intervention of science and technique. Scientific technique cuts reality into spheres, sectors, regions, and fragments, the better to take them up in accord with (presumedly) appropriate perspectives and methods, the aim of which is effective action through the employment of specific and adequate means.

Everyone undoubtedly knows that "man" and "history" are phenomena within cosmic nature and the limitless totality that opens out into the past and the future. No one is unaware of the fact that man and history cannot be wholly isolated. Nowhere is there found a man absolutely alone and isolated from any history; nowhere do we find a history that is entirely impersonal and empty of mankind. We are what we are as indissolubly historical and human. Men appear as the protagonists of becoming, wherein is manifested and whence withdraws the being-in-becoming of the totality of the world. As both individuals and collectivities men are the active and passive agents of this movement. At the same time there is a constant tendency to forget that human history does not exhaust the totality of being. Historical and social becoming, especially if understood in a narrow, reductive way, hardly coincides with the total openness that is the world, a totality never "total." This totality of nontotality which is the World, which is never simply an empirical aggregate, this open totality of fragments of World, cannot be reduced to the supposedly real world of the senses. True, there is but one *world*, but it is not as limited and accessible to grasp as the constituted, concretized world may be.

The epoch of subjectivity, the modern period, cannot but give privileged status to the anthropological, the historical, and the social, leaving human thought and action without some definite foundation. But *subjectivity* does not mean psychological subjectivity, an exclusively individual thing. Subjectivity here is the place and time of the revelation of the totality of world. It is "subjectivity" *in an objective way*.

The pairs of opposites, man/history, individual/society, subjectivity/objectivity, remain schematic and empty of content, in that we never encounter a single pole-element. When these pairs appear, it is we ourselves who introduce an artificial isolation of the elements that make up the whole, and it is the whole that allows them to be active members. But this does not mean that we possess the foundation, the common single root of whatever makes up the whole, taken either in its fullness or as completely dissected. Human history encompasses individual man as in an evolution that is everywhere and always within the context of society, as everyone knows. But the question, Is man the cause of his own history? or, Does history produce man? is meaningless, because it is artificial and abstract. What counts is to know how to understand simultaneously in a single movement both the globality of history as containing men and the greatness and poverty of the human being, the "maker" of history in concert with his fellows. Marx does indeed give predominance to one dimension and cuts it off from the rest, but he does not keep it isolated. He focuses upon one dominant aspect, analyzing one determinant factor for the whole of existence; he deepens and amplifies one reality, without for all that building up a concrete, total approach, one that would be more unitary than unifying. He proposes one reading of the book of history and recommends one particular kind of action. Nevertheless, the totality of the world holds other realities and other activities as well.

The Totality to which the Pre-Socratics were open, and which they took hold of for the first time in the history of the world by way of the *logos*, became a total system in the thought of Hegel. In that system *the true is the whole*. But, since Hegel, Totality has no longer been manifest *as such*. Regions and dimensions, fragments and aspects of Totality continue to be taken hold of by men, but the absence of ontological thought is a cruelly persistent phenomenon. The ground of all that is and the tie that binds everything into an All still elude man. Nature has become something against which one struggles, God is dead, and human Society shows awesome faults. Man and men strive to win themselves back from alienation, for neither individual existence nor human community can do without something that serves as a foundation. Human being tries desperately to be reconciled with itself, other men, and the World by socializing its drama, by mobilizing the powers of will and work. Yet it seems that the powers of *conceptual representation* and *reason*, of *theoretical knowledge* and *practical labor*, of *consciousness* and *technique* do not lead directly to the recapture of lost unity, of broken totality, of nature unnatured. The single truth of Totality no longer is to be seen, the unitary base is no longer visible, and for modern man the Whole takes on the features of Nothingness. Yet men continue to deploy a theoretical and practical activity that seems to encounter no limits, even though all this activity does very little to keep the question *Why?* from remaining unanswered—in fact its effect is quite the opposite. Historical dialectic wants to gain mastery of the secret of mankind's being and movement in order to face the issue of alienation directly and offer a remedy for the dissatisfaction that increases proportionately to the rhythm of the production of material goods. The sources of exploitation, frustration, and dissatisfaction come under attack by those whose aim is to transform negative, unhappy consciousness, as well as the wrenching dislocation and conflict between uncommunicating vessels that accompany it, into positive, revolutionary reflective self-possession (*prise de conscience*), leading to a harmony in which riddles are solved and a satisfactory order established.

At the very first, it is Marx, the belated Jewish prophet, who sets out on this trek. He takes the humanist tendency of Western thought and pushes it to its extreme consequences; he finds the language for expressing the evils of society and of man in order to propose a solution and a therapeutic treatment; and he does this on the basis of an analysis of given conditions as end products of a long process. In Marx's eyes, a laborious development, that of the forces of human history, has alienated man from his being, from the products of his labor, from his true nature, and from the world in its totality, and all this to the point where his true social nature is lost for the sake of a technicist civilization. Marx, in an analytic description of this unnatured and alienated state, violently questions it and radically condemns it, his intent being the preparation of a practical alternative that will overcome its many-sided alienation. His polemic analysis brings him to extremely negative conclusions about the *present* and foolishly conciliatory prospects for the *future*, whereas the problem of the *original past* remains shadowy.

Marx thinks he has found the secret of the tragedy in his analysis of human labor. Man's labor, the production of goods, human reproduction, and, still more, man's production of himself bind men to Nature even while constituting an alienated bond. Marx thinks he has found in the alienation of labor the riddle of the historical becoming of alienated mankind. Accordingly the reality of evil and unrest is most substantial in economic tragedy. Evil is projected, as it were, like a secularized shadow of original sin. This struggle for the production of life that men lead as workers with and against their fellow men, as well as against nature, is what suffocates living forces.

Is it possible that this "evil" whose secret Marx tries to probe might not have arisen? Is it not intrinsic to the very development of living forces? And, further, why are we unable to keep from thinking of "original sin"?

The God of Genesis bade Adam, the first man, not to eat of the tree of knowledge in the Garden of Eden that he was to cultivate and keep. Man was free to eat of any tree in the garden save the tree of the *knowledge* of *good* and *evil*: "Of the tree of the knowledge of good and evil you shall not eat, for in the day that you eat of it you shall die" (Gen. 2:17), the Eternal One said to man. His commandment was not respected, negativity worked its way, and man and woman became mortal, *mortal* but *knowing*; for when they had eaten the forbidden fruit "the eyes of both were opened" (Gen. 3:7). Henceforth generations coming after Adam and Eve must expect to see evil weigh upon their lives. Paradise is lost for men *now become* mortal and possessing knowledge by reason of original sin. For the Eternal God said to the first mortal man, "Because you have listened to the voice of your wife, and have eaten of the tree of which I commanded you, 'You shall not eat of it,' cursed is the ground because of you; in toil you shall eat of it all the days of your life; thorns and thistles it shall bring forth to you; and you shall eat the plants of the field. In the sweat of your face you shall eat bread til you return to the ground, for out of it you were taken; you are dust, and to dust you shall return" (Gen. 3:17-19). In this way Heaven condemned the first man and all his posterity to hard labor upon the earth, to be imprisoned in the necessity of *production*. Reproduction as well was cursed by the Eternal God, who said to the first mortal woman, "I will greatly multiply your pain in childbearing; in pain you shall bring forth children, yet your desire shall be for your husband, and he shall rule over you" (Gen. 3:16).

Both man, created by the Eternal God in His own image from the dust of the earth, and woman, drawn from man, his helper and companion (for "it is not good that the man should be alone" [Gen. 2:18]), were originally destined to be immortal and blessedly fruitful. Their mission was to multiply and *fill the earth, to subject it to their domination*, and *to rule over all other creatures*. After the fall and the expulsion from Paradise, they became mortal, destined to return to the earth, and condemned to produce and reproduce in difficulty, struggle, and suffering in order to wrest goods from it. And, since they were to struggle against nature to satisfy their basic needs, they were also to struggle among themselves to satisfy their desires. Man and woman obeyed the voice of negativity and ate of the tree of the knowledge of good and evil. They gained that knowledge; their eyes were opened, and they became indeed "like the gods" (Gen. 3:5). Is it not human, then, that, since mortal men were cast out of Paradise and condemned to work and suffer on the earth, they should want to become really "like the gods" and realize Paradise on earth? And is it not just as tempting for those whose evolution takes place on the earth of men to try to embark upon the conquest of the heaven of gods, in the spirit of the builders of Babel, who said, "Come, let us build ourselves a city, and a tower with its top in the heavens, and let us make a name for ourselves, lest we be scattered abroad upon the face of the whole earth" (Gen. 11:4)?

The knowledge that came with the birth of the first human beings, once introduced, was to continue its course. Adam *knew* Eve, who conceived and bore a child. Original sin was the result of a break, a dislocation in the one, divine, and "natural" world. The violent entry of knowledge marked this break, and the result was that men wish to use knowledge to overcome the wretchedness that affected the production of life.

Marx, the latter-day Jewish prophet, seems still to hear, even if from a great distance, the echo of the Old Testament.[3] But he wants mankind to take a path that leaves behind it original sin and their accompanying curses. Marx insists upon true knowledge, the recognition of the true social nature of man, the reflective consciousness of all his wretchedness, in order that by this very knowledge and consciousness man may be totally freed from the chains that shackle him to the stream of historical becoming. Original sin is put in question and denied. Labor is no longer to be alienation. Man is to become master of the earth without falling under the blows of any other master, human or divine. Marx assigns to man the task of conquering the earth and reconciling himself with himself, with others, and with the totality of being; and with this he, Marx, empties heaven of gods and tries to abolish the duality of heaven and earth.

Behind Marx stands a whole world whose origins are *Greek*, *Roman*, *Jewish*, *Christian*, *European*, and *Modern*. He wants his thought to be free of metaphysical presuppositions, but presuppositions, particularly if they are metaphysical, are not always easy to see. Even when he has no intention of doing the philosopher's work, his thinking has a philosophic side; he makes "concrete" something that comes to him from a long past. His explicit intention is that his system be scientific, because what counts for him most is a positive (not positiv*ist*, as is too often the case) scientific understanding of the process of the historical development of mankind, in order to transform the world through practice into one proportionate to human activity. But something that claims to be scientific does not draw only from sources in science. It could be said that Marx makes explicit on a scientific level his implicitly philosophical, that is, metaphysical, presuppositions, except that to describe his work as *scientific* is to use an inadequate term. More than science, it is technique that affects Marx's enterprise; indeed, perhaps (modern) technique is the internal motor of (modern) science.

Marx begins with a global consideration, an idea of the whole, moves to a work of analysis—economic, historical, political, sociological, ideological—and ends up once again with a perspective meant to be total. His thought (philosophic as it is) borrows a certain way of proceeding from science without, however, being satisfied with results of the scientific kind. Above all, Marx is a technician of the analysis of the alienation that the expansion of technique brings with it; but he wants as well to be a technician of the dealienation and transformation of world history. Technique, the secret of the modern era under various forms, also operates in Marx's work, and the objective of his effort is simply a dealienated and total deployment of the power of technique. The general direction of his thinking, the scientific construction of his work, and its practical consequences are permeated through and through with technique. We have to try to grasp the center of this thought, so inseparably philosophic, scientific, and technical, and see it as animating a method that leads to both the construction of a theory of the whole and the formation of a practical program.

To do this, we have to make every effort to grasp the essence of his thought, to understand his discourse in its totality, to discover the origin of movement in his thought, and to see its development as a process that is unitary, even though it involves stages. To prepare a future dialog with the discourse of Marx, we have to make contact with the living heart of his work, meditatively pursuing the methodic and systematic development of that work in its insertion in the course of the world. It is not easy to hear and understand the original sound of a voice. Communicating with a truth that is concealed in a construction is no easier. And to try to circumscribe the outermost limits for a thinking whose aim is conquest cannot but encounter difficulties.

Nevertheless we have to accept all these difficulties and run the risk that threatens this kind of enterprise, the risk of not really coming upon what it sets out to reach. We have accordingly made it our aim to make clear the methodic procedure and systematic content of Marx's thinking, realizing that for thought in progress method and doctrine are not separate things. We are undertaking neither *apology* nor *critique*. Attack and defense, necessary though they may be as military operations, too often turn senseless and end up as actions in the service of something unassailable and indefensible. Something far more meaningful can result if, instead of singing praises or launching into polemic, we prepare simply to follow the path taken by Marx's thought at grips with harsh reality and try to gain explicit understanding of the content of what is called his doctrine. To pursue and understand the course and movement of this thinking mean to discover its intention and follow it out to its ultimate consequences. By trying to hear the words of Marx as a coherent, consistent, and all-embracing discourse, we shall set ourselves as well to *make its truth shine out clearly*. To do this with a man's thinking one has to be open to it, because only in this way can its unresolved problems and limits, its internal difficulties, and its partiality show up in their proper light. One must be open as well to the radiance of the truth of a man's thinking, inasmuch as that truth, like any other, will hardly escape being superseded by something else. Scholastic commentary or clever criticism directed at ideas as constitutive and as constituted is and remains secondary. Ideas do not get "verified" or "falsified." They appear and develop organically; they arise as expressing and clarifying one of the core areas or one of the aspects of totality as an open phenomenon; by a process of transformation, they become the truth of actual reality and a real power, or they simply die in libraries because they failed to seize upon something overwhelmingly true. Even great ideas "die," but only after having exhausted their truth (or at least one of its dimensions) and made it reality, only after having become actually universal in seeding new thoughts and realities. Truly important ideas do "die," but their import is inscribed in the life of the world in its becoming.

In order to gain philosophic understanding of the metaphysical presuppositions of Marx's thought, the resounding significance of his doctrine, and the force of his method, in order to understand as well the aim of his whole enterprise, what it wants to accomplish, and where it wants to go, one must not forget that no understanding or interpretation exhausts the source from which those things come. Every thinking worthy of the name remains multidimensional and polyvalent, even and especially when we try to lay bare its core dimension; and only when we bring to the fore this core dimension do we succeed in making it speak. Thus we propose to bring *the central pivot* of Marx's thought to the light of understanding. Without reducing this thought to a *schema*, we shall try to question it regarding its proper intention in order that, once its own problematic as well as its problematic character are thereby made clear, it can be put into question. For Marx's thought contains negativity, which is movement by supersession; and its negativity awaits the hour for breaking out into the open. By putting into action the negativity that is implied in the very movement of Marxian thought, we loosen the structural frame that blocks negativity, we restore fluidity to the rigidity in Marxian thought. Without attacking it from an outside position, we proceed according to its own rhythm, but with our eyes open, realizing that we have to move and think beyond the point where that thought may have stopped.

If we return to the foundation of Marx's thought and follow it to its ultimate consequences, perhaps we shall find ourselves in the presence of something that both underlies and transcends it. To *present* a thinking, to *make it present* and *transparent*, means also to reveal the presence of the universality that encompasses its particularity. It means as well to assist the unfolding of its internal "contradictions," without for all that denouncing those contradictions in cavalier or scholastic fashion. No man's thinking ever succeeds in totally reducing all duality and plurality to unity, in monopolizing totality. Marx's thought is no exception. In truly making a thinking present, we already trace the path to what will supersede it in a more or less distant future. To *supersede* or *transcend* a thinking means to elucidate its truth, to understand its errance, to continue it in taking it up once more, while rising beyond it and going further. The movement of thought in history is a movement in which this kind of transcendence, this kind of dialectical elimination, becomes a reality; and through it is preserved the moment of truth proper to each great thought. The movement proper to history makes universal actualities of thoughts that are truly important, yet it transforms them. But, if a thought is instituted and then adapted in history, does this do more than transform it—does it betray it? But under what title do we pass judgment and claim it just? Obviously only one of many possibilities gets actualized, probably the one that is most forceful, the one most apt to transmute into "real" power. And so it is never the totality of significance in a thinking that is realized. Consequently, between the original thinking of Marx (Marxian thought) and Marx*ism* there are at once continuity and a break in continuity. Can we separate Marx from Marxism? Of course not. But neither can we identify them. Like Platonism and Cartesianism (*mutatis mutandis*), Marxism is bound to its founder by ties that manifest the power of continuity as well as of discontinuity. *Platonism, Christianity, Cartesianism, Marxism*, all follow a path blazed initially by a founder, even while moving away from his original intent. They each expand a movement generated by an initial thrust, modifying it, generalizing it, dulling it. Between Platonic thought and Platonism, between the Christian vision and Christianity, between the success-laden thought of Descartes and Cartesianism, and between the intent of Marx and Marxism there is a deep gulf that cannot be filled. In the end, no one can say what these *"isms"* mean. Whether they be polymorphic and elusive Proteuses, vague and flat "world visions," or mental or ethical attitudes, they pay for their effectiveness, their insertion into the world, the universal use of one aspect of their truth, by the loss of their first truth. The founding moment, the lightning flash of an idea, an intuition, a vision, probably cannot be maintained in its original intensity and purity. When it becomes a philosophic, religious, or political movement, what begins as a foundational insight takes on the character of a power seeking expansion; it follows a development that makes it change, and it becomes thus a formative force. The difference that separates it in its first appearance from what it finally becomes by way of its evolution must never be overlooked. Nevertheless, it has to be recognized that something within it has allowed and made possible this process of its expansion and reification.

The thought of Karl Marx has become Marxism, a theoretical and practical movement that not only comprehends the world but also takes hold of the world.[4] Marxism has shown itself to be a theory of history that works with resounding success; it is an instrument of sociological analysis possessing extreme effectiveness and a powerful weapon for political struggle. However, Marxism also has known, knows now, and will know difficulties within itself that will lead to its being superseded. In becoming Marxism— that is, systematized theory and practice and established method and doctrine—the original, creative thinking of Marx has externalized and alienated itself, has actualized itself in an objective and realist manner, and has given birth to something that has subsequently become autonomous. The Marxist development of Marxian thought fulfills an original intuition, realizes it, and makes it systematically and practically effective by extending and transforming it, that is, by extending one dimension of its intention and contracting what remains open and problematic, by transforming its aspiration into ideology and reality. Yet this kind of formative "deformation" was made possible by the initial formulation of Marxian thought. What became of it afterward only brought these initial elements to their fulfillment. That aspect of it that best lent itself to being taken up and pushed forward came to dominate. While not wishing either to detach Marx from Marxism or to fetter him tightly to it, we feel that to understand Marxism one has first of all to understand Marx. Let us learn to reestablish the link between Marx and Hegel and the whole philosophic tradition that preceded him. Let us see Marx's thought as one thought, without forgetting nonetheless that there are the young Marx and the other Marx. Let us think about the question of the tie that unites, and separates, the philosophic thinking of Marx and Marxism; for there is a unity there, and something in Marx has allowed Marxist development to take place. Let us try as well not to forget that a gulf, an abyss, separates creative geniuses from their disciples, more or less talented as they may be, and that one has to return to the source in order to better follow the stream. Philosophic thought constitutes the remote foundation for every scientific, technical, and practical theory that pretends to supersede philosophy, even while it follows from philosophy. One has, then, to try to understand by thought what philosophy becomes in giving birth to science and technique, in other words, when it "realizes" and "concretizes" itself, when it becomes practical. The philosophic basis for one body of thought gives rise as well to doctrines that in their development may turn quite autonomously away from it; and these doctrines generally intend to be a view of the Whole, a way of apprehending all that exists. Even when their individual truth takes in one aspect of the whole, they can hardly grasp the being itself of totality, the being-becoming of totality, because they throw only a partial light on one facet or another of the open whole which is the world. Expressing one reality found in Reality is never the same thing as really exhausting everything that is at work there. Neither the meaning of history nor the meaning of human life nor the course of the world can be exhausted. The problem implicated in all problems, the problem Marx wanted to solve, namely, man in his own history and at the heart of world history, remains open—let us call it a metaphysical and historical problem; and no one-sided approach can succeed in presenting it clearly, much less resolve it.

To understand the thought of Marx and make it transparent, to present it and make it a living presence, one has to find a guiding thread; and this is what we shall try to do, find *the* thread to follow. Our presentation ought not be attempted except on the basis of the letter and spirit of Marx's writings, the texts that make realities—men and things—speak. We shall try to see how the vision of man and of history and the program of salvation that Marx elaborates stand out against the background of broken Totality and dislocated Unity. We shall be taking all the texts into account without, however, putting them all on the same level. We shall pay special, though not exclusive, attention to the writings of Marx's youth (those that precede the *Communist Manifesto* of 1848) for the purpose of getting a better grasp of the origin, structure, and development of this thought; and we shall not forget for a moment that it is Marx himself who develops that thought along further lines, after which it takes on the form and content that Engels and the other Marxists give it. While not neglecting the total perspective or losing sight of the meaning of the ensemble of Marx's works, then, we shall be relying principally on (1) the Paris text entitled *The Economic and Philosophic Manuscripts of 1844*, the book of genesis of Marxian theory and its touchstone;[5] and (2) part 1 of *The German Ideology*. The many passages we shall quote will hopefully bring before us the authentic thought and language of Marx, the spirit and letter of his words.

Part I.

From Hegel to Marx

Marx the *thinker* is in a strict sense simply not understandable without the onto-theo-logy of Hegel.[1] It is a strange "dialog" that both joins them and sets them against each other. It is a dialog in which the dialectic—founded, thought, and then tortuously manipulated by absolute knowledge—is inverted and made to move on its feet, thereby somewhat losing its head. The dialectic seems actually to become world history, total praxis. Under the impulse given it by Marx, it undergoes a transmutation. History is no longer the place and time for absolute spirit to deploy itself, and it becomes the history of the development of technique. The World is no longer the world whose movement is the becoming of spirit; it is rather the world of human activity. Still, the problem of the "motor" and meaning of history remains an open one; for, whereas idealism fails in its effort to provide a basis for these things, antiidealism is deficient in not being bold enough to move its questioning beyond this obsessive phantom of an ultimate ground. Does the World, then, abide as a power *other* than man, one neither spiritual nor material, a question not yielding to a reductive answer, a horizon that will not let itself be completely conquered?

1.

FROM ABSOLUTE KNOWLEDGE

TO TOTAL PRAXIS

Marx's philosophic thought, aiming for the abolition of philosophy by its transformation into practical social reality, shapes itself within the horizon of the ontological metaphysics of Hegel.[1] In Hegel's hands, the whole tradition of Western philosophy from Heraclitus to Kant, the whole tradition of religion from the Jewish prophets to Protestantism, and the whole tradition of history and of art are brought back to life in thought; and Hegel does this with one central concern; to think the being-in-becoming of Totality. Philosophy for Hegel is reflective seizure in consciousness of the universal becoming that leads to divine, absolute *Spirit*. It is a movement that passes through the externalization of Spirit in *Nature* to the understanding of *History* in the concept; for it is in History thus comprehended that *Spirit* returns to itself and grasps itself in and by sensory subjectivity, psychological subjectivity, and finally rational subjectivity, in and by law, morality and ethics, art, religion, and philosophy. Philosophy tends toward the transmutation of itself into absolute knowledge of Totality by spirit; and Totality is truth. For Hegel, the truth of reality lies in the idea, so that the "rational" and the "real" coincide. He wants to translate the real into the form of thought, but the real (the true) is not even separable from thought; for the real is revealed in discourse and as such *is* thought. The determinations of consciousness, of (philosophical) science, and of knowledge are determinations of the essence of things as much as they are thoughts. The circular thought of Hegel is built upon the conception of identity as the identity of identity and nonidentity. Reality and idea, thought and thing are identical, given that identity implies negativity, mediation, and difference. Totality constitutes a reign of higher harmony and not that of nondistinction or of unification in nondifference. Thus, what is needed is a deeper understanding of these decisive words pronounced by Hegel in his *Encyclopedia of the Philosophic Sciences*: "... its [i.e., intelligence's] product (the thought) is the thing; it is a plain identity of subjective and objective. It knows that what is *thought*, *is*, and that what *is*, only *is* in so far as it is a thought."[2]

Philosophy, speculative thought, is inseparably logical-and-ontological; it is knowledge of being and knowledge of self, a knowing of world and a knowing of self. *Thought* is identical with *being*, *being* is *thought*. Thought is a force that says what is in becoming. The absolute-and-true is spirit but the absolute is in becoming, which is not change but the process of the revelation of the absolute (that is, of spirit). The truth of the totality is revealed in the unity of "being" and "nonbeing," which constitutes becoming; absolute spirit moves through a process of becoming toward absolute knowledge in (absolute) Subject. Hegel thinks in terms of "subject" rather than of "substance," although the "subject" he is speaking of is not easy to grasp.

Spirit manifests itself. It externalizes itself, alienates itself, and becomes reconciled and reintegrated with itself in the achievement of self-knowledge. Through many vicissitudes, including the overcoming of unhappy consciousness, man achieves reconciliation with universal historical destiny: history and man, mankind and the individual reach an accord which, though always in discord, transcends anarchic and tragic, romantic and senseless oppositions and conflicts. Self-consciousness and the World need no longer be separate, because they are one. The universal history of mankind, the history of the world, is the place wherein spirit is realized through time. Yet universal history is itself contained within the Totality of what is in becoming—which is the unity of being and nothingness; and this All is *the* truth which, implying its own negativity and temporality, endlessly leads the finite to the infinite (the absolute) by way of a perpetual process of supersession. Nothingness, negativity, and contradiction remain inherent in the being-in-becoming which characterizes both the totality of the world and every particular ontic reality.

Hegel's dialectical thought is both "conservative" and "revolutionary," if these terms can apply to it at all. Everything that is, is superseded, abolished, and overcome, while at the same time it is being maintained, preserved, and kept. The Hegelian *Aufhebung* indeed remains still enigmatic and resistant to any single interpretation. Marx is going to attack it from a certain angle, but will he really move beyond it, will he really supersede the Hegelian idea of supersession?

Those who came after Hegel would try to take issue with what he, Hegel, seemed to want to preserve and establish: *religion* and the *State*, on the one hand, *absolute knowledge* on the other. For Hegel, religion remained true; religion was spirit knowing itself and endowing knowledge with its own absolute power, thus raising itself by such knowledge to the pinnacle of rationality. God was revealed in and by history, and the State itself was a thing divine on earth. What was and would remain true in the total movement of Hegel's thought would soon be found isolated, becoming the object of attack by the tempest of criticism.

Hegel's thought is nothing less than polyvalent, plastic, and multidimensioned. Circular it is, without question, enclosing and imprisoning everything within a circle, and us as well; but it does not lend itself to being understood in schemata. We do not at all pretend to have drawn the main outlining features of that thought, or indicated its center. We have barely given a brief hint of the horizon against which Marx's critique will define itself. Hegel's philosophy seems not to allow capture in the seine of philosophic reflection; it seems not to permit philosophic comprehension or critique from outside itself, at the same time that it renders all interior "refutation" impossible. With his thinking, philosophy becomes problematic. For it, philosophy is an "inverted world"; for it, to philosophize is to "walk on one's head." But it is the world of naive consciousness that is in truth perverted, and naive consciousness, the good sense of the ordinary man, only imagines that it walks properly. What sense, then, can Marx's enterprise have, aiming as it does to invert this inverted world and set the Hegelian dialect back on its feet so that it may walk no more on its head? And what will the Marxian enterprise have in the end, given its advocacy of the passage beyond philosophy in and by its realization as praxis? Can total material praxis really replace absolute, real mind?

These questions are and will be all the more difficult to clarify (not resolve) given that Hegel himself seems to have wanted, by altogether different means and with a different aim, to take a path that would lead beyond philosophy. We have still not brought this purpose out clearly; indeed, perhaps we have not even begun to bring it to light. In the first pages of the preface to the *Phenomenology of Spirit*, Hegel writes, apropos of that love of wisdom which is philosophy: "The systematic development of truth in scientific form can alone be the true shape in which truth exists. To help to bring philosophy nearer to the form of science—that goal where it can lay aside the name of *love* of knowledge and be actual *knowledge*—that is what I have set before me."[3] This text can mean all kinds of things, but one thing it cannot mean is that philosophy, love of knowledge, must become science in the scientistic or positivist sense of the word. Knowledge (*Wissen*), the organic totality of systematic knowledge (*wissenschaftliches System*), and science (*Wissenschaft*) mean systematic and philosophical metaphysics, self-conscious knowledge, ontological knowledge. For Hegel, the highest form of science, true science, is philosophy; ontology, metaphysics, and the logic of spirit are identical. Hegel, accordingly, titles his ontological metaphysics the *Science of Logic*. The text just quoted says further that philosophy, in the Greek sense of the word φιλοσοφία, must give up its former essence and name to become real knowledge, that is, absolute knowledge of absolute idea which is the soul of Totality. Marx will take issue with this goal of absolute and total knowledge by a double opposition. On the one hand, against total absolute knowledge, he will put total productive praxis, and on the other, against absolute Subject, objective subjects, producers of material goods. What was a matter of absolute spirit and absolute idea will become for Marx material for labor and practical reality. Against the slow and painful elaboration of the Hegelian Idea working itself out through the phases of world history (the history of spirit), finally to reach reintegration and self-recognition in a knowledge superseding philosophy, there stands the Marxian vision of the history of the production of man and of the real world by way of concrete, determinate, material labor on the part of mankind aspiring to surmount its alienations.

We shall meet Hegel at all levels of the Marxian edifice, simply because he is behind the very act of its founding. We must try not to forget this. We must try as well not to forget that Hegel constitutes a definitive end, a grand culmination for that great stage of Western thought that begins with the Pre-Socratics, continues through Christianity, and finally, in establishing the *ego cogito*, the transcendental I, reaches consummation in absolute Subject, absolute Spirit. Hegel takes up the total thinking of total reality, enclosing all of it, both in its being and in its becoming, within a Circle from which one can escape only by breaking it. And that is the very thing Marx will do, if breaking a circle can be called escaping from it.

It may be that the world could not support the tension of thought at this high level or that this thought could not carry the world; in any case the philosophy of German Idealism collapsed, with Hegel's thought comprising the final act.[4] It was the hour for individual and piecemeal thinking to enter the scene. It was the hour for attempts to install men in a world that was no world.

Marx knew Hegel and yet misunderstood the essential dimension of his thought. He recognized his greatness, and yet Hegel is a figure both present and absent in Marx's work. Marx took up the sword against his genius adversary—who was already dead. He acknowledged in Hegel both the greatness of the global conception and the merit of many particular points. He regarded one aspect of the Hegelian dialectic with particular appreciation, namely, the recasting of stable, fixed ideas into dialectical form. In Marx's own words: "Hegel, for whom the modern world was also resolved into the world of abstract ideas, defines the task of the modern philosopher, in contrast to that of the ancient, as consisting of the following: instead of, like the Ancients, freeing himself from 'natural consciousness' and 'purging the individual of the immediate, sensuous method and making him into conceived and thinking substance' (into spirit), the modern philosopher should 'abolish firm, definite, fixed thoughts.' This, he adds, is accomplished by 'dialectics.' "[5] In fact, in the preface to the *Phenomenology*, Hegel calls for the surmounting of abstractness and the changing of static, determined, fixed thinking into fluid form in and by a dialectical movement within pure thought, within "this interior immediacy" that knows itself. Hegel recognized that it is difficult to make "fixed" thoughts "fluent." The dialectical movement that does this changes pure thoughts at the same time into *concepts*; they become in this way what they in truth are: "self-moving functions, circles . . . that of which their substance consists, spiritual entities." The determined, narrow self is likewise surmounted, consciousness and self-consciousness freed from the fixed self, and "pure certitude of oneself abstracts from self." "In virtue of this process pure thoughts become notions, concepts."[6]

In contrast, Marx wants to replace the force of thought and the movement of concepts by the forces of production and the real movement they possess. For the action of ideas he wants to substitute the activity of men: "*Ideas* can never lead beyond an old world system. Ideas cannot *carry anything out* at all. In order to carry out ideas, men are needed who dispose of a certain practical force."⁷ Perhaps we can begin here to glimpse how the dialog, or duel, between Hegel and Marx is hardly conducted on the same level and on the same terrain; Marx makes a reply to Hegel using weapons that do not correspond to his thought. Could it be otherwise? Was this not a situation in which some powerful force awaited the coming of its hour?

Men, inasmuch as they put into play a practical, real, material, and sensible power, make up the army Marx is undertaking to mobilize in this movement aiming for the revolutionary abolition of the world as it exists, the actual transcending of the present state of things. Against the effort to lay bare the logos of the phenomena of Spirit, which is what the *Phenomenology of Spirit* and the *Logic* are about, Marx sets up another effort and another labor.

> As Hegel here [in the *Phenomenology*] puts *self-consciousness* in the place of *man*, the *most varied* human reality appears only as a *definite* form, as a *determination of self-consciousness*. But a mere determination of self-consciousness is a *"pure category,"* a mere "thought" which I consequently also abolish in "pure" thought and overcome through pure thought. In Hegel's *Phenomenology* the *material, perceptible, objective* bases of the various ⟨alienated⟩ forms of human self-consciousness are *left as they are*. Thus the whole destructive work results in the *most conservative philosophy* because it thinks it has overcome the *objective world*, the sensuously real world, by merely transforming it into a "thing of thought," a mere *determination* or *self-consciousness*, and can therefore dissolve its opponent, which has become ethereal, in the *"ether of pure thought."*⁸

The *Phenomenology of Spirit*, the "science of the experience of consciousness," does conceive the phenomena of the appearance of spirit, revealed in and by history, as thoughts, ideas, concepts. The absolute does discover itself in this process, becoming as a result what it is in truth from the beginning, since "beginning" and "end" form one, the beginning tending toward its final accomplishment as toward its goal. In absolute knowledge, science, consciousness, and self-consciousness are reconciled, and absolute knowledge is the absolute knowing of absolute Spirit, of absolute Subject. The real world is not separable from the world of spirit, since the spirit of the world constitutes the essence and meaning of the history of the world; and this spirit, although alienated in becoming a stranger to itself, is reintegrated in self-recognition. Now, this central axis is what Marx will not leave in place; instead, he displaces thought as the axial principle in favor of the world, the world he calls real, objective, sensuous, and material.

The passage from Marx just quoted continues in that vein:

> Phenomenology is therefore quite logical when in the end it replaces human reality by *"Absolute Knowledge"*— *Knowledge*, because this is the only mode of existence of self-consciousness, because self-consciousness is considered as the only mode of existence of man; *absolute* knowledge for the very reason that self-consciousness knows *itself alone* and is no more disturbed by any objective world. Hegel makes man *the man of self-consciousness* instead of making self-consciousness the *self-consciousness of man*, of real man, man living in a real objective world and determined by that world. He stands the world *on its head* and can therefore dissolve *in the head* all the limitations which naturally remain in existence for *evil sensuousness*, for *real* man. Besides, everything which *betrays the limitations of general self-consciousness*—all sensuousness, reality, individuality of men and of their world— necessarily rates for him as a limit. The whole of *Phenomenology* is intended to prove that *self-consciousness* is the *only reality* and *all reality*.[9]

Is Marx getting ready to bury once and for all, and to pronounce his funeral oration over, a great epoch of the history of subjectivity now at its end with Hegel? Is he rejecting at the same time the dialectical form and the ontological content in the thought of Hegel, according to which *form* and *content* are identical? Hegel seems to have identified the totality of being with the totality of its revelation and the totality of our approaches to it. Marx's work will consist in dislodging this total totality by denouncing the alienations of both foundation and detail that lie in it and alienate us. In Hegel's thinking, spirit is alienated in nature, nature being the place for the externalization of idea and remaining always alienation. Nature has no history; it neither thinks nor knows itself. But nature alienates itself as well to the benefit of spirit which, in the movement of becoming history, "restores *subject*," reestablishes itself, finds itself again, internalizes itself, regains itself. Spirit finds itself after having been lost, and the process of its revelation, a process of alienation and recovery, consists in bringing mind to "*know* integrally what it *is*," to recognize the meaning of all that it is (which is identical with all that is). The result of this development is the reflective grasp of its preconditions, and what is in the beginning is what comes about in the course of the circular movement of self-knowledge. Marx simplifies to excess the dialectical movement of Hegel's thought when he reproaches him for treating truth as "an automaton that proves itself. Man must *follow* it. . . . the result of real development is nothing but the *truth proven*, i.e., brought to consciousness."[10]

However, what was gained through a conceptual and speculative effort on Hegel's part is in Marx's hands no longer in the speculative order. Thus, when Hegel said that truth lies in the adequation of knowledge with its object, he understood it speculatively and ontologically: "the true is the whole"; the whole is in becoming, on the way to what it is in truth, "the absolute alone being true, or the true alone being absolute"; truth is indeed "the movement of itself in itself"; it is true that "the spiritual alone is *actually real*"; "spirit is time"; the true and the nontrue are different even while being one. But these are all speculative propositions that cannot be understood except through the whole work of the dialectic—supposing still that they can be understood. In the last pages of the *Phenomenology*, correlative to the first, Hegel says: "While in the *Phenomenology of Mind* each moment is the distinction of knowledge and truth, and is the process by which that distinction is cancelled and transcended, Absolute Knowledge does not contain this distinction and supersession of distinction."[11] Obviously a short-sighted criticism, using the simple logic of common sense, could hardly refute Hegelian thought.

Marx has no intention of employing the weapons of speculative thought in the struggle against Hegel. He has no wish to engage in the exercise of speculative criticism, much as he might do so. His weakness here is his strength. He speaks another language and makes alienations of another kind speak out. In Hegel, the *externalization* and *alienation* (*Entäusserung, Entfremdung*) of spirit, of consciousness, of self, of concept, of idea, and of thought are necessary moments in the process of the unveiling of spirit in and by world history. They are nonetheless transcended in time through the process of becoming that moves toward the final result; absolute idea, knowledge that knows itself. The Hegelian conception of alienation and reconciliation, of externalization and reinternalization, is especially difficult to grasp. The last chapter of the *Phenomenology of Spirit*, "on *absolute* spirit," closes with this passage:

> Since its accomplishment consists in Spirit knowing what *it is*, in fully *comprehending* its substance, this knowledge means its *concentrating itself on itself (Insichgehen)*, a state in which Spirit leaves its external existence [*Dasein*] behind and gives its embodiment over to Recollection (*Erinnerung*). In thus concentrating itself on itself, Spirit is engulfed in the night of its own self-consciousness; its vanished existence is, however, conserved therein; and this superseded existence—the previous state, but born anew from the womb of knowledge—is the new stage of existence, a new world, and a new embodiment or mode of Spirit. Here it has to begin all over again at its immediacy, as freshly as before, and thence rise once more to the measure of its stature, as if, for it, all that preceded were lost, and as if it had learned nothing from the experience of the minds that preceded. But *re-collection* (*Er-innerung*) has conserved that experience, and is the inner being, and, in fact, the higher form of the substance. While, then, this phase of Spirit begins all over again its formative development, apparently starting solely from itself, yet at the same time it commences at a higher level. The realm of spirits developed in this way, and assuming definite shape in existence, constitutes a succession, where one detaches and sets loose the other, and each takes over from its predecessor the empire of the spiritual world.[12]

This text, which hardly lends itself to being readily understood in translation, goes on to culminate in an announcement of the *end*— though not the ending—of the succession of forms of mind that rule the world: "The goal of the process is the revelation of depth, and this is the *Absolute Notion*. This revelation consequently means superseding its 'depth,' is its *'extension'* or spatial embodiment, the negation of this inwardly self-centred (*insichseiend*) ego —a negativity which is its self-relinquishment, its externalization, or its substance: and this revelation is also its *temporal* embodiment, in that this externalization in its very nature relinquishes (externalizes) itself, and so exists at once in its spatial 'extension' as well as in its 'depth' of the self." Hegel reaches the final goal of infinite spirit in history (which is also "infinite") and pronounces the last words of the *Phenomenology*:

> *The goal*, which is Absolute Knowledge or Spirit knowing itself as Spirit, finds its pathway in the recollection of spiritual forms (*Geister*) as they are in themselves and as they accomplish the organization of their kingdom. Their conservation, looked at from the side of their free existence appearing in the form of contingency, is History; looked at from the side of their intellectually comprehended organization, it is the *Science of the ways in which knowledge appears*. Both together, or History (intellectually) comprehended (*begriffen*), form at once the recollection and the Golgotha of Absolute Spirit; the reality, the truth, the certainty of its throne, without which it were lifeless, solitary, and alone. Only
>> The chalice of this realm of spirits
>> Foams forth to God His own Infinitude.[13]

After this grandiose vision, based on an awesome conceptual and speculative effort, what was there left to do? No one can say whether this vision is turned solely toward the past. Marx has of course accused it of losing sight of real history, whose movement of becoming includes what is still to come. He takes his stand on the plane of *doing*, conceiving the totality of what *is* as a totality of what is *done*—which means he sees history with a different pair of eyes. "History is nothing but the succession of the separate generations, each of which exploits the materials, the capital funds, the productive forces handed down to it by all preceding generations, and thus, on the one hand, continues the traditional activity in completely changed circumstances and, on the other, modifies the old circumstances with a completely changed activity."[14] The kingdom of spirit is succeeded by the reign of practical labor.

Is time for Marx actually something other than this? He lives in and focuses upon a time different from that of Hegel. He means to open the way to future time, to what is to come. Hegel, even while identifying spirit and time, annihilates time for the sake of spirit. Spirit, alienating itself in space and time, that is, in nature and history, eliminates time by reintegrating itself: ". . . spirit necessarily appears in time, and it appears in time so long as it does not grasp its pure notion, i.e., so long as it does not annul time."[15] Marx hardly finds this elimination, this cessation, acceptable. The necessary becoming of real history, the ceaseless development of the forces and faculties of production, and the process by which nature becomes history owing to the practical labor of men do not admit of annihilation.

It was not only against the "spiritualism" and objective, absolute idealism of absolute Subject that Marx had to struggle: included as well were "materialism" and "positivism." We are not going to indicate here, even in broad lines, the history and fate of Hegelianism, especially the Left Hegelianism to which Marx belongs, though in opposition both to it and to Hegel himself. As we go along we shall be meeting all those figures that make their entrance only after the curtain has already fallen. For the time being let us simply say that the Left Hegelians—Ludwig Feuerbach, David Friedrich Strauss, Bruno Bauer, Max Stirner—engage in a movement of radical, positivist criticism of religion in an effort to naturalize and humanize the manner in which problems are posed. For them the only thing to be seen is the development of human nature on earth, but their conception of the realities involved is narrow and reductive, naïve and individualistic. It is a movement that feeds on the remnants of Hegel's thoughts, its representatives trying, like dwarfs, to mount the shoulders of the giant, imagining they will thus see farther. They manifest the very epitome of reflection that is one-sidedly, confusedly obsessed by a fixed theme, blind to anything that lies outside their focus. When Marx mentions them, he has only irony and ridicule for their pretense of having surpassed Hegel. "Their polemics against Hegel and against one another are confined to this—each extracts one side of the Hegelian system and turns this against the whole system as well as against the sides extracted by the others. To begin with they extracted pure unfalsified Hegelian categories such as 'substance' and 'self-consciousness,' later they desecrated these categories with more secular names such as 'species,' 'the Unique,' 'Man,' etc."[16] Marx's reproach to these secular critics is that they "have not left the terrain of philosophy," that is, of (speculative) abstraction. As a result, they cannot move beyond Hegel, bound as they are to him by a wrong kind of dependency.

Marx wants to unmask "these sheep who take themselves for wolves and are so taken by others" and take advantage of the "decomposition of the Hegelian philosophy," of the "putrescence of absolute spirit." Hegel had already pushed idealist and ideological understanding to its ultimate consequences, to its "purest expression." His was an order not of "real interests" but of "pure thoughts." He brought the work of speculative philosophy to perfection. "Hegel himself confesses, at the end of the *Geschichtsphilosophie*, that he 'considered the progress of the *concept* only' and has represented in history the 'true *theodicy.*' "[17] Ideologists who criticize him without leaving the terrain of abstract ideas secularize everything that was still deemed sacred, treating the world of religious and theological representations as an alienated world of the self-consciousness of individual man. They give themselves over to a rabid theoretical criticism all in words, their sole concern being individual man as a natural, material being. Anything that was assigned to God, having been stripped from man, is to be reintegrated by man. It is demystified self-consciousness, subjectivist, egoistic self-consciousness, upon which the task of the criticism of ideas by ideas lies. It is a criticism that therefore remains ideological and "theological" and "is seen in the final analysis to be nothing but the culmination and consequence of the old *philosophical*, and especially the *Hegelian, transcendentalism*, twisted into a *theological caricature*. This [is an] interesting example of historical justice, which now assigns to theology, ever philosophy's spot of infection, the further role of portraying in itself the negative dissolution of philosophy, i.e., the process of its decay ..."[18] A "historical Nemesis" brings it about that philosophy dies through its own capital sin. But the true death of philosophy can in no way be a matter of "negative decomposition." What has to be done is to abolish philosophy positively in order to transcend it in realizing it. This will be Marx's task.

In his critique of criticism Marx certainly treats Ludwig Feuerbach as a special case. He credits him with having inaugurated "*positive*, humanistic and naturalistic criticism."[19] He even considers Feuerbach to be the founder of this enterprise. Feuerbach's works conducted an *action*: "Feuerbach's writings [are] the only ones since Hegel's *Phenomenologie* and *Logik* to contain a real theoretical revolution."[20] "Feuerbach is ... in fact the true conqueror of the old philosophy."[21] His was a "great achievement" in three ways:

1. in having furnished "the proof that philosophy is nothing but religion rendered into thought and expounded by thought, and hence equally to be condemned as another form and manner of existence of the alienation of the essence of man";

2. in having established "*true materialism*" and "*real science*" by making "the social relationship of 'man to man' the fundamental principle of the theory";

3. in having stood against "the negation of the negation, which

claims to be the absolute positive, the self-supporting positive, positively based upon itself."[22]

There will be occasion later for returning to these elements that stand to Feuerbach's credit and also to the matter of the positive as founded positively upon itself. We shall have occasion as well, on the other side, to see in more detail what Marx reproaches Feuerbach with; for, when all is said and done, Marx's recognition of Feuerbach's merits is followed by his condemnation of him, as the *Theses on Feuerbach* and a number of other texts show. Feuerbach may well identify truth with perceptual reality (or materiality). He nonetheless misunderstands the primordial importance of real, sensuous action. Substituting love of man for love of God and faith in man for faith in God still does not get him beyond the narrow concept of the individual. He is naturalistic and humanistic, but his naturalism and humanism do not begin with, or lead to, the social reality of man, his communal essence. Anticipating Marx's criticism, let us look for a moment at the sixth and seventh theses on Feuerbach: "Feuerbach resolves the religious essence into the *human* essence. But the human essence is no abstraction inherent in each single individual. In its reality it is the ensemble of the social relations. Feuerbach, who does not enter upon a criticism of this real essence, is consequently compelled: 1. To abstract from the historical process and to fix the religious sentiment as something by itself, and to presuppose an abstract-*isolated*-human individual. 2. Essence, therefore, can be comprehended only as 'genus,' as an internal, dumb generality which *naturally* unites the many individuals."[23] It follows that Feuerbach is unable to see that "'the religious sentiment' is itself a social product, and that the abstract individual whom he analyses belongs in reality to a particular form of society."[24]

The significance and the import of what Marx is doing are quite different. Marx sets in motion an immense power of negativity, namely, negativity as it works in historical reality, and at the same time finds the right words for it. But, instead of leading to a synthesis, a negation of negation, in the present moment of time, this negativity results in a dramatic crisis, the historical crisis of the present in which man is found alienated from his true nature, from the products of his own labor, and from world history. And it is not man as the individual who is most of all alienated: it is men, all men. What is alienated is the *humanity* of Mankind. The Marxian view of alienation, as something soon to be transcended according to the optimistic prospect that Marx is opening up here, forms the horizon of all his philosophic and historical, anthropological and sociological thinking. Marx sets out to move beyond Hegel. Yet, having quit the terrain on which Hegel places himself, he meets his enemy only on the field of battle that he has chosen; because that is the one he is obliged to take with regard to Hegel.

The harmonious unity, the supreme synthesis with which Hegel ends up, that he finds at the end because it was already at work in the beginning, is in Marx's eyes a collection of dislocations whose gaping discontinuities are obvious. The restoration of subject, the reconciliation of man with fate and world history, the return of spirit to itself—in a word, the superior *synthesis* that Hegel had painfully elaborated—becomes for Marx a *thesis* to which he opposes the *antithesis*. To the Hegelian position a negation is set up that leads to a new negation of negation, to another and new synthesis. Hegel came to terms with the tragic without annihilating it. The tragic remained intrinsic to the being-becoming of the totality of the world, and Hegel knew that no historical human realization could abolish it. The true truth, total reality, the open totality of all that is in the past-present-future transcends any positive particular realization, however grand it may be. Ontic realities may express being-in-becoming, but they never exhaust it. Marx, for his part, begins—quite "concretely"—from a critical analysis of the present historical world, his objective being to unmask the true and real nature of alienation. Nonetheless he is not satisfied simply with tragic observations. His thought quickly moves on to unfold the prospect of salvation through the total reconciliation that in a future time will bring naturalism, humanism, and communism to full reality. Tragedy is in this way transcended by historical, social, practical, and material action on the part of concrete men. Drama will turn into dealienated action, bringing the (sensuous) meaning of real reality to realization on the level of universal history.

Marx links back up to Hegel and "extends" him in dialectically developing one moment of his, Hegel's, thought. He conducts the attack in one region of total reality, seizing upon one of its aspects, taking this region and this aspect to be the source for all essential reality and the principal figure of all that is. In Marx's eyes it is the historical, social world that, moved by the forces of production that set in motion the very agents of the movement of productive forces, constitutes the space and time of the drama. In his eyes the world is the totality of what is, as that totality is revealed by and in the productive activity of men. The (whole) World is indeed envisioned as a whole that nevertheless comprises *two* aspects: the essentially real *material* aspect and the secondary and derived aspect, the *spiritual* aspect. Alienation as well has a *double* nature. On the one hand, it is fundamentally *real*, *actual*, and *material* and, on the other, it is *ideological*, *superstructural*, and *epiphenomenal*. Hegel was the kind of metaphysician for whom *one single* totality organically encompassed all its dimensions and all its regions; thought, nature, and history obeyed the same rhythm; all duality was absorbed back into unity. Marx is chiefly the kind of thinker who seizes upon some one particular historical development, thereby giving one domain and one axis of investigation a privileged position with regard to the others, to their detriment. With Marx *metaphysics* turns into *social physics*, and philosophy is to be surmounted for the sake of its realization in an achieved naturalism-humanism, that is, communism. With Marx, negativity and alienation, which are both essential moments in the advance of Hegelian Spirit, become materially historical and social. The world of history is given one particular conceptualization, even though it is to be considered the whole world. One part of Hegel's system of speculative thought is pushed thus to its ultimate revolutionary, practical consequences. Hegel's absolute totality turns into something different: it is dislodged in order to be replaced, in the transcending of alienation, by a concrete totality, by *the* concrete, real totality which is the total expansion of human productivity.

Marx tries to develop a vision that is both global and concrete; but, though he wishes to be comprehensive enough to deal with the rhythm and meaning of the movement of mankind's history on the universal level, his attention focuses quite specially on bourgeois, capitalist, Western European society. He is only to a lesser extent interested in knowing if this vision really embraces all past history, for before all else he wants to take hold of the present and what in it is preparing the future for the world and history. And is it not the thrust of developments within bourgeois, capitalist, Western European society that is moving toward planetary expansion?

Whatever the violence in working out his plan, Marx's project has one goal: to find practical, concrete solutions to theoretical, abstract problems, to reach a real, effective solution to the only problems worthy of being solved and able to be solved, namely, problems that are real for a "real" sense-capability. There was an immense emptiness after Hegel completed his monumental enterprise of unifying logic, history, and encyclopedic knowledge within the Circle of totality by way of the indissoluble ties of sheer identity. This emptiness ruled simultaneously in the worlds of "thought" and of "reality," in that any attempt either to join or separate these two worlds remained empty of meaning. Marx set out to fill this emptiness. Animated by a strongly rationalistic Jewish prophetism, he unmasks the error and lies, the misery and alienation in the existing state of things, to prepare the way to a better future, one in which earthly happiness would be realized. In violent denunciation of the mystifications and illusions of man's consciousness, he takes up the battle against mysticism of any kind in the name of a real knowledge that would serve the praxis of production. Marx's vision of history, destined to have world-wide reverberations, is centered in what he calls the reality of human being, that is, the conception of human being as a being moved by natural drives and life needs and desires, tending toward the complete satisfaction of these drives, needs, and desires, and cruelly suffering because they are not satisfied, because they are alienated. This is how man's true nature is taken: it possesses an essence jointly and inseparably natural, human, and social, and yet it has always been alienated in history. Marx's thought, in its several guises (*naturalism, humanism, socialism*), is at once anthropological and historical. Its roots and development involve an analytic and reductive radicalism that takes everything back to data that are offered as positive, in order to make a radical denunciation of the given in the name of a positivity to be wholly realized in the future. *Man in history* is no longer deemed a metaphysical problem or metaphysically historical. He is rather a reality alienated from his true nature, which must be restored through the positive abolishing of alienation; this will be carried out in a practical way by the dismissal of anything that is an obstacle in the path, whether on earth or in "heaven."

Being sensitive only to what is offered in sensory perception, Marx, following his own rhythm of progress, takes a rather long route in exploring the vast panorama of a particular world and searching the hearts of its inhabitants. He is not at all interested in what is visible only to the eyes of the soul and spirit, or in the roads and paths that do not lead to a definite place, the vistas that do not have defined contours, in the secrets it is practically impossible to penetrate.

The positions Marx wins he takes by force. His separation from Hegel was a progressive one, and his early writings show the evolution of a young thinker working toward his own theoretical elaboration. These early writings, Marx's work before the *Communist Manifesto* (1848), contain the genesis of his thought. Of these texts, the Paris writings, known as the *Economic and Philosophic Manuscripts of 1844*, hold a place of absolute centrality and quite special importance; for here is found the comprehensive thinking by which the young Marx gained his own position against that of Hegel. In addition, the *1844 Manuscripts* have been and remain the richest in ideas of all Marxian and Marxist writings. Together with the *Contribution to the Critique of Hegel's Philosophy of Right: Introduction* (1844), the short "Theses on Feuerbach" (1845), and *The German Ideology* (1845), but surpassing them in import, the *1844 Manuscripts* show Marx's own thought in its first unfolding. At once philosophy, history, and anthropology, they show thought at work trying to grasp the drama of man, which is the tragedy of alienation, within the context of the horizon of history and society. Fundamental alienation, the alienation of labor in a system of private property, the alienation of man's social and political life, the alienation of man's very existence, and the alienation of religion and philosophy are all laid bare and brought back to the source of all alienations, the alienation of the production of life, economic alienation. Denunciation of these alienations is made precisely with a view toward their being abolished in a process by which, in a new world, socialism-communism would be brought to reality as achieved naturalism-humanism. A systematic economic theory, however, has not yet, on this level, carried the final victory. It is in Marx's mature writings, those beginning with the *Communist Manifesto*, that we find elaborated his theory of the economic and political evolution of history with its central axis in the development of the forces of production. It is there that we see the program of proletarian revolution spelled out and the method and theory of historical materialism built up as a systematic whole.

Marx's thought is a unitary thing. There are not two Marxes, the early and the late, the youthful and the mature. Thought is the sort of thing that undergoes development as a unitary process. Nevertheless, one must not forget that Marx's thought, even if unitary, shows two periods integrally bound together. Marx takes his beginning from the philosophy of Hegel, chiefly the *Phenomenology of Spirit*. This brings him to confrontations first with the Left Hegelians, then with the set of problems posed by the theories of political economy and of a so-called utopian socialism; and through these circumstances he hammers out the ideas by which he would offer a new and wholly practical solution for philosophical, political, human, and economic problems. It is in the course of this militant campaign that he constructs his scientific, economic, and historical systematics and composes the technical and political program of the communist revolution. The Paris *Manuscripts of 1844* contain the central point of Marx's thought, its philosophic core and the seed of a further scientific and technical elaboration; and it is this center with its blazing ramifications that becomes thereafter consolidated doctrine.

The early Marx searched into the meaning of human history and human life, denouncing the alienation of sensuous activity that lay in all the forms through which human life was externalized. Meaning (*sens*) was thus reduced to sensory meaning (*sens sensible*), and anything that disengaged itself from sensory meaning was nonsense. Marx then tried to settle the meaning of history and the evolution of society by determining the direction of the historical process on the basis of technical development. Accordingly, he rejected as (reactionary) absurdity whatever did not seem to follow the direction of the revolutionary action of the proletariat. Economic analysis held the "secret" of the total situation; grasp of economic dynamism yielded the "secret" of becoming.

Marx had a close interest in the history of mankind in its entirety, in that, for the first time in the history of the world, history was coming to be universal. He took up directly the issue of the destiny of History, interpreting that destiny in economic terms. For what constitutes the supreme power of world history, what "hovers over the earth like the fate of the ancients, and with an invisible hand allots fortune and misfortune to men, sets up empires and overthrows empires, causes nations to rise and to disappear,"[25] are the laws of world trade, the laws governing production and exchange, supply and demand, the laws of manufacture and commerce. Marx takes direct issue with this fate in order that mankind may free itself from it, in order that the historical and world activity of men may no longer be subjected to the alienating power of some world economy, in order that history may become truly universal and truly the work of men. No longer would there hover over them a fate alien to their manifold technical activity. We shall see later, however, that both revolutionary movement, in wholesale transformation of the world as its exists, and the organization of communist society are determined chiefly by the economic factor. And the question will be, Is this not the same power under another guise? Is this not the "same" fate that under new forms continues to hover over the whole earth, which is now caught in the net of economic communist organization on a worldwide scale? Will the elimination of the economic laws of capitalism, the transcending of primordial alienation and private property, the abolition of the capitalist world market open the horizon of a totally and actually different kind of world? Will these actions bring about for the first time an open world-form (*mondialité ouverte*)? Or will the new mankind have a computer heart that beats to the rhythm of the production of goods in endless variety, its pulse marking the productivity and planetary expansion of technique in its making everything over into the technical? Only a closely followed dialog with the thought of Marx can lead us step by step toward the answers to these questions.

Be that as it may, we must not lose sight of the fact that Marx, especially in his early period, knew how to discern certain problems in regard to the being of man in his gangrene-ridden humanity. As one who would propagandize for universal collectivization and radical socialization, he dwelt at times on the drama of man, on the failure of individuals in life and of the individual in the midst of society. Human existence itself, and not simply the social structures and stifling superstructures, was seen as an open sore. But, far from delaying before this vision, Marx hurried on to forge the weapon and instrument that would allow him to strike down the cause of suffering and, by violent intervention, bring root-deep healing to what was so ailing. One should not forget, moreover, that Marx's aim was born in the soil of Western European bourgeois, capitalist society and that he borrowed the weapon of critical analysis from that society and its theories in order to set off to war against it all. The world in which he lived was alien to him. Marx therefore denied the very reason for that world's existence, unmasked its myths and illusions, and rejected as lies the forms of self-consciousness of man now become stranger to himself. His purpose was to see that the rebirth of real truth, revolutionary reflective awareness, should guide men toward reconciliation with themselves and the world. The recognition of true reality in a new, demystified consciousness could be effective, however, only when joined to revolutionary action, sometimes to *serve* it, sometimes to *guide* it. The goal envisioned, once realized, would transform men's lives: every alienating distinction between private life and public life would be abolished, and universal history would become the history of the human deployment of universal, manifold technical activity. The individual will have thus healed his sickness, and society will no longer be based on the exploitation of man by man. Individual and Society will have ended their mutual opposition, and so also will mind and matter, subject and object, nature and history.

Nevertheless, in Marx's view of things, this reunification can be realized only through a social and socialistic movement that activates real, material forces. The last word, therefore, belongs to the social, the material, the real, the practical, the objective. But there are questions: Is dualism now radically overcome? Will thought and consciousness, theory and scientific knowing remain *different* and not merge completely into the movement of true reality? Will they be sacrificed to it or will they continue to hover over it? What will come of the claimed objective of a global unity encompassing simultaneously the truth of things alienation had separated? We shall for the moment keep these questions open, even if we are not at all sure we shall be able to answer them later.

Let us simply recall what Hegel wrote: "Mind [spirit] is all the greater the greater the opposition out of which it returns to itself."[26] But, when he said that, Hegel had in mind his own conceptual effort to get beyond representation in attaining the truth of the unity of thought present in things and of things present in thought: "Reason is to itself *all thingness*, even thingness that is *purely objective*; but reason is this *in the concept*, or the concept is only its truth."[27] But the thinking that comes after Hegel is incapable of maintaining the tension of this unity, a unity made by spirit, and, so, can it advocate anything other than the reverse of that unity?

It is not only the world Hegel lived in but also the Hegelian world (not to say simply the world) that becomes alien to Marx. For him the world has become uninhabitable, inasmuch as alienation has made all men uprooted beings. Marx's thought is inscribed in the world that has ceased to be a homeland or to contain homelands for modern man; he is without being, having no place whatsoever. Nonetheless, it is in this world that Marx speaks to give voice to a human, historical situation that is truly unbearable. He throws a vivid light on this age of the alienation of the subject to show men a path to exit. But is it possible for him not to be affected by the world that he rejects? Can he really advocate a world utterly different from the unworld he condemns so irrevocably? In the preface to the first German edition of *Capital*, after remarking that "the country that is more developed industrially only shows, to the less developed, the image of its own future," Marx has this to say a few lines later: "Alongside of modern evils, a whole series of inherited evils oppress us, arising from the passive survival of antiquated modes of production, with their inevitable train of social and political anachronisms. We suffer not only from the living, but from the dead. *Le mort saisit le vif!*"[28] The dead lay hold of the living! Marx is certain that the dead are not going to seize the living any longer in the world that will have abolished what is superseded. But is there a solid enough basis for his certainty and his hope?

The building of Marxian doctrine is a methodical development around the theme of man as endowed with a will capable of taking hold of the world by virtue of technique. The will, as radical demystifier charged with bringing about the reign of naturalism-humanism, has one aspiration, namely, to abolish the alienations that stand in the way of its action. But, after the transcending of whatever is an obstacle to its total activity, to its will for total activity, it recognizes no further limit external to itself. Consciousness, of course, is also *mobilized* in this work. But, once all its illusions are dissipated, consciousness, as serving action, thereby takes its place in the ranks of technical instruments. Yet, consciousness seems almost fated always to move beyond real action. All along our way, we keep coming upon this consciousness that, while reduced to being only an instrument, *at the same time* is regarded as going beyond the works performed by instruments. Consciousness will no longer be, of course, a celestial light or heavenly cloud. It will come down to earth, for Marx believes he has brought about the permanent collapse of a heaven now empty, its gods gone; and he can hardly allow that heaven continue to overshadow earth. Marx is determined that his anthropology and philosophy of history, as well as his program of salvation and his, shall we say, eschatological vision, be altogether real and deeply, radically immanent. The human collectivity, communist society—the generalized power of the human *ego*—becomes here the founder of whatever is and master of the planet; and this society achieves mastery of the totality by practical means, conscious of what it *does* and not floundering about directionless and astray (*dans l'errance*).

Absolute knowledge, speculative thought, and the (metaphysical) logos of the being-becoming of the totality of the world must be one and all superseded in and by total praxis, multilateral activity, real action. But total praxis, even in the form of an offered vision, remains nevertheless quite problematic. Will it be an activity that is only practical? Will it leave no place at all for any theoretical thought? Does it encompass thought *and* action, that nonetheless remain separated from each other? Or will it, precisely as praxis that is material, real, sensible, actual, effective, and objective, dominate and determine all thought, consciousness, and knowing? Will its totality contain a duality (between action and thought, between sensory and conceptual meaning) or will it establish unity? Total praxis involves what Totality? We shall meet these questions often on our itinerary. Let us say at this point, though, that Marx, setting out to war against Hegel's absolute knowledge, does not mean to end up establishing activity "in its dirty-judaical manifestation,"[29] as he puts it in the first thesis on Feuerbach. In his eyes, dealienated praxis is a far more comprehensive phenomenon, although its totality remains limited and one-sided, stained with objectivism and pragmatism.

Practical action seems to be the last word for Marx. There are *also*, however, the understanding and consciousness of this active praxis. All the mysteries and riddles, all the problems and questions that theoretical thought thinks it resolves in thought by thinking become questions pertaining to the practice that gives them a solution. Yet a question remains: Is the understanding of practice itself of the order of practice? "All social life is essentially *practical*. All mysteries which lead theory to mysticism find their rational solution in human practise *and*[30] in the comprehension of that practise,"[31] writes Marx in the eighth thesis on Feuerbach. Does the difference, then, persist? Do being *and* thought, action *and* understanding, theory *and* practice, sensory activity *and* its meaning *not* unite to form a single thing? Does not the fate of Western thought, which since Parmenides has been seeking the unity—even the identity—of being and thought, thought thinking being as thought, weigh on Marx also? For the whole of Western thought, since the end of the pre-Socratic dawn, is caught in the meshing of elements in which being and thought unite and separate, in which neither identity nor difference reaches full clarity in the light of a single, truly total foundation. Does the essentially practical praxis of Marx merely invert the order of the two orders, making real activity the judge of thought?

Part II.

Economic and Social Alienation

As Marx grasped what alienation meant in its concrete and cruel reality within human history, he moved progressively farther away from Hegel to nurture his own vision. Wishing to understand where universal history was going and where the historic destiny of mankind lay, Marx began with an analysis of present European society; for there he would meet the bitter reality of an alienation visible on every level of human life, beginning with its material basis in economic alienation, the alienation of labor.

2.

LABOR, THE DIVISION OF

LABOR, AND THE WORKERS

Man works to live. Unlike animals, he *works* to wrest the goods from Nature that will permit him to satisfy his natural needs. It is not something he does all alone but with other men, in a social frame. The essence of labor is social. It is the community, human society, always in some historic form, that struggles against nature to gain subsistence. Real, sensuous, material, and practical activity working some transformation is the very first given, behind which we cannot go. The essence of man is not an abstraction based on isolated individuals. It is constituted as human existence in and by the ensemble of social relations founded on labor and passing through historical development. The ultimate foundation of Marx's thought as the effort made to grasp all reality in social and historical terms lies in his belief that the root of human history can be found nowhere else than in the activity of man as immanent to his own history. Here is where we see Marx's humanistic radicalism in its foundation and origin. The history of human societies begins with the production of the means that permit men to satisfy their material needs. It is put this way in *The German Ideology*: "[Men] themselves begin to distinguish themselves from animals as soon as they begin to *produce* their means of subsistence, a step which is conditioned by their physical organization. By producing their means of subsistence men are indirectly producing their actual material life. The way in which men produce their means of subsistence depends first of all on the nature of the actual means of subsistence they find in existence already and have to reproduce."[1]

This production of material goods by labor not only allows men to *live*, it also manifests a specific *way of living*; for individuals show their existence in working and are determined by what they produce and the way they produce it. The satisfaction of needs, which is the aim of work, never comes to an end. "Primary" needs, once satisified, engender (produce) new needs that in turn demand satisfaction, and so on. Natural needs and human, social labor develop dialectically and progressively, and their ultimate limit is beyond seeing. It is not possible to find either the absolute beginning or the end of human history. Men are what we find at work, natural beings, who, as social, set themselves against Nature by force of the demands natural needs make upon them. Men are, therefore, "biological" beings led by their vital instincts (even though they become different from their ancestors by virtue of labor), and as men they are from the very outset social beings. The first things to note are the way the human body is organized (for it is the basecondition of man's productive activity) and man's resulting relationship to the rest of nature.

Nature, Humanity, and Society compose the historical being of man. The humanistic radicalism of Marx that goes on to become socialism-communism is based on a kind of naturalism: it is the organic, vital nature of man that drives him toward the satisfaction of essentially natural needs. This *naturalism* is nonetheless "*antinaturalistic*," since man sets himself against nature, fighting with it by labor, in order to realize his human fate. We find, then, a naturalism at the foundation of Marxian thought; but this same naturalism mutates into an antinaturalism on the question of the natural being of man; for it makes man's essence a productive, social phenomenon that in a radical exigency comes to manifest itself through its history. No transcendence and no *Logos* dwell originally in the Nature of man.

The question of the absolute beginning of human history remains unanswered. Marx thinks it is a question without any meaning, because it cannot be solved on the plane of sense experience. History has, "therefore," no absolute beginning. The problem arises only because we, products of history at some determinate stage in historical movement, pose it. What escapes logical grasp is declared to be nonexistent on the ontic level. In the *1844 Manuscripts*, Marx puts it this way: "Both the material of labor and man as the subject are the point of departure as well as the result of the movement.... Thus the *social* character is the general character of the whole movement: just as society itself produced *man as man*, so is society *produced* by him."[2] Still, Marx does not resolve the problem of the origin of historical mankind that easily;

and he says further: "Neither nature objectively nor nature subjectively is directly given in a form adequate to the *human* being. And as everything natural must have its *beginning,* man too has his act of origin—*history*—which, however, is for him a known history and hence as an act of origin it is a conscious self-transcending act of origin. History is the true natural history of man."³ The whole idea of an origin in creation is rejected by Marx; for to him the notion of creation is an extrapolation and projection back into the past of the obvious facts of the practical life of production. "Spontaneous generation" (*generatio aequivoca*) is all there is, both for Nature and for man.

We are coming here to see, perhaps, that Marx does not begin, as Hegel does, from the being-becoming of totality, that which, having become Nature, manifests, reveals, and grasps itself as Spirit in History. Refusing any foundation, he begins with the natural history of man, the "first" origin. In other words, his thought is in no way metaphysically ontological but is rather philosophically (and then scientifically) historical and anthropological. It begins with the appearance and development of labor and technique as subjugating Nature, Nature itself being what has led naturally to men who must work to live. But we have to ask, Is there here only *one* basis for all development, or are we, from the very beginning, involved in a *"dualism"* that sets *Nature* and *Technique* in opposition to each other? "Everything" derives from Nature, of course, since it is Nature that becomes productive human nature. Nevertheless, "everything" derives from *Technique* as well, because it is Technique that allows men to take hold of the natural world. The duality can perhaps be reduced in favor of either of the unities (and entities) that compose it, without thereby being abolished. Marx, who promised an explicitation of his point of departure, has not provided it. He did not give much thought to the source of the problematic of history and man, because the dramatic, historical human problems that lay then and there before his eyes absorbed all his energies.

Marx sees men at work, and he sees work as the externalization and manifestation of man that, as a reifying externalization, are his alienation. In violent opposition to Hegel's conception of work, he offers his own; and in so doing he passes from the sphere of metaphysics and the phenomenology of mind to that of historical physics and political economy. "The outstanding achievement of Hegel's *Phenomenology* and of its final outcome, the dialectic of negativity as the moving and generating principle, is thus first that Hegel conceives objectification as loss of the object, as ⟨alienation⟩ [*Entäusserung*] and as transcendence of this ⟨alienation⟩; that he thus grasps the essence of labor and comprehends objective man—true, because real man—as the outcome of man's *own labor*."[4] A few lines farther on, Marx goes on to criticize Hegel, to the effect that Hegel "grasped *Labor* as the *essence* of man—as man's essence in the act of proving itself: he sees only the positive, not the negative side of labor. Labor is man's *coming-to-be for himself* within ⟨externalization⟩ or as ⟨*externalized*⟩ man. The only labor which Hegel knows and recognizes is *abstractly mental* work."[5] Marx's thinking, aiming at concreteness and freedom from mystification, tries to show clearly both the positive side of labor (i.e., the manifestation of man in production) and its negative side, which is equally essential, since man loses his essence through alienating labor. Marxian dialectic does not forgive Hegelian dialectic its tendency to justify labor, particularly as labor is known in the society of that era. "Thus, by grasping the *positive* meaning of self-referred negation (although again in ⟨alienated⟩ fashion) Hegel grasps man's ⟨self-alienation⟩, the ⟨externalization⟩ of man's essence, man's loss of objectivity and his loss of realness as self-discovery, change of his nature, objectification and realization."[6]

As a revolutionary, Marx both emphasizes the negating function of labor in general as the moving principle of becoming and lays heavy stress on its negative aspect in its present actuality: the men *that he sees* do not affirm themselves in their work but deny and alienate themselves in this self-externalization. Yet all "labor" is not alienating. There was, perhaps, in the past a work that was more "realization" and there will be in the future a labor capable of reconciling man with the world. Nevertheless, in the historical present, work renders man an alien to himself and to his own products.

Economic activity is the activity that allows man to live and that makes him man. It is, indeed, the primordial and essential activity of man, but it is also the terrain on which man is alienated.

Economic activity is essentially social, for labor is everywhere and is always done by men, and each man performs a part of global social labor. A division of labor necessarily results, and this division is a cruel and tangible manifestation of economic and social alienation, of global alienation itself; for with the division of labor there is a division of *species man* (total and concrete man) as well, to the loss of his unitary essence. "The *division of labor* is the economic expression of the *social character of labor* within the ⟨alienation⟩. Or, since *labor* is only an expression of human activity within ⟨externalization⟩, of the living of life [*Lebensäusserung*] as the ⟨alienation⟩ of life [*Lebensentäusserung*], the *division of labor*, too, is therefore nothing else than the ⟨alienated, externalized⟩ positioning of human activity as *real activity of the species* or as *activity of man as a species being*."[7]

Marx reproaches political economists (Adam Smith, Ricardo, and other bourgeois theorists) with not having grasped the real nature of the division of labor and with not having seen that this division constitutes an alienated and externalized form of the species activity of men. Liberal, individualistic political economy, therefore, "atomizes" man, whereas the social, collectivistic political economy of Marx would not cut man off from society. Marx confronts Hegel, Smith, and Ricardo not for the purpose of providing a better history of philosophy—and philosophy of history—or a better systematic and historical exposition of political economy, but in order to introduce philosophical and historical *criticism* into philosophy and economy, criticism that would lead to a new "politics." Marx does not minutely trace for us the history of the division of labor from the dawn of prehistory up to modern times; instead, he conducts a critical attack only on certain main forms of the division of labor in society, with special concentration on the present misery that results from them.

Marx does, though, momentarily look toward the past when he writes: "The greatest division of material and mental labour is the separation of town and country. The antagonism between town and country begins with the transition from barbarism to civilisation, from tribe to State, from locality to nation, and runs through the whole history of civilisation to the present day."[8] The division of labor not only isolates every man from the community, it also cuts man in two: into someone who works with material and someone who works in the mind. By this division, some men are assigned the "specialty" of being the first sort of workers, others the second. The possibility of this rupture begins when men leave nature (barbarism) in order to gather together in towns (civilization). In leaving nature, man gives himself over to an antinatural labor and civilizes himself.

Contemporary civilization makes the division of labor unbearable. Man's act of production becomes an alien and external power that enslaves him. By the division of labor, the *exploitation of nature by men* is transformed into the *exploitation of men by men*. Productive labor and the consumption (and enjoyment) of its products fall to different individuals and different classes. Since town and country have become separated, leading to the opposition of farm labor and commercial and industrial activity, men have indeed continued to exploit nature, but they do it by exploiting men. Those who exploit nature to produce or make what men need are in turn exploited by nonworkers.

A break with nature, a break between individuals and the community, and a break between producer and consumer accompany the division of labor, which becomes in a way the principal and, as it were, single cause of alienation. Is not the division of labor, then, an externalized and alienated expression of human species activity? Men no longer *are* and no longer *know* what they *do*, and their activity is not whole but segmented. Each occupation is isolated and autonomous, each sphere of activity forms a sphere apart, and each man considers the domain in which he is "manifest" by self-externalization and self-alienation as *the true one*. And, although each being is necessarily linked to universality, he desperately compartmentalizes himself in *his* particularity. Global social power reduces individual beings (from whom it is derived) to impotence.

> The social power, i.e. the multiplied productive force, which arises through the co-operation of different individuals as it is determined by the division of labour, appears to these individuals, since their cooperation is not voluntary but has come about naturally, not as their own united power, but as an alien force existing outside them, of the origin and goal of which they are ignorant, which they thus cannot control, which on the contrary passes through a peculiar series of phases and stages independent of the will and action of man, nay even being the prime governor of these.[9]

Are, then, labor and the division of labor *evils*? Do they manifest themselves only negatively? Marx thinks that labor and the division of labor were facts of historical development in the past, permitting the progress of societies; yet in the present they no longer play a positive role. Less concerned in his study with the essence of labor than with its historical forms, Marx sees in the present division of labor a source of unbearable alienations. Yet he well knows that "this fixation of social activity, this consolidation of what we ourselves produce into an objective power above us, growing out of our control, thwarting our expectations, bringing to naught our calculations, is one of the chief factors in historical development up till now."[10]

In the past, it was *natural* that men be led to the division of labor. But in the future, by transcending the stifling framework of the division of labor, they will be able to give themselves *voluntarily* to social activities. Just as has private property (which we shall treat later), the division of labor has played a positive role in the past; but it becomes intolerable in the present, and in the future it can be abolished. Labor, the division of labor, and private property are indissolubly bound together. As for labor, we are told that "what was previously being *external* to oneself—man's externalization in the thing—has merely become the act of externalizing [*Entäusserung*], the process of alienating [*Veräusserung*]."[11] Only a communist future will be able to give labor its meaning. Private property and the division of labor[12] promoted the advance of mankind, but now they impede its evolution. One must never lose sight of the connection (which is more, really, than just a close bond) between the division of labor and property and between the different forms of the division of labor and the different forms of property. The different levels of the evolution of the division of labor coincide with the forms of property, being the expression of them. What is true for private property is true for the division of labor, namely, "on the one hand, that *human* life required *private property* for its realization, and on the other hand that it now requires the supersession of private property."[13]

So Marx works out an analysis and critique of the present in order to prepare a better future. He criticizes economic science, but his criticism is focused on economic reality. And, if he attacks objective social realities, it is to uncover therein the human being who is stifled by them. When he talks about labor or the division of labor, it is men who labor that he is thinking of, *workers*, not social things. The genealogy of the modern worker is but briefly set out; by a quick passage he moves from the slave of antiquity, through the medieval serf, to the modern, paid "free" worker, which is where he wants to get.

If the essence of man lies in his whole social activity, as practical realizing action, the worker is from the start an alienated being, doing nothing but work, and work for others. The development of productive forces has led to a social state where those who produce social wealth, namely, those who are workers par excellence, the proletarians, are "totally" blocked from the product of their labor. To live, miserably at that, they are obliged to sell not their labor but their labor force.[14]

Workers in all historical societies constitute the exploited class. They are bound directly to productive forces and struggle against the class of exploiters, who draw their power from the fact that they possess the means of production and that existing relations of production assure them domination. Workers, as well as those who possess the means of production, form a *class*, that is, a group of men bound together to the economy in some determinate way and possessing or being able to possess a consciousness of their class. Likewise, it is not only the worker, the laborer, the proletarian, who is alienated. The bourgeois and the capitalist are alienated as well. The class of proletarian workers is not formed by naturally existing poverty but by poverty that is artifically produced. The proletarized producer is himself a product of alienating economic process. Properly speaking, the proletariat is not a particular class. It has a universal character, it represents universality, since it is the social element of society, and its sufferings are universal. It is the last class, the class that leads society to society without classes.

According to Marx, the economic process of capitalism leads to a progressive concentration of all riches in the hands of those who possess the means of production. At the same time their number gradually diminishes, for they expropriate the less powerful. Correlatively, the decomposition of the middle class intensifies, and their members fall back into the ranks of the proletariat. Proletarians, by their life and struggles, bring the dissolution of the existing social order to completion and can march on toward social revolution; and from this will result the new socialist, communist society. Believing that he had discovered this dualistic polarization, Marx saw only the insurmountable opposition of two adversaries: the capitalist as possessor and the exploited laborer. Yet Marx was mistaken in matters both of concrete eventuality and of economics: the dualism has not intensified. On the contrary, it has lessened, and the middle classes have not been proletarianized. In Western Europe at least, as well as in America, the "pauperization" of society has not happened. The middle classes have increased, even absorbing a certain number of proletarian elements. The petty bourgeoisie does not seem disposed to die so easily. Besides, workers participate in the benefits of their exploiters, and the economic level of Western capitalist society is rising.

61 Labor, Division of Labor, and Workers

Workers, nevertheless, are and remain alienated from the products of their labor; and no political program that is simply social has succeeded in surmounting that alienation, even though the need of the worker is no longer the reduced one of "maintaining physical life at its most indispensable and most deplorable level." Marx sees in workers the protagonists of historical becoming "despite" their alienation and because of it; for they are the ones who use productive forces. The workers will be as well the conscious heroes of the new, total liberation—because of their total alienation. Those who animate productive forces have always been the *constructors* and *leaven of negativity*, which is what makes society evolve. They have been this even when they did not know it, since it was the oppressors who gave their forms to the matter that the oppressed furnished. But from now on they can know what they do and what they are: constructors, constructing and bringing to realization, forming the material world that they produce.

The famous preface to the *Contribution to the Critique of Political Economy* expresses the conception Marx had of the whole of things both as a young man and in his later years, and it formulates schematically, laconically, the whole vision of *historical materialism*. More than that, in the preface Marx tries to grasp the meaning of history (that of modern European history, more than of universal, past history) in order that political practice may be able to orient society in a new direction. The workers will be the ones who bring the (hidden and inverted) truth of history to realization. It is a fundamental Marxian and Marxist text to be read carefully:

> In the social production of their existence, men inevitably enter into definite relations, which are independent of their will, namely, relations of production appropriate to a given stage in the development of their material forces of production. The totality of these relations of production constitutes the economic structure of society, the real foundation, on which arises a legal and political superstructure and to which correspond definite forms of social consciousness. The mode of production of material life conditions the general process of social, political and intellectual life. It is not the consciousness of men that determines their existence, but their social existence that determines their consciousness. At a certain stage of development, the material productive forces of society come into conflict

with the existing relations of production or—this merely expresses the same thing in legal terms—with the property relations within the framework of which they have operated hitherto. From forms of development of the productive forces these relations turn into their fetters. Then begins an era of social revolution. The changes in the economic foundation lead sooner or later to the transformation of the whole immense superstructure. In studying such transformations it is always necessary to distinguish between the material transformation of the economic conditions of production, which can be determined with the precision of natural science, and the legal, political, religious, artistic or philosophic—in short, ideological forms in which men become conscious of this conflict and fight it out. Just as one does not judge an individual by what he thinks about himself, so one cannot judge such a period of transformation by its consciousness, but, on the contrary, this consciousness must be explained from the contradictions of material life, from the conflict existing between the social forces of production and the relations of production. No social order is ever destroyed before all the productive forces for which it is sufficient have been developed, and new superior relations of production never replace older ones before the material conditions for their existence have matured within the framework of the old society. *Mankind thus inevitably sets itself only such tasks as it is able to solve, since closer examination will always show that the problem itself arises only when the material conditions for its solution are already present or at least in the course of formation.*[15] In broad outline, the Asiatic, ancient, feudal and modern bourgeois modes of production may be designated as epochs marking progress in the economic development of society.[16]

Here Marx, in an overly broad outline, divides the whole of both prehistory and world history into four epochs, which in fact form only the "prehistory" of human society; for history properly speaking will begin with the new socialism in its innovative action. We have, then, five historical periods: Asiatic (Oriental), ancient (Greco-Roman and slave), feudal (Christian and Medieval), modern (bourgeois, capitalist, and Western), and socialist (proletarian, communist, and universal). The founder of historical materialism never wonders, however, whether this "Western" and "European" schema encompassed the totality, both spatial and temporal, of the becoming of mankind in history and whether it was always and everywhere the forces of production that instigated social changes. Marx concludes: "The bourgeois mode of production is the last antagonistic form of the social process of production—antagonistic not in the sense of individual antagonism but of an antagonism that emanates from the individuals' social conditions of existence—but the productive forces developing within bourgeois society create also the material conditions for a solution of this antagonism. The prehistory of human society accordingly closes with this social formation."[17]

This text is a summary of Marx's whole thinking, and, though in general his philosophic ideas were always laconic, summary, and lacking explicitness, it does give a clear statement of the plan and perspective of historical materialism, the theory that purposed to guide the revolutionary practice of workers. It brings together and subsumes all historical epochs under one schema, rigid and summary, perhaps, but illuminating. Rather than searching into the truth of historical particularities, it violently takes hold of "the" moving sense of the historical becoming that leads to Western capitalism, with workers being given the task of completely overthrowing that epoch, in order to lead humanity to true universal history. Marxian *materialism* is *historical*. (The founder of Marxism never spoke of *dialectical materialism*.) It is based on the idea of the primacy of economic process as structurally determining the whole development of the superstructure. This historical materialism cares little about knowing whether it in fact explains the past, present, or future becoming of world history and whether it is applicable to any culture whatever. It describes *the actual state of affairs* existing in bourgeois, capitalist Europe and has no concern for the riddles posed by Indian or Chinese history. In its view, the truth of the present-day world is the *actual* truth of the planet. Occidentalism thus becomes universal, and the workers of the entire world receive the task of uniting to realize the global social destiny of the world. Marx does not ask whether this *existing state of affairs*, in which economic technique causally determines all the rest, is not a quite special reality, the result and product of a particular metaphysics as realized, of a certain reading of the world (first Greek, then Christian, and finally modern European) that gives a position of privilege to *technē*, the idea of *Creation*, and *practical reason*. Marx's historical dialectics is one-sided: producers produce products and the mode of production of material life determines politics, religion, and philosophy, while these things themselves are unable to produce or generate anything. Marx does not confront Hegel in his totality. He opposes his historical "idealism," but this new position has no explicitly ontological import; nowhere is it a question of *matter* and *spirit* on the level of being in its totality. Marx resolutely turns his back on the kind of illumination metaphysics would offer with regard to a history and intends to ignore all "abstract" ontology. He wants workers to concretely appropriate the products of their labor through a real, material history.

The tragedy of history reaches a culmination after each great economic-historical period without achieving a really positive solution. The true heroes remain unknown and despised. Henceforward, what the producers of social wealth do negatively is to be done positively. Workers can and must make themselves negators of negation, liberating productive forces and preventing constraints and obstacles from arising.

Though the emphasis in Marx's thought is philosophical in the *1844 Manuscripts*, economic in the preface to the *Contribution to the Critique of Political Economy* (1859), and political in the *Communist Manifesto* (1848), his thought, even while evolving, always starts from the same center, namely, productive labor, which in its development conditions the whole of historical movement. His eyes are always set on the present and its alienations, on what generates the necessity for the revolutionary transformation of society by workers whose actions will realize in praxis what is now part philosophic, part scientific, and part practical theory. The first lines of the first chapter of the *Communist Manifesto* constitute an important text for Marxism that we must take into consideration; for, since the "dogmatic" texts of Marx are few and scattered, they have to be constantly kept together in mind. The chapter entitled "Bourgeois and Proletarians" begins by telling us: "The history of all hitherto existing society is the history of class struggles.[18] Freeman and slave, patrician and plebeian, lord and serf, guild-master and journeyman, in a word, oppressor and oppressed, stood in constant opposition to one another, carried on an uninterrupted, now hidden, now open fight, a fight that each time ended either in a revolutionary reconstitution of society at large or in the common ruin of the contending classes."[19]

Once again, Marx's historical vision, much more Western than abstractly universal in surveying the three great epochs (Greco-Roman *Antiquity*, the feudal *Middle Ages*, and bourgeois, capitalist *Modernity*), asserts and admits the reality and possibility of a catastrophe-laden resolution of class struggles: *the mutual perishing of the two classes in conflict.* Tragedy does not, then, always lead to a progressive solution, something the Marxists too often forget.

Marx once again concludes: "The modern bourgeois society that has sprouted from the ruins of feudal society has not done away with class antagonisms. It has but established new classes, new conditions of oppression, new forms of struggle in place of the old ones. Our epoch, the epoch of the bourgeoisie, possesses, however, this distinctive feature: it has simplified the class antagonisms. Society as a whole is splitting up more and more into two great hostile camps, into two great classes directly facing each other: Bourgeoisie and Proletariat."[20]

Marx keeps constantly in view this fundamental dualism: *bourgeois-capitalists*, the exploiters and oppressors, and *worker-proletarians*, the exploited and oppressed. Starting from this dualism, even while wishing to be not dualistic but dialectical, he continues to think in a dualistic, even Manichean, way. The material life of men *and* their thoughts, the economic structure *and* ideological superstructure are illuminated in terms of a radical opposition between the true *and* the nontrue, light *and* darkness, good *and* evil. The only outcome that leads to a surpassing of the antagonisms, that is, to their abolition, is the total and absolute victory of basic realities over the epiphenomena. Yet it is not easy to admit the universal truth of this vision or that "the history of all hitherto existing society is the history of class struggles." This truth and this conflict essentially characterize Western Modernity, the bourgeois, capitalist phase of the becoming of societies. For it is not true that the slaves, the plebeians, and the serfs led a struggle against free men, the patricians and the barons, a struggle that ended *by its internal dialectic* "in a revolutionary reconstitution of society at large." Historical becoming was not so univocally determined by the development of productive forces and by the revolt of men who worked them. Without mentioning the prehistory of primitive societies, the empires and peoples of the Orient and Asia, we can certainly state that neither ancient Greece nor the Roman Empire nor the Christian Middle Ages died under the blows of slaves fighting against free men, or plebeians struggling against patricians, or serfs battling barons. The passage from one historical stage to the next did not result from the victory of the exploited over the exploiters but from an internal exhaustion and the appearance of a new "third power." Dualistic antagonism was abolished and superseded by a third power that abolished and superseded the two parties in conflict. The Romans triumphed over the Greeks, and the barbarians crushed the Greco-Roman world which was no longer able to sustain itself. And the Middle Ages came to an end owing to the development of the burgher (bourgeois), "independently" of the struggle between baron and serf. Shall we therefore exclude the possibility that the present antagonism, between capitalist and proletarian, be abolished and superseded without a definitive victory for one over the other but with the development of a third solution, arising, of course, from within?

In any case, Marx concentrates all his attention on the present and not on what he calls in one place "pretended universal history." The present with its impairments absorbs all his energies, and it is because he thus took a one-sided and violent hold on the world that he was able to grasp it. Labor, the division of labor, private property, and capital haunt his vision, and he continues to search these alienating realities for the opening to the future.

3.

PRIVATE PROPERTY,

CAPITAL, AND MONEY

Economy is the backbone of historical becoming, and political economy with a historical method is science par excellence. Labor is the principle of economy and of the science of political economy. It is also the essence of private property. Political economy is itself a product of real economic movement and, above all, of private ownership. Yet, far from recognizing man in his productive activity, (bourgeois) political economy is the logical realization of the denial of man.

The division of labor and private property go hand in hand. If the division of the city and the country was the first major form of the division of labor, the first major form of private ownership was landed property. After it came industrial capital, the objective form of private property. The important thing, now, in regard to private property is to analyze its alienating essence. "This *material*, immediately perceptible private property is the material perceptible expression of *alienated human life*. Its movement—production and consumption—is the *perceptible* revelation of the movement of all production until now, i.e., the realization or the reality of man. Religion, family, state, law, morality, science, art, etc., are only *particular* modes of production, and fall under its general law."[1]

Private ownership alienates the species productive activity of man and prevents him from showing his true universality. Linked to the division of labor, it stabilizes, consolidates, fragments, and individualizes what is essentially communal. Private ownership introduces the particular into the heart of the universal, generalizes what is thus particular, and alienates and subjugates man; by always individualizing what is by its social essentiality common to all, it prevents man from being man (a natural, human, social being) and locks him into relations of *having*. Under the sway of private ownership, man ends up by becoming an alien object in his own eyes, a subject who *is* not but who *has* (who above all *has not*) and who is *had*.

The history of labor and the development of productive forces are bound up with the history of the division of labor and with the relations of production, that is, the forms of ownership. As a result, each form of productive labor has corresponding to it a mode of the exploitation of labor. What men take from nature is taken from them by other men. Ownership, that is, expropriation, has a whole history, and the history of labor (agricultural, commercial, industrial), which is the history of the appropriation of material goods, the history of the expropriation of workers, conditions all History. In the last pages of part one of *The German Ideology*, Marx gives us a brief history of ownership in a kind of foreshortened sketch of property as landed, communal, feudal, and, finally, modern.

The *first* form of ownership is *tribal*. Tribal ownership corresponds to an undeveloped degree of production, based on hunting, fishing, animal raising, elementary agriculture, and landed property. The division of labor is here but little developed, being confined to a further extension of the primitive division of labor found in the family. This period is characterized by community property and patriarchal rule, with a social organization consisting of patriarchal heads as chiefs over both the simple members of the tribe and the slaves. We do not know what period of history tribal ownership is to be identified with, perhaps that of prehistoric, primitive *barbarism* (or even that of certain *Asian* societies). Marx does not tell us, and we cannot infer it with certainty.

The *second* form is that of *communal ownership* in *ancient Greece and Rome*. This ownership comes about from the unification of several tribes into a city, either by agreement or by conquest. In it the opposition between city and country is heightened, and slavery is perpetuated and strengthened. The division of labor develops as well as movable private property and, later, immovable private property, which however remains subordinated to communal ownership. For "the citizens hold power over their laboring slaves only in their community, and on this account alone, therefore, they are bound to the form of communal ownership. It is the communal private property which compels the active citizens to remain in this spontaneously derived form of association over against their slaves."[2] Nonetheless, private property continues to develop, becoming concentrated more and more in fewer hands.

The *third* form of ownership is the *feudal property* of the Middle Ages (the period of "unreason realized"). Unlike Antiquity, in which the city with its territory was the starting point, the Middle Ages starts out from the country. Feudal property, estate property, results, on the one hand, from the breakup of the Roman Empire by victorious barbarians and, on the other, from the influence of Germanic military organization. The slave gives way now to the serf, the small tiller of land. Corresponding to the feudal organization of landownership, in the cities arises guild ownership, the feudal organization of trades. Weak, crude agriculture and artisan industry are characteristic of this order. Its estates hierarchy consists of princes, the nobility, the clergy, and the peasants, in the country, and, in the city, the masters, journeymen, apprentices, and soon the day-laborer populace. Slowly but surely this form of ownership, essentially property ownership involving secondarily a little capital, will lead to the next one.

> The necessity for association against the organised robber-nobility, the need for communal covered markets in an age when the industrialist was at the same time a merchant, the growing competition of the escaped serfs swarming into the rising towns, the feudal structure of the whole country: these combined to bring about the *guilds*. The gradually accumulated small capital of individual craftsmen and their stable numbers, as against the growing population, evolved the relation of journeyman and apprentice, which brought into being in the towns a hierarchy similar to that in the country.[3]

Modern property (that of modern times) is the *fourth* form of ownership. Essentially urban, it shifts in character progressively from manufactural property to industrial capital. *Estates* are abolished or, rather, transformed into clearly antagonistic *classes*. This period marks the apotheosis of private property as such. Here the necessary development of labor has freed industry and capital and turned the slave and serf into free workers. "The *real* course of development . . . results in the necessary victory of the *capitalist* over the *landowner*—that is to say, of developed over undeveloped immature private property—just as in general, movement must triumph over immobility—open, self-conscious baseness over hidden, unconscious baseness; *greed* over *self-indulgence*; the avowedly restless, adroit self-interest of *enlightenment* over the parochial, worldwise, naive, idle and deluded *self-interest of superstition*; and *money* over the other forms of private property."[4]

These four major stages in the history of ownership correspond, *mutatis mutandis*, to the great periods of economic and social history rapidly reviewed in the preface to *A Contribution to the Critique of Political Economy* and in the first paragraphs of the *Communist Manifesto*. The development of productive forces, labor, the division of labor, the modes and relations of production, class struggles, and the forms of ownership are together bound to the same historical dialectic. The "first" period, *"prehistoric," "primitive," "barbarian,"* that of *tribal property* in a *half-patriarchal, half-slave-owning* system of rule, is left in shadow. Marx does not shed much light either on the *"Oriental"* or *"Asian"* period. His treatment becomes impassioned when he turns to the three major phases of Western history: *Antiquity* with its *slavery*, the *feudal Middle Ages*, and, above all, *capitalist modern times*. At the end of this process, labor is found to be totally alienated, and private property, the alienator, has reached full realization in the form of capital. Here, in capital, we find the protagonist of economic and social tragedy, to the point where laborers—free workers—come to occupy the whole historical scene in order to inaugurate universal socialism-communism. The dialectic of Western history is thus built up as the dialectic of universal history. Universal history has not been—it will be.

The periods of the development of productive forces constitute so many eras of alienation. Every new step in its progressive achievement freed productive forces and further alienated the producers, but it had to be this way. The social life of men, in order to realize itself, needed the division of labor, the formation of classes and their conflict, and private property and the development of capital. Henceforward, in order to put a decisive end to alienation, what is needed is the abolition of all these alienating realities, especially private ownership. Capital, the economic, social, and historical victor, is now itself to be wholly vanquished by the victorious workers.

Capital has to be understood in its real nature before it can be annihilated. Capital has gained complete victory by the transformation of all private property into industrial capital. Marx often returns to the topic of the separation of town and country, the true dawn of the epic of history for a mankind emancipating itself from nature. "The separation of town and country can be understood as the separation of capital and landed property, as the beginning of the existence and development of capital independent of landed property—the beginning of property having its basis only in labour and exchange."[5]

Capital is just this, productive (and produced) property with its basis in labor and exchange. Capitalist profit which nourishes capital is not theft, as Proudhon claims. Capitalist profit constitutes a certain purchase of the labor force of workers, since workers sell not their labor but their labor force. We are not going to describe here Marx's economic theories, properly speaking, those concerning the formation of capital and its mode of functioning. All the historical, economic, and technical problems studied in detail in the massive *Capital* will not be taken up here. At the level on which we are situated, that of the economic and social alienation of the worker's labor, capital enters as the consummate form of private property, and capitalism is seen as the last stage in the history of class struggle, the stage that, through its internal contradictions, leads to its supersession in socialism. The production of material goods comes about more and more in a social way, and the private ownership of the means of production finds its special expression in capital; and in these developments lie antagonistic relationships which contain the seeds of the abolition of capitalism by a social form that, harmonizing social production and the socialized means of production, no longer tolerates the struggle between Labor and Capital.

Capital rests on social labor and yet alienates it and sets itself up as autonomous with regard to it. "But labor, the subjective essence of private property as exclusion of property, and capital, objective labor as exclusion of labor, constitute *private property* as its developed state of contradiction—hence a dynamic relationship moving to its resolution."[6] The contradiction resulting from *social* labor and its *individual* mode of expropriation (and from the mode of appropriation of its products) is so alienating that it develops the dialectic that will abolish it.

Workers, themselves the products of capitalism, produce capital, and they will be its gravediggers as well. It is the worker who, by his alienated labor, "produces" capital, and it is capital that transforms man into the worker, thereby reducing him to being nothing but a worker. Man the worker, at once producer, product, and simple commodity, is now the fatal terminal point of the movement of history and technology. At the same time, the worker becomes living capital with needs. If he does not work, selling his labor force as a commodity, he loses his interests and thereby his very existence. The value of the worker, as capital, rises according to supply and demand, and the value of his life lowers to its farthest limit. The whole life of the worker is only the supply of a commodity. Reduced to being only a worker, his human properties exist only to the degree that they are for him alien capital that can relate back to him. Thus, owing to capitalism, the workers make a living by losing life, themselves producing the forces that alienate them and of which they are the by-products.

It is certainly true that modern capitalism has caused certain modifications in the life of workers and that a surplus of gains has made it a little more accommodating to the workers. Nevertheless, Marx's intention was to grasp the deeper truth that governs the empirical truth of capitalism. The colors may have changed from black to gray without the workers' for all that becoming dealienated. What we have called Marx's humanistic radicalism energetically shows itself in his conception of capital. Capital is indeed the great protagonist of the present-day drama. It dominates and crushes both those who produce it, thanks to their social labor, *and* those who enjoy it, individually or in small groups. However, it is men who are involved in this experience. Wishing to break all fetishism of the so-called objective, the passionate objectivism of the man who authors *Capital* points constantly to human subjects in subjection. His vision and prophecy replace relations between things by human relations. His cool passion scientifically analyzes and denounces the existing state of affairs, with a view to freeing these human relationships as reified in and by capitalism. Hatred, romanticism, science, passion, and action are put into play in condemnation of this alienating reality that makes human being a piece of merchandise. When the capitalist process comes to an end, man, himself a product of the whole movement of production, ends up becoming a worker decomposed into vital need and salary, capital and commodity.

This whole cool passion and this prophetic rage turn also to another particular reality of the reified world. Logic and feeling take up the battle against the *res* par excellence, *money*. Money, having the capability of buying anything, capable of appropriating anything, has become the object par excellence. Its essence is simply the universality of its properties, and its being can be seen

73 Private Property, Capital, and Money

as a being-able-to that is unlimited in its power. Money is the alienator because it is, and especially because it has become, the go-between for the needs of human life on the one side and the objects of these needs on the other, the pandering intermediary standing between the life of man and the existence of other men.

Reviving the passion for purification shown by the Jewish prophets and in alliance with the great ("bourgeois") poets, Marx utters a violent denunciation of the corrupting power of money, citing Shakespeare and Goethe. He makes his own the invectives of Timon of Athens, who, with the pronouncement "All is oblique; there's nothing level in our cursed natures . . .", sets himself to dig the earth in search of roots. But it is gold he finds instead, and he begins to curse, vowing to have done with the damnable precious metal by burying it once and for all:

> Gold? Yellow, glittering, precious gold? No, gods,
> I am no idle votarist. Roots, you clear Heavens!
> Thus much of this will make black white, foul fair,
> Wrong right, base noble, old young, coward valiant.
> Ha, you gods! Why this? What this, you gods? Why this
> Will lug your priests and servants from your sides,
> Pluck stout men's pillows from below their heads.
> This yellow slave
> Will knit and break religions, bless the accursed,
> Make the hoar leprosy adored, place thieves,
> And give them title, knee, and approbation
> With senators on the bench. This is it
> That makes the wappened widow wed again—
> She, whom the spital house and ulcerous sores
> Would cast the gorge at—this embalms and spices
> To the April day again. Come, damned earth,
> Thou common whore of mankind, that put'st odds
> Among the rout of nations, I will make thee
> Do thy right nature. . . . Thou'rt quick,
> But yet I'll bury thee.

And a little further, Timon addresses money in these words:

> Thou visible god,
> That solder'st close impossibilities
> And makest them kiss! That speak'st with every tongue
> To every purpose! O thou touch of hearts!
> Think thy slave man rebels, and by thy virtue

> Set them into confounding odds, that beasts
> May have the world in empire!⁷

Marx also quotes with full approval the words Mephistopheles speaks in Faust's study:

> What, man! confound it, hands and feet
> And head and backside, all are yours!
> And what we take while life is sweet,
> Is that to be declared not ours?
> Six stallions, say, I can afford,
> Is not their strength my property?
> I tear along, a sporting lord,
> As if their legs belonged to me.⁸

Money is alienating and cursed because it reverses the (natural?) properties of things, reconciles contraries, puts the humanity of men into subjection, and prostitutes everything it buys. The properties of money become the properties of him who, having it as the "universal property," can appropriate everything. Summarizing these passages from Shakespeare and Goethe, Marx writes: "The overturning and confounding of all human and natural qualities, the fraternization of impossibilities—the *divine* power of money— lies in its *character* as men's ⟨alienated⟩, ⟨externalizing⟩ and self-disposing *species nature*. Money is the ⟨externalized⟩ [alienated] *ability of mankind*."⁹

If a man wishes to satisfy a need and has money, he can *realize* his desires, whereas the man who has only needs and no money can only *imagine* the satisfaction of his desires. The first man is thus in contact with the reality of the external world, while the second has to take refuge in internal thought, which itself develops within alienating externality. Moreover, the absence of money determines as well the absence of needs. Money can accordingly transform the representation of a need into objective realization, and the reality of need into mere subjective imagination. Since human labor, through alienation, produces values that are used and exchanged, money has become that which is exchanged for all things and that for which all things are exchanged.

Long before Timon of Athens, Mephistopheles, and Marx, Heraclitus had already grasped, in all the splendor of its universality, the "dialectical" process of the conversion of all things into money and of money into all things. We find Fragment 90 saying this: "There is an exchange: all things for Fire and Fire for all

75 Private Property, Capital, and Money

things, like goods for gold and gold for goods."[10]

In the first volume of *Capital*, in a chapter entitled "Money, or the Circulation of Commodities," Marx, discussing the metamorphoses of commodity, quotes Heraclitus directly, seeing his fragment as illustrative of the process he is studying, namely, the transformation of commodity into money and the retransformation of money back into commodity, totally capable thus of being bought and sold.[11] Nevertheless, what is particularly interesting to the author of the *Manuscripts of 1844* is the alienating power of money, the fact that it provokes the general confusion in our upside-down, perverted world. Not only can money be exchanged for any property, thus being able to buy as well the opposite of any property, it can also be exchanged for the whole of the human world of objects. The real needs of men and their real satisfaction are subservient to money. The need for money becomes the sole, true need that the science of needs and of their satisfaction (political economy) produces and recognizes, by the very fact that real movement (economy) only produces the need for money.

THE MACHINE, INDUSTRY,

AND TECHNICIST CIVILIZATION

The upside-down, perverted world that Marx attacks is not only dominated by the division of labor and of capital and money: in it as well the economic and social alienation of human being is brought to consummation by the reign of the machine, industry, and technicist civilization in general. Man, product of nature and technique, is distinguished from the other animals by his practical activity, which is expended to satisfy his compelling natural needs. "Life involves before everything else eating and drinking, a habitation, clothing and many other things. The first historical act is thus the production of the means to satisfy these needs, the production of material life itself. And indeed this is an historical act, a fundamental condition of all history, which today, as thousands of years ago, must daily and hourly be fulfilled merely in order to sustain human life."[1] What allows man, then, to posit himself as man and oppose himself to Nature, in order to wrest from Nature his goods, is the *tool*. Marx subscribes to the American, *zoo-technological* definition of man made by Benjamin Franklin: "Man is a toolmaking animal."[2] The use and making of tools and the correlative development of productive forces and instruments of production form the real guiding clue to the historical becoming of mankind, generating—and being generated by—an infinite dialectic. For, as Marx says in naming the next historical condition, "the second point is that the satisfaction of the first need (the action of satisfying, and the instrument of satisfaction which has been acquired) leads to new needs; and this production of new needs is the first historical act."[3]

A reciprocal action goes on between natural needs and the instruments of satisfaction, between new needs and new instruments; and yet one cannot reduce everything either to an initial, progressing dialectic of needs or to a primordial dialectic of the evolution of productive techniques. Need determines the instrument that leads to the satisfaction of the need, and the available means of production engender (produce) new needs. The action of these two realities is certainly reciprocal, but does it not rest on a dualism, that of natural need on one side and technique on the other? A unitary basis is not yet attained, since Marx remains less taken by the search for ultimate historico-anthropological *foundations* than by the process of the *development* of technique. His point of departure is the initial relationship men have with Nature. The natural and social relationship of man with (historicized) Nature is a relationship of struggle, and the natural history of man is a product of this struggle. Man fashions tools, and his history is "fashioned" by way of the development of the weapons of struggle; history itself, in the course of its becoming, fashions the weapons that determine its becoming.

The founder of historical materialism is always and everywhere looking for the true, the real agency of movement in the historical development of mankind. He devotes himself to inquiring into the process of alienation, finally tracing out the lines of the perspective that sees the transcendence of alienation in a future universal reconciliation. His *intent* is to make his point of departure (namely, living individuals with a determined bodily organization, struggling socially against nature, with the aid of tools, in order to produce their livelihood) free of any metaphysics. He always sees the manifestations of human nature as taking place in history (which is, of course, man's history) and never presupposes a "state of nature" that would be the origin of man's historical nature. Primitive men hardly interest him. He knows that every age creates its own ethnology and "produces its own primitive men."

Marx's thought is essentially historical. It looks to economic development for the sense of the direction of global historical movement. His thought is historical in two ways: it tries to understand every phenomenon in its history (and in History), and it concentrates particularly on the present moment of history, the present historical situation; for it is the past that produces the present and the present that prepares the future. This is how the past-present-future is joined together in the dimension of historical time.

Time as historical in this way presses the prophet of socialism to make haste, to avoid delaying too long in historical studies. The present state of alienation calls for remedy, and so his principal adversary is that alienation which is produced by general mechanization in capitalism.

As long as men used mostly natural instruments of production, water, for example, they remained subject to nature, whereas the means of production that civilization creates (and that thereupon re-create it as a technicist civilization) help men to oppose nature more effectively. But, while exploiting nature, men have let themselves be exploited by other men; and they remain, thus, subordinated to what they have themselves produced. The development of the instruments of production leads necessarily to the creation and development of the machine, for the labor that presupposes the machine will show itself to be the one most suitable for development.

The machine is the last term so far in the development and constant, progressive improvement of instruments of production. The long path of historical becoming for humanity leads from the use and making of the first extremely crude tools to the reign of powerful, highly perfected machinery. The machine is in a certain way the synthesis of all instruments. It contains them in that it makes synthetically what they make analytically. Yet man has continued to be alienated progressively in and by his labor. The age of the machine completes this alienation, and man, having himself produced the machine, now finds himself to be a mere cog in the immense machine and machinery of capitalism. Marxian positivism, in admiration of the evolution of productive forces, turns into passionate romanticism when faced with the machines as inhuman and alienating. While necessary to the development of human societies, the machine nevertheless crushes men; yet it crushes them not simply as the machine but by way of the relationships that the workers have with it.

These inhuman relationships that bind men to the machine make the essence of man mechanical. Today, Marx observes, the machine adapts itself to man's weakness in order to make weak man into a machine. Man has thus become a slave to the machine, just as he is likewise a slave of divided labor, private property, capital, money, industry, and the whole of technicist civilization. The constant progress of the division of labor and the continual simplification of machine and mechanical labor transform the child into a worker and the worker into a child. The development of productive forces, leading to the reign of capitalist machines, has not brought the worker to maturity but to infantilism and debilitation. The wheel of history crushes those who set it turning.

Since real productive forces, in satisfying the material needs of men and in endlessly creating new ones, are the internal motor of historical development, it is their rhythm that determines the rhythm of the development of society as a whole. Any disruption of the rhythm of functioning of these forces, especially on the complex level of general mechanization, becomes a generalized social disruption. It is true that *Economy* is not *Society*; but economy does form the moving agency in the functioning of society and its development. Marx seems sometimes to "identify" *the ensemble of relations of production* (which is obviously more than pure economy) with *the totality of society*, thus identifying a "part" with a "whole." If it is defined too narrowly, economic movement loses its great importance; defined too broadly, it ends up by including everything. In his intent to bring out the import and breadth of economic movement, Marx gives in a little too much to the second tendency. When engaged in concrete criticism of all the dramatic economic realities that alienate man, he attacks as well society as a whole, without distinguishing between "economic evils" and "social evils." But are all these things that externalize and alienate—productive forces as autonomizing and alienating the workers, labor transmuting into commodity, the division of labor segmenting the very being of the worker and of society, private property emancipating itself from the community, capital, money, and the machine subjugating individuals and the proletarian class, the social classes themselves becoming autonomous—are all these things economic, principally economic, or globally social? Is it only the internal dialectic of the development of productive forces that has led mankind to the stage of capitalism, general mechanization, industrialism, and technicism? Is not the development of productive forces also a *product*, something that reaches beyond the strict framework of economy and the ensemble of relations of production?

To the question, What has made possible this monstrous development of mechanization and industrialism, this state of affairs that characterizes only a few centuries of the total history of mankind? Marx does not give a complete answer. He thinks that the being of men is dependent upon the material conditions of production and that men *are* as they *show* themselves in fact. As a result one cannot go beyond their real manifestations, that is, economic manifestations, and one has to stay with the tangible data of the process that produces goods for consumption. ". . . a certain mode of production, or industrial stage, is always combined with a certain mode of co-operation, or social stage, and this mode of co-operation is itself a 'productive force.' Further, ... the multitude of productive forces accessible to men determines the nature of society, hence ... the 'history of humanity' [single quotation marks by Marx] must always be studied and treated in relation to the history of industry and exchange."[4] And the same statement is made even more strikingly elsewhere: "The history of *industry* and the established *objective* existence of industry are *the open book of man's essential powers*, the exposure to the senses of human *psychology*."[5]

Industry is the most complete form of labor. This enormous productive machinery allows man to set himself up in effective, victorious opposition to nature. The crucial separation of town and country gave rise to the separation of production from trade, which in turn fostered the development of manufacturing. The spread of commerce, navigation, and manufacturing speeded the accumulation of movable capital and gave birth to big industry. It was big industry that completed the victory of the town over the country, that generated a mass of productive forces, that generalized competition, that established the means of communication and the world market, that gained control of commerce, and that transformed all capital into industrial capital and led to the development of the financial system and the centralization of capital. This mother-goddess of the modern world "destroyed as far as possible ideology, religion, morality, etc., and where it could not do this, made them into a palpable lie. It produced world history for the first time, insofar as it made all civilised nations and every individual member of them dependent for the satisfaction of their wants on the whole world, thus destroying the former natural exclusiveness of separate nations."[6]

The commitment of industrialism to the systematic and the automatic therefore dissolves every natural relationship and reduces to nothing what little there still is of the "natural" in labor, the division of labor, and the social life of men. Mechanization, industrialization, and technicism are doubtlessly immense conquests in the historical development of mankind, being the core of *all* conquests. Yet they lead to bourgeois and capitalist civilization in its consummate, unlivable form. All these conquests have as well intensified and extended the drama of alienation, pushing the external character of man's practical activity and the externalization of his being to their ultimate consequences. Everything seems to have become strange, alien, hostile, and alienating, at the very moment when industry transforms history into universal history. The nature that gave life and soul to men seems to have been definitively vanquished, but it is within this state of things that vanquished men shall engage in the greatest battle against their vanquishers, the industrial capitalists; and this battle shall be based upon the very contradiction that exists between the prodigious but one-sided development of industrial productive forces and the structure of private property.

What Marx the romantic finds "deplorable" in the process of industrialization, of denaturalization, Marx the positivist admires. The regime of industry puts a positive mark on the whole of civilization. Industry has universalized economic life and unified it with social life, something that previous ruling orders, such as the barbarian, the Asiatic, the ancient, and the feudal, could not bring about. What distinguishes one economic and historical age from another is less a matter of what it produces or makes than the manner in which it does it. And so the means of production in the industrial age define the whole period of bourgeois and capitalist technique. The process of becoming through which technique has passed constitutes the gradual, progressive movement toward industrial mechanization; and the elementary tool by which man tried to satisfy his needs has been transformed, in the course of the ages of civilization, into modern technique. Nevertheless, though destined by its very nature (a nature simultaneously natural and antinatural) to tie into nature, exploiting it for the sake of the whole of human society, technique has not yet realized its destiny. Technicist civilization has made life and labor *unbearable*. It has reduced the play of human activity to the role of the worker free to sell his labor force to those who possess the instruments of production as private property; it has heightened as far as can be done the contradiction between productive *forces* and the *forms* of the organization of labor and property; it has consequently made broader and broader the *base* of social life while making narrower and narrower the *closed circle* of those who determine the relations of production. In doing all this, technicist civilization is an impediment to the full and harmonious development of productive forces, stifling at once industrial workers and the true creative, social possibilities of industry itself. Life, the city, and the industrious activity of men in industry no longer have any justification within the framework of alienating technicist civilization. Marx, often enough nostalgic over the past, always horrified by the present, and full of hope for the future, writes this about industrial machinism and technicist civilization in general: "[Big industry] took from the division of labour the last semblance of its natural character. It destroyed natural growth in general, as far as this is possible while labour exists, and resolved all natural relationships into money relationships. In the place of naturally grown towns it created the modern, large industrial cities which have sprung up overnight."[7]

These human lives and these cities, technical in character, technical in origin, alienated and alienating, are the obverse of the coin of progress. Men no longer enjoy the products of their labor, since the worker receives only what is indispensable for him to perpetuate his physical life and continue to sell his labor force. Those who hold the means of production throw out their products as so much bait to draw the money of others, creating and awakening needs and desires, too often artificial, in order thereafter to satisfy them. Real needs are far from being really satisfied, while a crowd of artificial needs is artificially produced and artificially satisfied. "This ⟨alienation⟩ manifests itself in part in that it produces sophistication of needs and of their means on the one hand, and a bestial barbarization, a complete, unrefined, abstract simplicity of need, on the other."[8] Everything has become crude, uniform, automatic, and mechanical. The reign of quantity, abstract quantity, spreads and transforms the earth of men into a civilized desert. Nevertheless, the way the rich evolve in this desert is not like the way the poor do. "The meaning which production has in relation to the rich is seen *revealed* in the meaning which it has for the poor. At the top the manifestation is always refined, veiled, ambiguous—a sham: lower, it is rough, straightforward, frank—the real thing. The worker's *crude* need is a far greater source of gain than the *refined* of the rich."[9]

Industrial, technicist civilization thus develops in the heart of a *crude barbarism in need and labor*: labor is exploited more than ever before, and needs, offered satisfaction that is but gross and artificial, remain unsatisfied. This overcivilized, crude barbarism is a terrain of a double speculation: the "pillars" of civilization speculate at the same time on the sophistication of needs and on their crudeness, a crudeness artificially produced so that what is supplied as enjoyment are a giddiness and an illusory, merely apparent satisfaction of real, rich need. Bourgeois, capitalist civilization, consequently, dissimulates to men all the true wealth of the material world, mechanizes needs and the productive means for satisfying needs, and substitutes for the natural, social, real human world an artificial, alienating world technified to the extreme, alien and hostile to those who live in it and have built it. "The base pestilential breath of civilization" emitted in the capitalist stage of historical becoming indicates the degree of rottenness that the process of mankind has in it attained, and one may well ask if things can go further still in the same direction. Capitalist technicism poisons and alienates everything, and only the negativity that is implicated in its very essence will be able to furnish the antidote that can reconcile men with a social and human civilization and technique.

Once technicist alienation is overcome, technique will be able to develop in a manner that is integral and nonalienating if it is kept under the control of the whole of the human community. The comprehensive planning of technical production should prevent it from generating alienation and disorder.

Capitalist society, by generalizing labor and positing the basis for integral technological development, has brought about the preparation of what will abolish it.[10] It has *universalized* labor, making it maximally *alienating*, and set up the practical reality of *abstract* labor operating in a total *indifference*. Technicist labor is present no longer under a particular form but imposes itself on everyone in the universality of its abstraction; it no longer is intimately bound to the individual. "The fact that the specific kind of labour is irrelevant presupposes a highly developed complex of actually existing kinds of labour, none of which is any more the all-important one. The most general abstractions arise on the whole only when concrete development is most profuse, so that a specific quality is seen to be common to many phenomena, or common to all."[11] This is the situation found realized in the United States and, for other reasons, in Russia. Here are Marx's own words: "This state of affairs is most pronounced in the United States, the most modern form of bourgeois society. The abstract category 'labour,' 'labour as such,' labour *sans phrase*, the point of departure of modern economics, thus becomes a practical fact only there. The simplest abstraction, which plays a decisive role in modern political economy, an abstraction which expresses an ancient relation existing in all social formations, nevertheless appears to be actually true in this abstract form only as a category of the most modern society."[12] But could this characteristic of society in its most modern form, of industrialization in its most evolved state, of mechanization in its reign of conquest, of civilization in its most technicist role, also be found in a technically underdeveloped society? The passage just quoted continues:

> It might be said that phenomena which are historical products in the United States—*e.g.*, the irrelevance of the particular type of labour—appear to be among the Russians, for instance, naturally developed predispositions. But in the first place, there is an enormous difference between barbarians having a predisposition which makes it possible to employ them in various tasks, and civilised people who apply themselves to various tasks. As regards the Russians, moreover, their indifference to the particular kind of labour performed is in practice matched by their traditional habit of clinging fast to a very definite kind of labour from which they are extricated only by external influences.[13]

The extreme mechanization and automatization of labor, the transformation of every reality and process into components of an industrial mechanism, the abstract and automated technicism that has been developed to the utmost by the most modern societies and invades the technically underdeveloped countries, all this moves in the direction of its own negation. This state of affairs—the transformation of all men into workers free to sell their labor force and the development of a total indifference toward the mode of labor—can and must lead to the freeing of all workers, to the superseding of both traditional and modern labor. Yet as a liberation of workers it coincides with the liberation of productive forces, for capitalist technique is not only alienat*ing*, it is *itself* alienat*ed*.

Part III.

Political Alienation

According to Marx, the political rests on the economic and is determined by it as a domain of superstructure. Politics is the *form* that organizes the productive forces of economy, the real material of society. Nevertheless politics also distorts the logic of economic development; it is a frozen form of becoming. Political alienation constitutes the expression (itself alienated) of economic alienation. Thus politics and the State appear as powers that are both alienated and alienating.

5.

CIVIL SOCIETY AND THE STATE

The anatomy of economic society provides the instrument for the analysis and study of the real genesis of civil society. Political society is the sphere of total society that has become quasi-autonomous by reason of alienation, and its development is conditioned by the development of productive forces. The objective reality of Society rests on human forces in their activation of the productive process, while these same forces get their organization in and by the forms of political society of which the State is the most powerful expression.

In producing their material life, men (themselves natural social products) enter into relationship with the means of production, producing relations of production; and in this they become bound to each other. The relationship they have to productive forces is determined by the evolution of these forces and by the modalities of their development. These relations of production into which men enter are conditioned by men's needs and by the mode of production, and they exist in reality prior to political and legal relationships. Political forms, in turn, exercise a return action upon economic forces, an action which is thus real but holds a second-order position. "The form of intercourse determined by the existing productive forces at all previous historical stages, and in its turn determining these, is *civil society (die bürgerliche Gesellschaft)*. . . . Already here we see how this civil society is the true source and theatre of all history, and how absurd is the conception of history held hitherto, which neglects the real relationships and confines itself to high-sounding dramas of princes and states."[1]

The State is only a reaction on the part of civil society to the class conflict within it. Civil society is the theater wherein is played the confrontation of real protagonists: the exploited and oppressed class—the class that produces social wealth—and the class of the exploiter and oppressor—the class that holds productive forces in legal possession as property. The State is the instrument and weapon of the power of the dominant class, and it constitutes the armature for the superstructural. Its nature has to be grasped through its historical development right up to the present and into the future, which may indeed be its nonfuture.

The separation of town and country not only resulted in and precipitated the development of the division of labor, private movable property, and the conditions that lead to the formation of capital and industry—all of which is "counter to nature"—but it brought about as well the need for an administrative and political setup. Administration and political arrangements, which are thus historical creations deriving from the economic sphere and not some eternal realities, became subsequently autonomous, working in diametric opposition to the logic of the development of productive forces. The relations which the forms of political organization maintain with the forces of economic production are the source of conflict; civil society is not only the birthplace of history *(le foyer de l'histoire)* but also a hearthstone of endless conflagration *(un perpétuel foyer d'incendie)*. Actual contradiction between productive forces (and those who work with them) and the forms of relationship (forms of the organization of labor and property, legal and political forms) breaks out in the heart of civil society and will shatter it.

Formally, civil society sees itself as the guarantor of justice and liberty and the fair administrator of society as a whole. Yet things do not work out that way in real, material fact; they work out just the opposite. The personal liberty of which civil society is deemed the guarantor exists in fact only for those individuals who belong to the dominating class; it is hardly universal. Within society, taken both from the economic and the civil point of view, are, on the one hand, the totality of productive forces put into action by the labor of the majority of individuals (a class that is not, properly speaking, a class, since it constitutes the total mass of society and represents Society) and, on the other, the set of social relations that serve the interests of a minority of individuals (the class that possesses the means of production and dominates political institutions). At the same time, the totality of productive forces appears to be independent and detached from individuals, to form a world of its own alien to the individual, whereas these forces are real forces only thanks to the labor of individuals bound together socially. The concrete form of the totality of real productive forces stands opposed to the abstract form of individuals from whom these forces are detached. An equal appearance of independence and detachment from individuals accrues to the ensemble of social relations that form the political world in its autonomy and alienness to the individual, whereas these relations are relations of men. The concrete form of the administrative sphere and of the State stands opposed to individual men, here turned into abstractions. And this is why political alienation is the alienated expression of economic alienation.

The State, consequently, emancipates itself in relation to society and leads a quasi-independent existence, all the while fighting on two fronts: against its internal enemies, the proletarians and in general those who are exploited, and against its external enemies, other national societies. External politics remain subordinate to interior politics.

> Civil society embraces the whole material intercourse of individuals within a definite stage of the development of productive forces. It embraces the whole commercial and industrial life of a given stage and, insofar, transcends the State and the nation, though, on the other hand again, it must assert itself in its foreign relations as nationality, and inwardly must organise itself as State. The word "civil" society [*bürgerliche Gesellschaft*] emerged in the eighteenth century, when property relationships had already extricated themselves from the ancient and medieval communal society. Civil society as such only develops with the bourgeoisie; the social organisation evolving directly out of production and commerce, which in all ages forms the basis of the State and of the rest of the idealistic superstructure, has, however, always been designated by the same name.[2]

Organization in various spheres—the social, the legal, that of the State, the political, and the administrative—develops, then, from production and economic life. It is a superstructure rising up from a foundation in the real. This is the way it has been for all time but especially since the capitalist and the bourgeois era. But *has* it always been like that, or has not Marx once again projected and generalized the truth of the reality of one historical epoch (which is tending to become universal) upon the whole of history? It seems that we can answer in the affirmative to both questions. In his survey simultaneously of the social and historical evolution of mankind and of all the collisions that have taken place in the course of its becoming, Marx never stops affirming the primacy of the economic over the political. The development of productive forces and their conflict with the relations of production are regarded as capable of explaining social evolutions and revolutions. Nonetheless, the economic dialectic is not the only one (or, to put it another way, economic logic is not *principally* the one) that accomplished the passage of *prehistory* into history, that gave birth to the life and vicissitudes of the *Oriental* empires and *Asian* peoples, that led to the *Greco-Roman* world, that provoked the *barbarian* invasions, that founded *feudal* and *medieval* society, and that brought us to the *modern* period. Class struggle within each society alone does not explain historical becoming, no more than the collisions between productive forces and relations of production. Economic *matter* and *content* and the modes of production are not so easily separated from political *forms* and *organizations*. *Economic forces* and *political power* are not bound together as real base to idealistic superstructure. The great noneconomic events of history are not simply cases of organization imposing a structure upon an existing organism; they are themselves constitutive and themselves organic.

The passages from the preface to *A Contribution to the Critique of Political Economy* and from the *Communist Manifesto* that we have already cited, as well as many other texts in the *Manuscripts of 1844* and *The German Ideology* (not to mention other writings of Marx's that are even more categorical), insist upon this famous contradiction between productive forces and relations of production. Productive forces form the bases and are handled by the workers, while the forms of intercourse constitute the superstructure, organized in and by the Law and the State to serve the interests of the exploiters. This contradiction and the class struggle that it generates appear through the whole history of mankind; and all evolutions, revolutions, and collisions are considered from the angle of "internal" economy and politics, *with no third power or solution having any hold on the fundamental dualism.* The greatness and the limit of Marx's thought lie in this perspective, that, starting from the existing state of things, universalizes it into the past and into the future. What he sees going on before his eyes and the outlook contemporary history offers become total reality. What is true for a few centuries, what has animated the present movement of universal history, what is true now and will be true, is found also to have been true. All conflicts reduce to a common denominator: "This contradiction between the productive forces and the form of intercourse, which, as we say, has occurred several times in past history, without, however, endangering the basis, necessarily on each occasion burst out in a revolution, taking on at the time various subsidiary forms, such as all-embracing collisions, collisions of various classes, contradictions of consciousness, battle of ideas, etc., political conflict, etc. . . . Thus all collisions in history have their origin, according to our view, in the contradiction between the productive forces and the form of intercourse."[3]

Marx fixes his gaze on the present contradiction between the development of technique, which is put to work by the proletarians, and administrative, legal, and political forms, which assure the bourgeoisie its economic and political domination; and as a result he narrows considerably the meaning of the political. The distinction he makes between economic and social changes, structural transformations, and simple political changes is, of course, consummately true; but politics in general does not simply exercise an action that reflects back upon economy and the struggle between States. It is at least as important and determinative as class struggle itself is within the State. One can, to be sure, look to find the economic basis for the fight between States and see them as becoming autonomous entities that seem to be quasi-independent in relation to economic movement. Nevertheless, States in the past have played a constitutive and primary role, and even within post-Marxian thought and politics we are witness to the revenge taken by politics on the sphere of economy.

As for the political alienation of the present period, Marx is tireless in his negation critique, one might say even his anarchic critique, of civil society and all institutions. In criticizing the State, with a view to abolishing it, he concentrates his attack particularly on the bourgeois State, which is the state apparatus of these latter centuries.

> By the mere fact that it is a *class* and no longer an *estate*, the bourgeoisie is forced to organize itself no longer locally, but nationally, and to give a general form to its mean average interest. Through the emancipation of private property from the community, the State has become a separate entity, beside and outside civil society; but it is nothing more than the form of organisation which the bourgeois necessarily adopt both for internal and external purposes, for the mutual guarantee of their property and

> interests.... The State exists only for the sake of private property.... Since the State is the form in which the individuals of a ruling class assert their common interests, and in which the whole civil society of an epoch is epitomised, it follows that the State mediates in the formation of all common institutions and that the institutions receive a political form. Hence the illusion that law is based on the will, and indeed on the will divorced from its real basis—on *free* will. Similarly, justice is in its turn reduced to the actual laws.[4]

Neither the political sphere nor the State, neither law nor institutions have their own history; they develop owing to and by economic movement, and it is precisely the fact that they become autonomous in alienation that gives them the false appearance of a specific and independent existence. All these superstructures, constituting the "structure" of civil society, alienate themselves from the sphere of the economy while alienating the men that they contain. The framework that they erect maintains things and men in their place, preventing their development and flowering. The whole of "becoming" in any of these domains—the political, that of the State, the administrative, the institutional, the legal—takes place over the heads of individuals and consecrates alienation. Marx tears the mask off the lie of Law and Laws: juridic law is only the deformed expression of economic laws; it derives from them and mystifies them. Law is simply the law of the dominating class and assures its rights. Within civil society, which is the organized and "artificial" form of real society, that is, economic society, the State is ruler, organizing the law and the lawful status of the existing state of things. Organic life, the real society of men who work and live in common, falls under the blows of the State and its Law and suffocates. Abandoning the terrain of the social, the State develops its "own" logic and betrays universal interests; it watches out only for the interests of those who hold possessions. Social classes, the apparatus of State, and political institutions and powers emancipate themselves from the community, set themselves up as autonomous with respect to individual men, and then enslave them, even while still giving the illusion of being independent of them. Just as illusory is the law, inasmuch as it in fact forces men back into alienated reality, thus functioning as a power for oppression and repression. The fulminations of the law, far from being the expression of the general will, express "legally" the special domination of one given class. The world is thus upside-down and perverted, everything seeming to follow the right path when it is exactly the contrary which is in fact true.

In the light of what we have just seen, we can better understand the radical critique Marx makes of Hegel's political philosophy. Taking his point of departure in the *Phenomenology of Spirit* and the *Philosophy of Right*,[5] Marx applies his critique of Hegel's philosophy of history, the State, and law principally in his "A Contribution to the Critique of Hegel's 'Philosophy of Right': Introduction,"[6] his *Critique of Hegel's "Philosophy of Right,"*[7] and in the *Manuscripts of 1844*. Marx's critique finds its beginning in the achievements of Hegel's political thought, in its conception of the essence of modern society and history, and these receive full recognition. Marx does not reject Hegel; he criticizes and extends him, and in this way he overturns him. He accuses Hegel of having grasped the ideal essence, the ideal genesis and development of the apparatus, function, and power of the State and of right, not their actual truth, their real, material genesis and history. Marx takes Hegel as holding that whatever is alienated is so in its ideal form and not through its true reality. The State and right (law) remain spiritual entities, since spirit alone is the true being of the State. In this way the political alienation of man within civil society and its State, as autonomous realities, is maintained and justified. Marx does not forgive his great teacher for having grasped dialectical movement, for having seen negativity at work, for having instituted critique that works by a negating action, only to dissolve all that in the justification of existing reality, a justification that, being merely spiritualizing, idealist, and mystical, degenerates into mystification. Hegel stands accused of masking over contradiction as such as well as contradiction in particular manifestations, that is, the contradictions between economic life and civil society, between civil society and the State, and between the real functioning of the state apparatus and juridical laws that justify such action on an ideal level. Hegel is accused of justifying the empirical reality of all that exists, both in its reality and in its deceitful and idealistic justification. Hegel's *criticism* is purely *apparent* and its *positivism false*, says Marx.[8]

Marx appreciates both the breadth and depth of Hegel's political vision, and he does not subscribe to the summary, depreciatory judgments laid against it. Marx sees the role that negativity plays in Hegel's thought, and so he relentlessly criticizes him. He constantly reproaches him for not denying apparent being, so as to assert true being, and for asserting and confirming apparent being (alienated being) in and by the justification of alienation. In Hegel's thought the abolishing and transcending of *economic society* lead to *civil society*, the abolishing and transcending of civil society to the *State*, and the abolishing and transcending of the State to *universal history*. Yet Marx, in thus interpreting the becoming of Spirit in history according to Hegel, sees in this process of abolishing and transcending only the *Aufhebung* of being as thought by, and grasped in, consciousness. In other words, the act of abolishing remains in the realm of ideas, with no effect upon reality. It is a dialectic that believes itself to have really conquered its object, while leaving it existing in reality and justifying it there. Correlatively, in the eyes of this dialectic and this phenomenology, reality becomes an ideal element, entering into consideration only in its abstract reality.

Real, alienating political existence remains thus dissimulated, being revealed only in thought and philosophy. Political philosophy can in no way, however, offer compensation for the reality of political drama. Men who live in a society organized by the State are citizens only in the abstract, and that situation is demonstrably shown in *thought* rather than in *actual reality*. Given that political alienation is only the alienated expression of economic, and basically social, alienation, the alienation of political thought is a third degree of alienation. Thus Marx can write: "Hence . . . my true political existence is my existence within the *philosophy of right* . . . my true human existence, my existence in *philosophy*. Likewise the true existence of . . . the state . . . is the *philosophy* of . . . the state . . ."[9] As a result, the existence of political alienation is maintained *in practice* and "transcended" only *in theory*. The State is and remains the statutory assertion of alienation, maintained by the whole apparatus pertaining to it, institutions, laws, and politics. "The man who has recognized that he is leading an alienated life in politics, law, etc., is leading his true human life in this alienated life as such. Self-affirmation, in *contradiction* with itself—in contradiction both with the knowledge of and with the essential being of the object—is thus true *knowledge* and *life*."[10] For even the theoretical existence of the citizen and the purely theoretical "reality" of social justice are not grasped by thought and knowledge; theoretical knowledge is not an adequate expression of practical reality.

Hegel expressed in philosophical language the untrue truth of civil society, the State, and the politics of his era when he raised real existence to truth, transformed the rational into the real and the real into the rational, and saw the latter incarnated in the former. In this way becoming escapes the force of negativity, the thrust of time as historical and revolutionary, which lies within given positive actuality. Hegel has *said* what *is*; his word justifies and yet masks the true reality of what is. By the mere fact that something is, it is justified and, at the same time, it is rational justification that accords it its being. Hegel is the truthful and untruthful interpreter of the modern bourgeois State, yet he fails to translate its actual truth into a language equally real. At least, this is what Marx thinks.

Marx himself, on the other hand, wants to expose all the contradictions and inconsistencies that lie in social and political life; he wants to prevent them from achieving a conciliatory balance. His analytical dialectic works by dissociation, for he sees dissociation as the result of the fragmentation of society as a single whole into partial, particular societies that are simultaneously alienated and alienating. Contrary to Hegel, Marx says that it is economic society, the productive forces and the relations of production, that determines civil society. Civil society is not in opposition to the political State, which moreover would be situated still "higher" than civil society in order to rule over social and historical development as a whole. The State is the armature of civil society and its weapon against both internal enemies (the workers) and, in a secondary way, external enemies. The whole of economic life constitutes the basis and terrain of the social. Civil society, as organized and organizing form, is conditioned by the movement of economy both as technological and as human, and it forms the theater for class struggle. It is by this class struggle, expressing specific relationships to production and to its means, that the political reality of the State and of the administration both of personnel and *especially* of goods is constituted. Marx stands in violent opposition to any idea of the State as constituting an autonomous organism of management, one that would develop a strong bureaucratic apparatus and civil service whose functions are social in some self-styled universal way. Statism, centralism, and bureaucratism simply organize, centralize, and institutionalize social and political alienation. Administration and the civil service exercise no communal or universal function but, rather, express particular and determined reality and interests. Those who administer the affairs of State administer the affairs of a dominating class that has "confused" its business with the total social interest. In his *Critique of Hegel's "Philosophy of Right,"* Marx writes: "Hegel gives us an empirical description of the bureaucracy, partly as it actually is, and partly according to the opinion which it has of itself. . . . Hegel proceeds from the separation of the state and civil society, from the separation of the particular interests and the absolutely universal; and indeed the bureaucracy is founded on this separation. . . . Hegel develops no content of the bureaucracy, but merely some general indications of its formal organization; and indeed the bureaucracy is merely the formalism of a content which lies outside the bureaucracy itself. The Corporations are the materialism of the bureaucracy, and the bureaucracy is the spiritualism of the Corporations."[11]

Hegel, however, in his attempt to describe and justify existing reality by thinking it philosophically to its foundation in Spirit and Reason, is not so naïvely "intellectualist" (*spiritualiste*) or "idealist" as the Marxists, more than Marx himself, claim. Hegel's political philosophy has an extraordinarily *realistic* outlook, and through all the mediations and reconciliations he knows how to keep in view the tragic contradictions that lie inherent in all human history. Civil society stands opposed to economic society, while the State is in opposition to civil society, since it emerges over and above the political and civil life of concrete individuals. Political reality does not lend itself to being entirely unified and harmonized, and contradictions remain at work despite all mediations. But Marx, not wishing simply to stop before the supreme reality of contradictions, denounces them. His aim is not simply to supersede them for the sake of some total unity: he wants them eliminated.

Marx explodes the contradictions that exist between the *form* of civil society and the State, a form that is bureaucratic, and their actual *content*; between the *idea* which is thought to give life to or represent political life and the implacable and sordid historical *reality* of it; between the deceitful *spirit* of institutions and their calculating *materiality*; between the pretension of *universality* on the part of the civil service and its severely narrow *particularity*; between the *fictional totality* of the State and the *organic totality of men*; between the *citizen* and the *man*; between *public life* and *species life*. All that emerges beyond actual society is denounced by Marx in the name of historical realism and materialism. But, with his intention being thus to abolish all contradictions and realize a totalist unity, which is what the negating critique of civil society and the State aims for, is not Marx himself exhibiting idealism? This romantic longing for unity, this dream of a realized totality, this communitarian anarchism that animates the Marxian theory of alienation and its various forms and provides the term of comparison for their being deemed precisely *alienations*, is it not intensely colored with idealism? And does not Hegel here cut the figure of the realist, recognizing and accepting mediations and alienations as he does? The movement of Marx's dialectic, however, is also realistic and materialistic, especially in its aspect of analytic critique; it undertakes the unremitting dismantlement of the political, social, and economic mechanism that stifles natural and organic development. But does not any natural and organic development necessarily become alienated in social organization, whatsoever it may be? Can the enormous development of productive forces, the organization of labor, the control of technique, and the distribution of wealth be administered by society as a whole and in quasi-anarchic fashion? Can politics as such be superseded? Or is it abolished by being generalized (rather than by being brought to completion)?

Marx, who founded the movement that tried and still tries to bring his theory to realization, is less concerned with knowing what could or is going to be than with knowing what *is*. He breaks down what is to show all its faults, *as if* those faults could also not exist.[12] The political State is thus broken down and explained *materially* by civil society, which itself is a *spiritualization* of economic process, since society taken as a totality (whose essential nature Marx has not explicated very well, any more than he has that of species man) does not reach—certainly has not up to the present reached—the simultaneous abolition of "spiritualism" and "materialism." Founder of the movement that invokes his name, Marx never tires of making assertions like the following:

> The fact is, therefore, that definite individuals who are productively active in a definite way enter into these definite social and political relations. Empirical observation must in each separate instance bring out empirically, and without any mystification and speculation, the connection of the social and political structure with production. The social structure and the State are continually evolving out of the life-process of definite individuals, but of individuals not as they may appear in their own or other people's imagination, but as they *really* are; i.e., as they operate, produce materially, and hence as they work under definite material limits, presuppositions and conditions indepen-

dent of their will.[13]

As a structure erected on the real foundation of productive forces and relations of production, both of which are activated by men and not things, the State assumes an alienated form that at the same time acts to alienate, thus becoming the organization of an illusory community independent of the real universal interest. The State not only is not what it claims to be, the reality of the universal, it also derives its raison d'être from the contradiction that exists between the common interest (which is unsatisfied) and particular interests (the dominating interests). Modern man is fragmented through the processes of reification, by which both the natural and the human are eliminated. He becomes worker, salary winner, economic man, political animal, citizen, civil servant, and so on, all separately, never in the form of species man (which perhaps he has never been anyway), that is, man as a totality working, living, and organizing community life with his fellow men. Man is thus split in two: he has a *public life* and a *private life* which are incommunicable and contradictory.

The oriental, Asian empires, the city-states of Greece, Rome, Christianity and the Middle Ages, the bourgeoisie—none of them has succeeded in fully solving the problem of political life. In the despotic states of Asia, one man alone is free, the despot, and all the "subjects" are subjected to him; the political State, if it can be called that, is in the hands of one particular individual. In the ancient city-states, the community became a "truth" by the fact that it allowed the flowering of the citizen, but only free men were citizens. Nevertheless, the private and public lives of the man who was not a slave did not form two worlds foreign to each other; a man was fully a citizen of a community of citizen-men. As a reality, this was naturally only the reality of the idealistic political man and was not actually true: Greco-Roman antiquity, being an order in which slavery was intrinsic, was never able to make public matters an affair of all men.

Did the conditions that permitted the possible arrival of actual democracy result from Christianity and not from the development of productive forces? Marx's thought is not very precise on this important point. One passage in the *Critique of Hegel's Philosophy of the State* says this: "Just as it is not religion that creates man but man who creates religion, so it is not the constitution that creates the people but the people which creates the constitution.

In a certain respect democracy is to all other forms of the state as Christianity is to all other religions. Christianity is religion κατ' ἐξοχήν ["par excellence"], the essence of religion, deified man under the form of a particular religion. In the same way democracy is the essence of every political constitution, socialized man under the form of a particular constitution of the state. It stands related to other constitutions as the genus to its species."[14] Marx of course is not asserting that the Christian imperative of equality and freedom for *all* individuals (and no longer just for *one*, as in the Orient, or just for *some*, as in Greece and Rome) is at the origin of democracy; at least he is not asserting this explicitly. Nonetheless, he does tell us that Christianity is the most fully achieved form of religion and that democracy, here related to Christianity by Marx himself, is the essence of all political constitutions. This does not prevent one from thinking that Christianity and democracy have not actually realized their essence, that they have realized it in particular and partial fashion through new forms of alienation, and that their promises remained formal.

> It was most difficult to form the political state, the constitution, out of the various moments of the life of the people. It was developed as the universal reason in opposition to the other spheres, i.e., as something opposed to them. The historical task then consisted in their revindication. But the particular spheres, in doing that, are not conscious of the fact that their private essence declines in relation to the opposite essence of the constitution, or political state, and that its opposite existence is nothing but the affirmation of their own alienation. The political constitution was until now the religious sphere, the religion of popular life, the heaven of its universality in opposition to the earthly existence of its actuality.[15]

In the Middle Ages the life of the people and the life of the State were identical, but real men, while not being slaves, were far from free. "It was therefore the democracy of unfreedom, accomplished alienation. . . . real dualism."[16] The dualism becomes *abstract* with the *modern age*. Everything now becomes abstract: private life, public life, the State as such, the whole of the political. If the community among men of antiquity was still a "truth," among modern men it has become an idealistic "lie"; and modern man, the bourgeois, is a realist only inasmuch as he is a friend of money and commerce. The whole of political life becomes equally formal, forming and deforming any material content, government action takes over, and the State is appropriated by administration. Even the political regime that most pretends to universality, namely, democracy (and not republicanism), does not succeed in de-alienating political life. Democracy is indeed in principle "content and form" and "the unresolved mystery of all constitutions"; it is true that its "formal principle is simultaneously the material principle" because it is "the true unity of the universal and the particular."[17] However, since "it is evident that all forms of the state have democracy for their truth and for that reason are false to the extent that they are not democracy,"[18] therefore, lofty as it is for Marx, democracy remains in his eyes abstract and formal, bourgeois and particular. Political life rules and stifles the true life of the people.

The progressive development of administration, the civil service, and the bureaucracy further aggravates political alienation. The bureaucracy increases "State formalism," intending to be the self-consciousness and will of the State and of Society as a whole, but it is only a tissue of practical illusions; it is the "illusion of the State." It offers the formal for content and content for the formal. An autonomous entity, it is the (vicious) circle from which no one can escape. "The state formalism, which the bureaucracy is, is the state as formalism."[19] Institutions are empty of substance. Human subjectivities cannot objectify themselves through political structures. The stifling atmosphere of politics weighs heavily upon the earth of men. Man, a *social* animal rather than a *political* one, according to Marx, cannot realize himself, either in the sordid reality of civil society and the State or in the ideological or political attempts to abolish his alienation. Remedy for the evil has not been provided by the philosophers of the Enlightenment, by the liberal thinkers, by the French Revolution, by Kantian morality, by Hegel's political ethics, or by the dreams of utopian socialists. The real material life of men is still betrayed by idealistic and spiritualizing forms and abstractions that disguise the meaning of species life in society. Political powers reduce those they govern to impotence. Established state power is not a living organism but a bloodless organization, a system that, instead of integrating its elements into an ordered whole, merely makes them work mechanically like wheels and gears. With his forceful negating critique of the past and present, and open to the actual preparation of the future as salvation though without positive statements about it, Marx tears off the mask from the protagonists of political drama and demolishes its machinery. Political tragedy is disguised to those who live its true meaning, but, when its secret is perceived, it becomes comedy.

> The bureaucracy must thus defend the imaginary universality of particular interest, i.e., the Corporation mind, in order to defend the imaginary particularity of the universal interests, i.e., its own mind. . . . The bureaucratic mind is through and through a Jesuitical, theological mind. The bureaucrats are the Jesuits and theologians of the state. The bureaucracy is *la république prêtre*. . . . The mind of the bureaucracy is the formal mind of the state. It therefore makes the formal mind of the state, or the real mindlessness of the state, a categorical imperative. The bureaucracy asserts itself to be the final end of the state. Because the bureaucracy makes its formal aims its content, it comes into conflict everywhere with the real aims.[20]

The passion with which Marx thunders against the modern world is that of the Jewish prophetic figure, but his passion is armed with the weapon of critical, negation-structured analysis. It is directed against the modern world, the bourgeois world—which in fact is not a world—but it certainly throws flashes of light upon other, equally modern, worlds. And indeed his artillery is aimed not only at the political philosophy of the last great philosopher: its target is all of modernity, from the Renaissance to the nineteenth century, this whole period of history—in all its manifestations in the realms both of things and of spirit—that has pushed alienation to its ultimate limits. Ongoing technicism and bureaucratism have seized human social reality, their formalism has set itself up as a terribly real power and become their very content, the aims of bureaucracy have become confused with the aims of the State and those of the bureaucratic State with those of global society, political life has lost its senses—given all this, what must be done before all else is to describe and denounce the whole situation in order to be able finally to dealienate human history. Marx finds mask after mask to wrench off:

> Its [the bureaucracy's] hierarchy is a hierarchy of knowledge. The highest point entrusts the understanding of particulars to the lower echelons, whereas these, on the other hand, credit the highest with an understanding in regard to the universal; and thus they deceive one another.

> The bureaucracy is the imaginary state alongside the real state; it is the spiritualism of the state. As a result everything has a double meaning, one real and one bureaucratic, just as knowledge is double, one real and one bureaucratic (and the same with the will). A real thing, however, is treated according to its bureaucratic essence, according to its otherworldly, spiritual essence. The bureaucracy has the being of the state, the spiritual being of society, in its possession; it is its private property. The general spirit of the bureaucracy is the secret, the mystery, preserved inwardly by means of the hierarchy and externally as a closed corporation.... Accordingly authority is the principle of its knowledge and being, and the deification of authority is its mentality. But at the very heart of the bureaucracy this spiritualism turns into a crass materialism, the materialism of passive obedience, of trust in authority, the mechanism of ossified and formalistic behavior, of fixed principles, conceptions, and traditions. As far as the individual bureaucrat is concerned, the end of the state becomes his private end: a pursuit of higher posts, the building of a career. In the first place, he considers his real life to be purely material, for the spirit of this life has its separate existence in the bureaucracy. Thus the bureaucrat must make life as materialistic as possible.... Real knowledge appears to be devoid of content just as real life appears to be dead, for this imaginary knowledge and life pass for what is real and essential.[21]

Marx, then, drains the substance out of one function, *the* function, of modern politics, which is itself empty of substance. The civil service exists and functions in its administration of the State, civil society, and the whole of society, but its being is empty. This seems to be Marx's thinking in regard to any civil service, for any such officialdom, when considered precisely as officialdom, becomes autonomous, separate, and alienated from the species life of individuals living in society. The same condemnation applies to all institutions as oppressive realities empty of life and meaning. Anything which is in essence universal becomes, when it is particularized, alienated and alienating; men cannot live and act by delegation. Marx sets crass *materialism* and deceitful *spiritualism* back to back as inseparable and does not delineate any systematic theory of the nonalienating State; nor does he even ask how such a State would be possible. He performs his pitiless critique of what exists now as his preparation for the future, a future in which bureaucracy is to be abolished and the general interest made to coincide with particular interests, the formal becoming congruent with the real beyond any opposition. Aspiring to this kind of situation for things and men, without a State, Marx wants man to realize his species life to its full extent and to cease being two things, man on one side and citizen on the other. Total intimate life would thus be personal and public, without the two spheres constituting quasi-independent regions, since there would be but *one single* terrain for the one yet many-sided manifestation of human activity. All this is what Marx wants. He wants to universalize the economic structure of society and abolish the superstructure of politics and the State. He wants man and men to reach full realization as subjects and to rule in community form over objects. But is the actual realization of this human aim fully possible?

Marx, taking up and perpetuating Western tradition, with its basis in the human *ego* as agent, that is, in the will and its power, pushes that tradition to its ultimate consequence and rejects one particular phase of Western history, the bourgeois, capitalist phase. For it is within this phase especially that state bureaucracy rose as master of society, "in its will to do everything, i.e., in its making will the *causa prima*, for it is pure active existence, which receives its content from without."[22] The State thus becomes the all-powerful subject that as absolute, that is, as separated from its basis, treats everything as its object. Marx, in contrast, *wants* the *will of men* as agents to be coextensive with *the general social order* and *collective power*, the agent-ego thus being universalized and the whole of human history as a single phenomenon becoming the Subject who transforms, produces, and administers objects. Western tradition is thus pushed to its ultimate consequences in being extended universally; rationalist, technicist individualism is transmuted into communism, the foundations of which remain technique, *ratio*, and will; and communism finally leads to anarchy. Man, once worker, wage earner, proletarian, bourgeois, civil servant, public man and private man, theoretician, practician, man subjugated by labor and the division of labor, by capital and class distinction, by institutions and the constructions of the spirit, will become total man in a total society, and his being will be inscribed in the rhythm of the historical becoming of the Universe.

However, even though civil society and the State break down by virtue of the contradictions of their own logic, they must be assaulted by the subjects of those contradictions, men themselves. The apparent community, the State and civil society, the "community" based on the state of the economy, on the right of the State, and on the state of bourgeois law, may well become independent, may well become foreign to individual men and constitute itself over their heads; nevertheless it remains torn by class struggle. To the domination of the possessor class corresponds the struggle of the oppressed class for its total enfranchisement. ". . . all struggles within the State, the struggle between democracy, aristocracy, and monarchy, the struggle for the franchise, etc., etc., are merely the illusory forms in which the real struggles of the different classes are fought out among one another. . . . Further . . . every class which is struggling for mastery, even when its domination, as is the case with the proletariat, postulates the abolition of the old form of society in its entirety and of domination itself, must first conquer for itself political power in order to represent its interest in turn as the general interest, which immediately it is forced to do."[23] The political State is therefore only the summary of the political struggles that express, in travesty, economic interests. The socialization of production, distribution, and consumption, the superseding of the division of labor, the abolition of private property and capital, the elimination of the bourgeois and the worker as such, the universalization of human labor, the destruction of bourgeois state apparatus—in short, economic and political dealienation—begin with the *dictatorship of the proletariat*, which in turn will abolish itself. In this way, through a new particular form of "domination," humanity will advance toward integral communism, a society without classes and without the State, where the "political" fuses with the "economic" and the "social" with the "individual" and where there is no action of a government directing the action of the collectivity.

The whole analytic vision of current political alienation is to lead to the superseding of that alienation in a future which, for the first time in the history of the world, will not be based on alienation. Marx, almost at one and the same time philosopher (theorist and visionary), scientist (sociologist and economist), and politician (technician of practice), never raised in his own mind much question about the possibility of this nonalienation becoming a reality. That is what gives him both his depth and his limits. Since his aim is to annihilate reality as it is, he takes hold of things in their present existence—seeing them, of course, from his particular vantage point, yet at least that vantage point reveals an aspect of the whole of ontic reality; and by his vision of mediation as radically transformative and revolutionary he thinks he is opening up *the* perspective for the historical salvation of mankind. Questions about this prevision of the future, a prevision that he always refrains from elaborating in a priori systematic form, give him little concern. His thought can be characterized as at the same time too *theoretical* ("idealistic," romantic, anarchist, etc.) and too *practical* (naïvely "realistic," down to earth, etc.), too *open* and speculative (surveying universal history in order to dealienate it in thought) and too *fixed* and dogmatic (reducing everything to economy and class struggle). But we gain nothing of the essential in thinking we can pillory Marx's thought with designations of this kind. That thought remains still actively fertile and extremely problematic, if not ambiguous, and not simply dialectical; it can sustain several interpretations, in the realms both of thought and of reality. The political problematic of this thought is far from being univocal, and, in a world that sees the political as if it were its global destiny, the questions posed by Marx's political philosophy remain open; for politics may not be so easily reduced to the economic, nor can the dualism between the social and the political disappear so effortlessly, nor may the formal be dissociated from the material, from the real content of which it is the form.

Marx condemns politics and the State as such in the name of humanized and totalized economy and society. Nevertheless, the *economical*, the *social* (which for Marx is the total, and whose foundation is the technologically and humanly economic), and the *political* (which is infinitely more than formal) cannot be so schematically distinguished. Marx writes in explicit terms that *the emancipation of the proletariat will be above all a political action*[24] and will thus bring the last of the classes into the play and engagement of politics.[25] Politics even seems to take the lead over economy in the course of the process that transformed the thought and intent of Marx into Marxism. Already in "What Is To Be Done?"[26] Lenin speaks of the threefold perspective of proletarian struggle—economic, political, and theoretical (we always meet this tripartite schema)—and of the necessity of subordinating the economic to the political in the course of one's fight. In actuality, political power, though linked of course to economy and technique, seems still to be more in evidence than economic factors; the history of the past hundred years shows the conflict of Nations (which are a kind of reality much too neglected by Marx) and of States as combining with the conflict of Classes, to the point of even dominating the latter. For the will to power, at grips with technique, comes to gain the mastery of economy; and politics, in deployment of its will to power, develops its own dialectic.

It is not our purpose to apply Marx's thought to meta-Marxism and Marxist reality. We wish to let Marx himself speak and then comprehend his words, filled as they are with questions, in the manner of all great words. The link with reality for which one invokes Marx poses a serious problem which cannot be hastily settled in some univocal way. Would it be possible to apply the words of the Gospel to that reality which intended to be Christian—a question one asks only while keeping in mind, of course, the differences that separate the two visions and the two realizations?[27] Thought always takes flight, and reality always keeps its own weight, without, for all that, thought being unreal or nonreal. Faced with the questions of thought and the problems of reality, *man and mankind set themselves such tasks as they are unable to solve*, for not everything is of the realm of the given and the practical. Thought can at certain moments grasp that which is and blaze a path to becoming, but being-in-becoming once more slips away. So it is with political thought, the term *political* being taken in its historical, formative sense. The West saw the birth of Plato's *Republic*, St. Augustine's *City of God*, and Marx's *Communist Manifesto* in the course of its history, which is now tending to a universal, planetary becoming. Nevertheless, no thinker has ever yet penetrated the riddle of this History and brought it to light. Nor did Marx, which does not necessarily make him "utopian." He directed his efforts to deciphering reality, lending his ear to the voices of human tragedy that cried out from earth to heaven, though he held no belief in a heaven. Could he have spoken or done otherwise? Thought thinks that which is, and political action constructs and destroys cities and empires; yet that does not mean that thought is ineffective.

Marx means to surmount speculative thought once and for all, in order to bring the ideological, abstract "truth" of theory to realization in practical energy. The great heroes of ideological alienation, nevertheless, help us often enough to understand not only Marx's thought but also the economic and political realizations of practical Marxism. Plato, chief of all idealist metaphysicians, recognizes in his *Republic* that the practical, political application of true thinking is not simply taken for granted; for, he asks, a few lines after speaking of the philosopher-kings and the coinciding of philosophy with politics, "is it the nature of things that action should partake of exact truth less than speech . . . ?"[28]

As for Aristotle, whom Marx does not hesitate to call a "giant thinker,"[29] he quite calmly asserts that "the end of theoretical knowledge is truth, while that of practical knowledge is action (ἔργον)."[30]

Part IV.

Human Alienation

Human beings are the visible and yet invisible protagonists of becoming. They are the subjects and at the same time the objects of history, those who are responsible for the development of a technique while being developed by it. And yet human beings are alienated in relation to themselves and their essence. They lose their true existence in the struggle for subsistence, becoming strangers to themselves. Man as Marx envisages him has a nature that is essentially historical, and it is this (historical) nature of man that is externalized and alienated in the course of the becoming of mankind. Man is the being of all those beings and of all those realities through which he manifests himself. His essence is that of a universality, of a community of possibilities. He is therefore the species being (*Gattungswesen*) and the community being (*Gemeinwesen*) that is alienated through life in all its dimensions—the economic, the political, the familial, the human. The abolition of alienation will consequently give birth to man as a totality, and man will *become that* which he truly *is*; his *nature* will at last become *human*—because that is what it essentially is.

THE RELATIONSHIP OF THE SEXES AND THE FAMILY

Man is the species being who acts according to his nature against Nature in order to satisfy his needs. In his transformation of what opposes him he transforms himself; thus "history is nothing but a continuous transformation of human nature."[1] Every positive satisfaction leads to new needs, which in turn seek satisfaction, and so on. The life of man, whose essence is communal, and thus we should say the life of men, is therefore a perpetual realization and superseding of what men have and what men are. We have already seen how Marx takes up two of the three primordial conditions of history.[2] *The first historical condition* is "the production of the means to satisfy these needs," that is, "eating and drinking, a habitation, clothing and many other things." The second historical condition consists in this, that "the satisfaction of the first need (the action of satisfying, and the instrument of satisfaction which has been acquired) leads to new needs." To these "two" conditions and premises for all history, which together comprise "the first historical act," a "third" precondition is linked, a *third type of natural-social relationship*;[3] and yet one must take these three "factors" as a single whole. "The third circumstance which, from the very outset, enters into historical development, is that men, who daily remake their own life, begin to make other men, to propagate their kind: the relation between man and woman, parents and children, the *family*. The family, which to begin with is the only social relationship, becomes later, when increased needs create new social relations and the increased population new needs, a subordinate one (except in Germany) . . ."[4] Three lines further, Marx makes explicit how these three things link together: "These three aspects of social activity are not of course to be taken as three different stages, but just as three aspects or, to make it clear to the Germans, three 'moments,' which have existed simultaneously since the dawn of history and the first men, and which still assert themselves in history today."[5]

The history of mankind, which is simultaneously natural and social, and the conditions and relationships that generate it imply, therefore, a double production: the production of material goods by *labor* and the production of human life by *procreation*. To the *instruments of production*, then, correspond in a way the *organs of reproduction*. Men are, consequently, bound among themselves and with Nature by both natural and social ties.

That hardly means that Marx admits a duality of forces for the development of the historical nature of man. Hunger does not exist on the one hand and love on the other. Nor do sexual forces develop alongside productive forces.[6] Procreation, that is, the reproduction of the human species, is an essential condition of historical becoming, but it remains above all—though not exclusively—a natural function. The motor agency of the genesis and development of *history*, which is more than just *evolution*, lies in the utilization, creation, and development of productive forces, which determine human relationships in general as well as the form and human content of sexual relationships in particular. Men are historical beings from the very beginning and thus are more than simply natural. The development of the relations between the two sexes and of the family is, as a result, conditioned by the development of productive forces and relations of production.

Just as the division of human labor in production puts the possessors and oppressors in opposition to the exploited and oppressed, so the "labor" of human reproduction and the relationships within which it lies embedded separate master and slave within the family. Marx even goes so far as to think that the division of labor "was originally nothing but the division of labour in the sexual act."[7]

It seems that from the very beginning there was a *conflict of the sexes*, a conflict that has continued right up to our own day, meeting with and being determined by the conflict of classes. From the union of the two mutually attracting "contraries," the male and female sexes, in the conjugation of their ultimate differences, is born human being, the product of the unity-opposition of contraries. Yet human beings are not born by virtue of Nature alone; they are born as well by virtue of two human beings living in a Society. And society is characterized by the division of labor and the exploitation of man by man. "With the division of labour, in which all these contradictions are implicit, and which in turn is based on the natural division of labour in the family and the separation of society into individual families opposed to one another, is given simultaneously the *distribution*, and indeed the *unequal* distribution, both quantitative and qualitative, of labour and its products, hence property: the nucleus, the first form of which lies in the family, where wife and children are the slaves of the husband. This latent slavery in the family, though still very crude, is the first property . . . "[8]

Moreover, the progressive development of private property conditions the development of the forms of family life. Private property and the division of labor isolate the family within the community and make it a private instead of a communal reality. The movement of private property directly determines a separated domestic economy. Civil society and the State have the family as their presupposed foundation. Since the relations of private property are what bind the illusory community and its members to the natural world of objects, that is to say, bind them to it in alienation, it appears that marriage is "certainly a *form of exclusive private property*."[9] Private property, marriage, and the family are the practical foundations on which the domination of the bourgeoisie rises. The institution of the (bourgeois) family rests on quite empirical questions, namely, economic matters; and the very body of the family is the state of its fortune. What induces people to marriage is generally a combination of boredom and monetary interests, not the autonomous validity of the family. Thus, although the dissolution of the family is realized in practice and dismemberment becomes more and more common, the institution remains, because its existence is made necessary by its connection with the mode of production and civil society. Moreover, the dissolution of the family has been theoretically proclaimed by socialists in France and England and by French novelists; even German philosophers have taken note of it, Marx adds in irony. Two things, now, are meant by dissolution of the family: a *real dissolution* that actually takes place, since the bourgeois themselves are the protagonists of this dissolution, and a *necessary dissolution*, that is, the conscious recognition of this state of things and the abolition of the family in its bourgeois form through the proletarian movement of dealienation.

The family continues to exist and remains relatively intact in theory, because in practice it is one of the foundations on which the bourgeoisie erects its domination. Hypocritical by his very essence, the bourgeois man does not particularly conform to the institution, but the institution is in general maintained. Marriage and the family are negated, both in theory and in practice, on the level of individuals, but they are maintained, in theory and in practice, on the level of society. The boredom and emptiness of alienated existence, the interests of money, the theoretical criticism of marriage and the family, and the dismemberment they in practice undergo do not lead to *actual* dissolution and superseding for the personnel of bourgeois capitalist business and industry. The general hypocrisy of bourgeois existence builds a sacrosanct world for the relationship between the two sexes, which, existing as it does in the form of dehumanized reification, is a dirty thing. And this same hypocrisy, based as it is on straightforward empirical conditions, finds ways of escape and emergency exits, especially in adultery and prostitution.

Nonetheless, to the degree that man thinks of, and puts a value on, woman as prey and servant to pleasure, he expresses all his own infinite degradation and alienation. In extra-marital sexual relations, in particular in prostitution, this alienation of the true social nature of man manifests itself in a manner still more infamous, for prostitution is based on a commercial relationship that makes the one who prostitutes even more alienated and alienating than the one who is prostituted. Money, possessing the power to appropriate anything by buying it, can thus buy "love" as well. Itself the universal prostitute, the universal go-between among human needs, money confounds all the natural properties of species and communal human being, overturns them, and exchanges them for their opposites. Money "makes contradictions embrace."[10]

In contrast, according to Marx in the *Manuscripts of 1844*:

> Assume *man* to be *man* and his relationship to the world to be a human one: then you can exchange love only for love, trust for trust, etc. . . . Every one of your relations to man and nature must be a *specific expression*, corresponding to the object of your will, of your *real individual* life. If you love without evoking love in return—that is, if your loving as loving does not produce reciprocal love; if through a *living expression* of yourself as a loving person [and not in alienation and externalization] you do not make yourself a *loved person*, then your love is impotent—a misfortune [a misfortune inherent, then, in the human condition and no longer in the particular form of the social order].[11]

To enable the relationship between the two sexes to be dealienated, marriage in its bourgeois form—and all that conditions it, surrounds it, and accompanies it—must change in both form and content. For the family is only one of the particular modes of production and falls under its general laws. Thus it is that the abolition of private property and of a *separate economic life* cannot be separated from the abolition of the family.[12] Positive, effective abolition of private property means, according to Marx, appropriation of true human life (as a species life and in totality) and the radical abolition of all alienation. Man in the family will thus find again his human existence, and it will be indivisibly individual and communal. The question here concerns the *elimination*, the annihilating, of private property, not its *generalization* and extension. Consequently, the abolition of marriage in no way means the communal status of women, as if they were to be common property, as seems to be the intent of every "crude and thoughtless communism."[13] Marx condemns this kind of communism as simply a universalizing of capitalism, as containing the denial of the personality of man, and as fixing his place by way of a leveling process contrary to the true nature of man. Transcending human alienation, in the erotic domain among others, will lead man to regain "his *human*, i.e., *social* existence."[14]

Marx has, it is true, written that "the highest function of the body is sexual activity";[15] but Marx in no way thinks of man's sexual activity as a purely animal function. Genuine physical love cannot be reduced to the material act of semen secretion. Marx uncovers the deforming and constricting alienation that human sexual activity suffers in the stifling regime of private property. Far from wishing to take the particular as setting the rule for all cases by generalizing private property, Marx wants the concrete to reflect the universal. And so he says this:

> In the approach to *woman* as the spoil and handmaid of communal lust is expressed the infinite degradation in which man exists for himself, for the secret of this approach has its *unambiguous*, decisive, *plain* and undisguised expression in the relation of *man* to *woman* and in the manner in which the *direct* and *natural* species relationship is conceived. This direct, natural, and necessary relation of person to person is the *relation of man to woman*. In this *natural* species relationship man's relation to nature is immediately his relation to man, just as his relation

> to man is immediately his relation to nature—his own *natural* destination. In this relationship, therefore, is *sensuously manifested*, reduced to an observable *fact*, the extent to which the human essence has become nature to man, or to which nature to him has become the human essence of man. From this relationship one can therefore judge man's whole level of development. From the character of this relationship follows how much *man as a species being*, as *man*, has come to be himself and to comprehend himself; the relation of man to woman is the *most natural* relation of human being to human being. It therefore reveals the extent to which man's *natural* behavior has become *human*, or the extent to which the *human* essence in him has become a *natural* essence—the extent to which his *human nature* has come to be *nature to him*. In this relationship is revealed, too, the extent to which man's *need* has become a *human* need; the extent to which, therefore, the *other* person as a person has become for him a need—the extent to which he in his individual existence is at the same time a social being.[16]

Marx, who wrote love letters and poems[17] to the girl he ardently loved from his youth and who became his wife, faithful companion, and mother of his children, does not dissociate true, that is to say, human, sexuality from love (as passionate love)—just as he does not dissociate more generally nature and man, man and mankind, individual and society, subjectivity and objectivity. He protests against those who wish to reduce the force of passion in love, attacking at the same time both critical bourgeois materialists and humanity-lacking speculative idealists. "Love is a passion, and nothing is more dangerous for the calm of knowledge than passion. . . . How could absolute subjectivity, the *actus purus*, 'pure' Criticism, not see in love its *bête noire*, that Satan incarnate, in love, which first teaches man to believe in the objective world outside himself, which not only makes man an object, but the object a man!"[18]

The development of the passion of love—as related to "the profound, sensitive, most expressive object of love"[19]—cannot in fact be constructed a priori, since it is a real, material development that from the outset takes place in the world of the senses, affecting concrete individuals. And, just because of that, there can be no theoretical and speculative construction regarding love. It seems that the internal development of the passion love, that is, its origin and its aim, cannot be studied dialectically, nor can programs be outlined for it. Rather, what principally matters is that whatever may impede the flowering of love and the full, rich manifestation of human sexuality be dealienated through the abolition of bourgeois marriage and the bourgeois family, leaving entirely open the problem of what they become in the future. Marx implies that, once these two things in their bourgeois form are eliminated, men and women will form couples freely and as equals. Woman will no longer be an instrument for pleasure or (re)production, and children will be educated chiefly by the community. Woman will no longer be dependent upon man, the couple will no longer depend upon the civil and economic situation, and children will no longer depend upon their parents. But, to repeat, that does not mean that women will be a community item and desire a socially generalized phenomenon, which is what crude communism wishes, to the horror of the bourgeois mind. Sometime after Marx's youthful philosophic writings, the *Communist Manifesto* offers one more explanation on this score:

> The bourgeois sees in his wife a mere instrument of production. He hears that the instruments of production are to be exploited in common, and, naturally, can come to no other conclusion than that the lot of being common to all will likewise fall to the women. He has not even a suspicion that the real point aimed at is to do away with the status of women as mere instruments of production. For the rest, nothing is more ridiculous than the virtuous indignation of our bourgeois at the community of women which, they pretend, is to be openly and officially established by the Communists. The Communists have no need to introduce community of women; it has existed almost from time immemorial.... At the most, what the Communists might possibly be reproached with is that they desire to introduce, in substitution for a hypocritically concealed, an openly legalized community of women. For the rest, it is self-evident that the abolition of the present system of production must bring with it the abolition of the community of women springing from that system, i.e., of prostitution both public and private.[20]

Marx's grasp of the relationship between the two sexes, of love, marriage, and the family, is, once again, much more powerful in its analysis and rejection than in its positive assertion and prescription. It begins as always from the fundamental fact—Marx's fundamental idea—that men produce their life and produce each other, so that human being is thus simultaneously, one might say, producer and product. In this product-rich process of production and reproduction men are always, and have always been, at grips with a historical nature, and their very history is natural. What they are and what they do in the course of their becoming gets denaturalized, dehumanized, alienated, and alienating; and it thus alienates, dehumanizes, and denaturalizes them. Everything becomes rigid, fixed, and thinglike (reified) in the very course of its progressing, universalizing evolution. *Work* and *love* suffer the same fate, and it is highly significant that Marx associates *sexuality* with *technique*, going so far as to see in the *sexual act* (of reproduction) the first manifestation of the *division of labor*. The development of technique is therefore the secret of the whole natural history of men (which is also nature-opposing history) and that which illuminates the becoming of love. Technique, moving farther and farther from Nature the better to dominate it, removes man as well farther and farther from his own nature. His nature is, doubtlessly, nature-opposing by essence, but in the course of human history, the history of alienation, it loses even the connection to that which it opposes: it becomes autonomous and empties itself of all life.

Because technique is on the same ground with exchange and commerce, it is consistent to see a form of commerce in the relationship between the two sexes. This relationship has always been one of *commerce*, but it is above all on the level of modern bourgeois society that, according to Marx, the alienating character of this commercialization has become blatantly obvious. Marx thus tries at the same time to discern the essence of love—still not realized after the abolition of the cause of all alienation, namely, private property—and to keep in view the realities of sexuality and the family in the bourgeois era, as both alienated and alienating. On the one hand he lauds love, physical, passionate, spiritual, and, "therefore," total love; on the other he describes all the alienated forms love has taken. Still, love is from the beginning referred to the division of labor and production, since the sexual act rests on one form of the division of labor. One wonders, then, what in fact it could mean that love will fully unfold after the division of social labor is transcended. In not dissociating sexual relations from love, has Marx grasped the import of what is called love? If he does not see the work of Christianity in the development of individual love, what can the free unfolding of individual love mean for him in a non-Christian, resolutely atheistic world, one that will universalize the individual? The human society that will succeed capitalism will have superseded conflict between both classes and sexes; man and woman will have become equals. But this society will also have abolished private property and all the forms of individualization that accompany it. Yet we ask, is love universalized still love? Once a human being is no longer "producer" and "worker," in what regard will he or she still be reproducer and even lover? What will he or she make of the "object" of "love"? What will become not only of *love* but of the *sexual act* as well, which is the original form of the division of labor, after the radical abolition of all division of labor, for this is the step through which the enterprise of human dealienation passes? With a new mode of technical and economic *production* will we see a corresponding new mode of *reproduction*?

Marx's whole thought concerning sexuality and love (which he does not dissociate, though he does not identify them either), concerning marriage and the family with its children, rests on the idea of the man-animal as making tools and reproducing. With this global, total, "species-level" idea as his base, Marx traces out the alienations that man in actuality undergoes, and then he points toward a salvation. While doing this, Marx remains nonetheless bound to modern, bourgeois, capitalist society; it is this society that he dissects, but it is also this society that provides him the weapons in his fight for "a movement beyond." Thus it is that he carries on the whole critique exercised in the course of the two last centuries in regard to social forms of love and family institutions. He himself refers to eighteenth-century criticism and the French Revolution in their shaking of the foundations of the family.[21] Marx continues, radicalizes, carries to its ultimate consequences, and universalizes the radicalism of modernity. Chains of confinement must be utterly done away with in order for human being to be fully realized, in order for its natural and social essence to be totally objectified, without the support of any metaphysical transcendence. So his aim is above all to abolish bourgeois marriage and the bourgeois family in order for love to manifest itself in its universality as love, "that makes man an object and the object a man." Within this world of Subjects and Objects, the modern world, one particular form—the bourgeois—is denied, but the deeper direction of the process remains the same: human subjects are to take hold of everything, universalize it, and themselves coincide with objects. But, in protesting against crude, mechanistic communism, which only generalizes private property instead of radically transcending it, does Marx really supersede that kind of thinking and offer a vision of the relationship of the two sexes *and* of love that institutes a new kind of tie? For, even though he rejects the thesis of the community of women, his own "thesis" of the dealienation and dereification of the relationship between the two sexes remains within the axis of a thinking that universalizes that which is; he avoids, it is true, a false communalism, but without succeeding in wholly freeing himself from the weight of what he denies. As a result, he denies what breaks down, marriage and the family, and generalizes what was a constitutive element throughout alienation, namely, love as passion for the object. Marx lays a shadow over his whole glorification of love, and that same love seems unable to dissipate it.

7.

BEING, MAKING, AND HAVING

Marxian thought comes to grips with objects both as themselves objects of alienation and as objects alienating man. Though remaining close to objects, it struggles to liberate them in liberating man. It is an effort that can be characterized as at the same time romanticism and realism, if there is any meaning in speaking of "isms." Marx is simultaneously a visionary and a practical man, a prophet and a technician.

Marx's central preoccupation is the being of man, his essence (*Wesen*), his true historical and social nature. Nevertheless, human being is manifested through making,[1] and making results in having. In making, man alienates himself, and his true essence is revealed only negatively, in and by alienation; it remains itself beyond reach, since all history up until now is but the development of alienation. Thus man is such that he manifests himself in his social activity, but all his activity makes him a stranger to himself, to things, and to the world. Man's being is therefore something never yet expressed in the fullness of all its possibilities. Only in deciphering the history of this alienation can one grasp this species essence; it remains positively out of reach to any empirical observation. Marx refuses to do anything else but read the book of experience, but he still must decipher it, read between the lines, and scrutinize something which is visible only in deformed images.

To be, man must make, and his basic activity, productive labor done in common with other men, alienates him from his being. We have already had occasion to see the whole Marxian theory of labor. Making results in having, and it is this having that seems to alienate in an even more fundamental way the relationship of men with all that is. Labor, the division of labor, and above all private property alienate the being of man from his true social human nature, making him seek the consolidation of his alienated being in having, in possession. Instead of being realized in the course of historical becoming, the human subject only denies itself in coming to be more and more alien to its true essence. The production of social wealth impoverishes men, the external world is falsified to them, and what they do not have pushes them to the development of wealth in an internal world. The real needs of human beings impel them to work, to making, and their work acquires fixed form in having; yet the hold that possession and ownership have on human affairs alienates human beings, just as it alienates the objects of their activity. Want, in its turn, induces men to seek satisfactions in an imaginary realm cut off from the real.

> Private property has made us so stupid and partial that an object[2] is *ours* only if we have it, if it exists for us as capital. . . . Hence *all* the physical and spiritual senses[3] have been replaced by the simple alienation of them *all*, the sense of *having*. Human nature had to be reduced to this absolute poverty so that it could give birth to its inner wealth. . . . the most beautiful music has *no* meaning for the unmusical ear—is no object for it, because my object can only be the confirmation of one of my essential capacities. . . . The care-laden, needy man has no mind for the most beautiful play. The dealer in minerals sees only their market value but not their beauty and special nature; he has no mineralogical sensitivity. Hence the objectification of the human essence, both theoretically and practically, is necessary to *humanize* man's *senses* and also create a *human sense* corresponding to the entire wealth of humanity and nature.[4]

But what is Man, this natural being that is also human (*menschliches Naturwesen*), this community and species being whose essence is expressed by a making that hitherto has been an alienating action? What is the meaning of the human senses, which should place man in contact with the totality of the world?

The being-becoming of Totality is grasped—or rather thought to be grasped—by Man. Nonontology transmutes into anthropology, but this *anthropology* remains *negative*; it tells us everything that man is not. In Marx's thought, men enter into relationship with the World by the *senses*, which in turn tie men as subjects to entities, that is, to other men and objects. Yet the natural human senses are physical and material *and* "metaphysical" and "spiritual,"[5] practical *and* theoretical. Seeing, hearing, smelling, tasting, touching, thinking (*Denken*), observing (*Anschauen*), feeling (*Empfinden*), wanting, acting, and loving all constitute the manifestation of human reality (though in alienation) and of modes of appropriation of total reality (though this appropriation is an essential nonappropriation). The totality of the senses and their meaningful activity, then, ought to insert man in the totality of what is, were there no alienation; and after the total radical superseding of alienation they will have to do just that. *None* of the senses, whether physical (*physische*) or spiritual (*geistige*), is reducible in any way to sensations, for it is the sense of objects that presents itself, or disguises itself, to human subjects. By his own essence, man could appropriate entities and his own universal being in a universal way, if it were not for alienation that separates sensation from meaning, objects from subjects, the particular from the universal, the partial from the total, making from being, having as sanctioned possession from being as the process of being made in making something else. The total, unitary, multidimensional reality inherent in men as human allows them to appropriate in a nonpossessive way both objects and this same human reality; and *appropriate* here means to actualize fully the subjective and objective possibilities in a being, to act as maker in the movement of becoming, beyond any hindrance from some static, possessive fixation.

Being (i.e., human being), making, and having, subjectivity, reality, and objectivity constitute a strange knot of problems. Man is manifested in work, but this manifestation is an externality; he externalizes himself in an alienation of his being. In that man ceases to be man in order to become worker, especially in a world where labor is divided, the act of making is alienated within necessary productive activity. A human being enters into contact with things; yet in taking hold of them in having he does not have them, and he gives them up.[6] Man is the subject of historical becoming and he is "also" subject to this becoming: man the subject produced by his industry in objects. But what do these terms, *subject* and *object*, mean in Marx? Does he give them a clear meaning?

Man is not an absolute "subject," nor is he an object. Wherein reside, then, his subjectivity and his objectivity? What is the actual reality of man (*der wirkliche Mensch*)? According to Marx, man seems to be "the subjectivity of *objective* essential powers (*gegenständlicher Wesenskräfte*), whose action, therefore, must also be something *objective*. An objective being (*gegenständliche Wesen*) acts objectively, and he would not act objectively if the objective (*das Gegenständliche*) did not reside in the very nature of his being. He creates or establishes only *objects*, *because* he is established by objects—because at bottom he is *nature*."[7] This is Marx's point of departure: man the being that is immediately natural, the subjective-objective being, real, acting being, the being who makes; and it is from here that he sets out on his analysis of the alienation that man undergoes. Man is less a subject or an object than he is *activity on the sensuous, perceptible level (une activité sensible)*. It is owing to the human and social *nature* of man that "in the act of establishing [objects] this objective being does not fall from his state of 'pure activity' into a *creating of the object*; on the contrary, his *objective* product only confirms his *objective* activity, establishing his activity as the activity of an objective, natural being."[8]

Productive labor, this creation of objects, is unquestionably the realization of the essential, substantial, and objective forces of man. Yet the manifestation of his life is the alienation of his life; its concretization is its abstraction. Man's essential reality *becomes* an alien reality. The natural "and" social world constitutes the essence of man, but his ties to the world in general make him a stranger to the world and to himself; everything leaves him upon taking form in a having. Making in consort with others becomes, then, a matter of possessing and being possessed. Man *becomes* thereby a simple *object*, setting value for both beings and things in terms of objects. Man's being is directed to realities no longer by friendship and love for them but simply in order to possess them. Historical becoming is both realization *and* alienation; and through it man both "fulfills" *and* betrays his naturally community-structured essence and falls into a commerce-minded egoism. Ours is a world in which the becoming of being has been betrayed by making, inasmuch as making is utterly engulfed in having. Man is impoverished in creating the whole, immense world of wealth, to the point of becoming an *I* who *has* a *Me*, as one might say: I have my ... me. He "is" that being that by labor only aspires to make *its own* what exists and what is made.

Nevertheless, human being attains the only true and real objectification when, far from transforming itself and all things into objects, it objectifies itself humanly, that is, naturally "and" socially, knowing how to see human society and its works in all their objectness. This alone is how it realizes its individuality communally, instead of alienating it. The abolition of private property and of the whole world conditioned upon it will permit the full objectification of the objective forces that activate the human subject. This abolition must go together with the abolition of the condition that has made all men workers. "Man is not lost in his object only when the object becomes for him a *human* object or objective man. This is possible only when the object becomes for him a *social* object, he himself for himself a social being, just as society becomes a being for him in this object. . . . it is only when the objective world becomes everywhere for man in society the world of man's essential powers—human reality, and for that reason the reality of his *own* essential powers—that all objects become for him the *objectication of himself*, become objects which confirm and realize his individuality, become *his* objects: that is, *man himself* becomes the object."[9] This ambiguous, multivalent text seems to say that, by the development of human being (i.e., of man, the indissolubly natural and social, individual and communal being) in sensuous, material, real, actual objects, the wealth of (subjective) human reality—the wealth of possibilities for realization in man's essential forces—can become for men, for the society of individuals, a field wherein are manifested and recognized the forces of human being, for these are what become reality in the action of making. The being of man belongs to sensuous material nature, and man can only appropriate nature materially. *Appropriation* here does not mean the possession of natural or produced objects, that is, alienation or *expropriation* as of alienating powers of ownership. The proper materiality of each human being exists only as an individualization of total human materiality; and from that it follows that human being denies itself and alienates itself if it wishes to possess other human beings as objects. The content of man, *his true reality* (*wahre Wirklichkeit*), is constituted by his objective essence, which is in no way separate from external materiality. It is alienation that separates content from form, the subjective from the objective, the interior from the external, matter from spirit.

The materiality of nature constitutes man. Nature is, in a way, the first object, and the forces of man as substance and subject become objective reality only in objects, in natural, produced objects. Marx does not transcend the phase of philosophy that modern philosophy has lived in since Descartes: *the philosophy of subjectivity*. It is not by generalizing subjectivity in society and making it objective that one truly supersedes subjectivity. It is of the essence of this philosophy not to have a clear view of the issue of the subjective and the objective, which is the very thing it must question, in view of its lack of adequate grounding. Marx begins with man, a subject who is objective. The being of man is real, and his aspiration is to act for realization, to achieve realization, and to be realistic. But what does *reality* mean here? Is it anything other than sensuous objectness? It does not seem so. What is real, what is *truly* real, Marx says often enough, is that which exists objectively in a sensuous way. Man the subject is, consequently, a being that is real as object.

Marx is operating with a whole particular metaphysical conception of reality, namely, reality as objectness for the senses and subject to empirical grasp; and he in no way succeeds in emancipating himself from this realistic objectivism, just as he does not get free from the hold of the philosophy of subjectivity or from the idea of objective subjects. Yet he fights against bad objectivity, that is, against reification. The reality of man is stripped off by his fall into the realm of reification; each man and all men lost their reality and their humanity, and even things are alienated. Fetishist "thing-ification" (*chosification*) alienates the objectness of reality, shatters the relationship frameworks that unite man with nature, man with men, and men with things, and transforms those relationships into relations between things, into relations that are thus falsely objective in an inverted realism. The social relation of man to man is, in its very communal essence, the basic principle of the true practice that can be realized through a dealienated making. Naturality, sociality, humanity, materiality, objectivity, and reality—that is, the essence of what is—have up to now been alienated in, respectively, the form of a denaturalization, a desocialization, a dehumanization, a dematerialization, a "thing-ification," a deobjectification even while everything has been transformed into an object, and a derealization. Superseding reification can, nevertheless, allow the essence of man to be realized and to deploy itself fully in a making that leads no more to a having. But what Marx does not see, and cannot see, is that perhaps *thingism* (*chosisme*), *reification*, *objectivism* (of the "bad" sort), and *realism* are the necessary and inevitable consequences of the whole movement to which his own thought belongs, wherein *objectivity*, *reality*, and *sense experience* have privileged status though they are powers that, even while privileged, are cut off from the rest and remain without foundation. He does not and cannot ask the crucial question: How can a consequence and a set of consequences be actively surpassed without consideration of *that of which* they are the consequences?

The radical surpassing of alienation, Marx thinks, will allow man for the first time in the universal history of mankind to realize his needs, all his needs, fully, without sinking into reification, without being dominated by the negative forces of dispossession and frustration. The need of human being is infinitely rich, and man's own internal necessity tirelessly drives him to external realization. Man is not only rich in needs; poverty, not in the sense of the artificial poverty produced by the accumulation of wealth in the hands of possessors, but as human poverty, can have a human and social significance, for it is part of those needs that direct man toward beings and things. In a society which will have alienation, it can and should constitute the positive link uniting human being and the ultimate wealth—other human beings. Poverty can thus acquire meaning, as that human need of man for man.

Man's need, his active passion, pushes him toward the objects of the material world. But the "object" of human activity, its goal, its objective, has been alienated; it has foundered in "thing-ification," reification, fetishism; real objectification has not, has never yet truly, taken place. Man is the subject constituted by objectness, yet he becomes alien to himself and to the world in taking himself for a simple (reified) object and in wishing to possess objects. A fundamental alienation separates subject and object and is the source of this dualistic alienation. Radical, total alienation, presiding over all special alienations, is revealed in all particular realities, that is, in the different fields of activity, which amount to the fields of man's alienation. The forces inherent in the substance of human being become reified, and man lives in a world of objects—alien objects. But the "objective," possessive appropriation into which he is thus led is false and abstract, that is, *abstracted* from true reality as objective but not reified. That *I have* and that *I want to have*, that the *other has* and that *I* also *want to have*—this whole dialectic between (subjective) desires and (falsely objective) havings, this whole circle of being, making, and having, is simply infernal, that is, antinatural, inhuman, and antisocial. "Alienation is manifested not only in the fact that *my* means of life belong to *someone else*, that *my* desire is the inaccessible possession of *another*, but also in the fact that everything is itself *different* from itself—that my activity is *something else* and that . . . all is under the sway of *inhuman* power."[10] But may not this *making* that leads to *having* (though a not-having), this spread of the reign of antinatural, antisocial, *inhuman* powers, this whole thing of "alienation," be more essential to human history than Marx thinks? Is it not owing to a making that goes counter to nature that history is built up? Marx thinks that, after the surpassing of alienation, making will bring being to realization. Having will be abolished, and all things will be and will be made in their plenitude. What is abstracted from reality will be replaced by the concrete in its totality and in totality as such. But is not all making "alienating" and does it not remain so, especially when it takes the form of our enormous modern technical machinery? Can this power mediating between human needs and that which exists and can be produced, this dispensing agency of the means of existence, this awful μηχανή ("contrivance, device, mechanism"), divest itself of otherness and "inhumanity"? Once again, Marx does not turn his eyes upon this problem, for his gaze is directed elsewhere. He is completely taken up by something whose existence he wants to bring to an end, namely, the tyranny of labor as labor and private property as possession-dispossession.

The victory of having over making and being is seen most clearly in the phenomenon of *thrift*. Thrift is the consummate way in which the (reified) object lays hold of man's being, it is the power that deprives man of all that belongs to him in the name of private property, it is the force that blocks man from spending and being spent. For "political economy, this science of *wealth*, is therefore simultaneously the science of renunciation, of want, of *saving*.... This science of marvelous industry [and, above all, the real economic movement that founds it, one must read here] is simultaneously the science of *asceticism*, and its true ideal is the *ascetic but extortionate* miser and the *ascetic* but *productive* slave. Its moral ideal is the *worker*..."[11] In man's deeper being lies the movement that brings wealth into being, in that he is himself driven by the wealth of his needs; but in becoming *worker*, or even *capitalist*, he is alienated both in relation to his activity and in relation to the product of his activity. He has but one desire: to have, or even to put aside in order to be able to have. He thus renounces at one and the same time his own being and the objectifications of his making; he even renounces direct possession, seeking only to save, that is, to get rich while making himself poorer. This constricting of human life is seen most particularly in the worker, the proletarian, even though alienation has transformed all men into simple workers or "laborers." The worker must have just enough to live on in order to be able to produce and reproduce, and he must have the will to live only in order to have. Surrounded by his own works, man is essentially dispossessed of what belongs to him in a mode that is not one of having; he must either accumulate and hoard capital, if he is rich, or "economize," if he is poor, whereas

131 Being, Making, and Having

there is never any economizing of troubles in his favor. The economic theory of saving is, on the level of the science of political economy, only the alienated expression of the real alienating function of thrift within the movement of economy as a productive, expropriative process. It is not only the economic life of man that thrift affects: his whole existence is made poorer and smaller. "The less you eat, drink and buy books; the less you go to the theater, the dance hall, the public house; the less you think, love, theorize, paint, fence, etc., the more you *save*—the greater becomes your treasure which neither moths nor dust will devour—your *capital*. The less you *are*, the less you express your own life, the greater is your ⟨*alienated*⟩ life [*entäussertes Lebans*], the more you *have*, the greater is the store of your ⟨alienated⟩ being [*entfremdeten Wesen*]."[12]

Making and producing, human activity, and technique have led men into the heart of the most evolved form of alienation, this alienation of the modern world that has so queerly constricted man's being, forcing it into one reduction after another. We must understand clearly Marx's generous criticism of thrift. Beyond any particular economic modality of saving, that criticism is directed at the thing itself; it strives for the negation of the luxury, extravagance, and wealth that are spread throughout the modern world; it aims for the very foundation of the human aspect (which as things stand is inhuman) in those powers that reduce man to impotence. ". . . extravagance and thrift, luxury and privation, wealth and poverty are equal. And you must not only stint the immediate gratification of your senses, as by stinting yourself on food, etc: you must also spare yourself all sharing of general interest, all sympathy, all trust, etc.; if you want to be economical, if you do not want to be ruined by illusions. You must make everything that is yours *saleable*, i.e., useful."[13]

The difficult thing to get hold of in Marx's analysis of human alienation is the nature of this entity, man, who alienates himself. For alienation to exist, someone or something must get alienated. One can ask, Just what is the human essence that becomes alienated, since there have never yet been men who were *not* alienated? Marx defines the nature (*Natur*) of man, his being (*Sein*), and his essence (*Wesen*) in rather summary fashion. His wish is rather that man be recognized as human, as a natural and social being, and that he be realized; he wishes that, through the social community and the historical process, men may fill all their needs, both natural (food, drink, clothing, habitation, reproduction, etc.) as well as spiritual (*geistigen*). Marx's starting position is, therefore, a metaphysical idea of man, an idea that he specifies only negatively, laying down the requirement that it be brought to realization in true reality. Man the human species being must, after dealienation, bring his natural nature to accomplishment; and he must do this in Society, since man is in essence a natural-human-social being, while engaged through Technique in a struggle against Nature. Yet, in all the world's history, man has only alienated himself more and more. Man externalizes himself in manifesting himself, and he alienates himself in working. By virtue of the total sovereignty of private property, "what was previously being *external* to oneself (*Sich-Äusserlichsein*)—man's externalization in the thing (*reale Entäusserung*)—has merely become the act of externalizing (*Tat der Entäusserung*)—the process of alienation (*Veräusserung*)."[14] It is *as if* a certain *interiority* in man is alienated in externalizing itself, externalization itself being alienation. Marx says little, however, about what is lost in being externalized. Marx is faithful to the spirit of the whole metaphysics of subjectivity; for, though metaphysics and subjectivity may take different forms, they proceed from one same foundation that is not easily overcome. And so Marx keeps his focus on man as led by his objective needs to manifest himself in production in order to realize his natural *and* spiritual needs, while in so doing he is alienated in the resultant realities; for they *deprive* both him and things of their *being*, and within them everything becomes external, alien, and hostile. In this way it is true for man that "the manifestation of his life (*Lebensäusserung*) is the externalization of his life (*Lebensentäusserung*), that his realization (*Verwirklichung*) is his loss of reality (*Entwirklichung*), is an *alien* reality."[15]

At the core of all the different dimensions of alienation (*economic* alienation, the basically determinant one, *political* alienation, in the state and bureaucracy, *ideological* alienation, in religion, art and philosophy) lies *human alienation*, properly speaking, the alienation of human being, to which the whole of being and man's own being as well are become alien and inimical. It is thus man that Marxian humanism wishes to dealienate, abolishing everything that prevents man—the social animal of reason—from satisfying his vital, social, and spiritual needs: in short, his *human* needs. For this social animal which is man is endowed with life forces and is active under the stimulus of his needs. The action of *human subjectivity* works on sensuous, real *objects*, and this subjectivity is itself real and sensuous. The instincts, or drives, constitute the natural motor agency that sets going the development of those forces that produce new objects.

> The *objects* of his instincts[16] exist outside him, as *objects* independent of him; yet these objects are *objects* that he *needs*—essential *objects*, indispensable to the manifestation (*Betätigung*) and confirmation (*Bestätigung*) of his essential powers. To say that man is a *corporeal*, living, real, sensuous, objective being (*Wesen*) full of natural vigor is to say that he has *real, sensuous, objects* as the objects of his being or of his life, or that he can only *express* (*äussern*) his life in real, sensuous objects. *To be* objective, natural and sensuous, and at the same time to have object, nature and sense outside oneself, or oneself to be object, nature and sense for a third party, is one and the same thing.[17]

Man, according to Marx, can manifest his life only in sensuous, real, material, "external" objects. Yet, in manifesting his life, he alienates both it and himself. Prophet of a humanism to be consummated in naturalism-socialism, Marx certainly thinks that man manifests his life in alienation because of the basic alienation on the economic level. He does not question the whole metaphysical conception of "subjectivity" and "objectivity," whose significance transcends one particular phase of social history. He does indeed wish to get beyond the problematic in which subject is opposed to objects, but he does not succeed fully in making that passage; he continues to transform that which man encounters into sensuous, material object.

It is always an objective power that drives man, instituting the action by which an object is energetically sought, even though that active power takes the form of suffering and passion. It is man's passion that inspires his action. "To be *sensuous* [i.e., to be real] is to be an object of sense, to be a *sensuous* object, and thus to have sensuous objects outside oneself—objects of one's sentience. To be sentient is to *suffer*. Man as an objective, sensuous being is therefore a *suffering* being—and because he feels what he suffers, a *passionate* being. Passion is the essential force of man energetically bent on its object."[18] Man's action is fed by his congenital suffering, by his nature that experiences passion and the passions. Yet man is not first passion and then action. As sensuous activity, he is at the outset a being of active passion. Need and suffering move in him as action seeking the overcoming of need and suffering; his action is itself passionate. Passion in the two meanings of the word seems to characterize man as such. Within the era of alienation, the action side of passion remains alienating and alienated; but, after the abolition of alienation, the action-passion of men will deploy its true forces. One must not confuse here man's true *passion* and that *unrest* that comes over him, that psychological and social *anxiety* that is a sign of the alienation of his existence. For the unsettling anxiety that takes hold of bourgeois man and makes him vulnerable to all kinds of idealistic and spiritualistic—that is, nonobjective and unreal—mystiques and mystifications is far from being an essential trait of man's historical nature. It is rather the inevitable accompaniment of the labor and boredom, of the emptiness and alienation that follow from the regime of bourgeois capitalism that keeps man in a state of inferiority and drives him to retreat into his "interiority." The "unrest" of the proletarian, on the other hand, is a sharp and violent distress; it makes him a revolutionary, inciting him to a fight to the death and inspiring him to action aimed at the true appropriation of the external world. This unrest is a victoriously militant passion.

Far from wanting to reduce to nothing all trace of human suffering, unhappiness, and "poverty," Marx thinks that in alienation these realities are prevented from taking on their true human meaning. The economic and social regime that he is fighting to the death alienates suffering, unhappiness, passion, "poverty," and human needs, all equally as much, cutting them all off from their essence and their objects. Yet meaningful suffering, the unhappiness inherent in man's being, and his active passion deserve to be de-alienated, for they hold a rich human content; they too constitute sources of action. But action must not obscure things as it does now. For political economy and bourgeois psychology know how to do justice to neither the true needs of life nor those of rest, to neither action nor passion. The need for rest in man is also alienated, and the mad race for property keeps this need from becoming an enriching human need.

All the manifestations of human being and all the negativities that it implies are, consequently, alienated. Men *are* as they are *manifested*, such is Marx's idea, but their manifestation is alienation. Where, then, is their being? Man is an *objective subjectivity*, sensuous, real, material—but *objects reify him*, reality unrealizes him, and the world of the senses makes no sense. What, then, can true objectivity and reality be, and what is the meaning of sense-experience? Man is defined as a sense activity, as a being that produces and is produced—but labor is *self-externalization* and production is dispossession. What is the making, then, that will impart being?

We do not expect to find an answer to these questions; Marx works with his ideas, without stopping to clarify his own operative concepts. He conducts an attack from the angle of analytic critique and negativity-structured polemic. He offers only a very brief definition of man; he asserts certain things about his being, posits his essence, and goes on to talk at length about the alienation of man, the ruined manifestations of his being, the betrayals of his essence. Marx, the apostle of human observation as empirical, objective, and free of every metaphysical or philosophical presupposition, describes men as he sees them, yes, but does not seem to allow that nonalienated man has yet existed. The notion of *true, real, species man* that he uses as the measure of alienation is a highly metaphysical idea; it precedes and transcends all sensuous, objective, real, empirical, natural, social, etc., experience. And it is with this metaphysical (and anthropologico-historical) idea that Marx will attack other metaphysical conceptions of man, indeed, any (metaphysical) conception of man—especially that of Hegel. By the simple fact that the very conception of man is one of the forms of alienation, that man is further alienated in being interpreted in some way, this fight against the anthropological consequences of a philosophy seems to him to be absolutely necessary.

At this level, then, Marx rises up against his "predecessor" of genius, the Hegel in particular of the *Phenomenology of Spirit*, which is "the true point of origin and the secret of the Hegelian philosophy,"[19] because he thinks that Hegel, the last philosopher, continues the alienation of man, justifying it in his thought, in the hypocrisy of his moral theory, and in the lie of his philosophy. The being of man for Hegel, Marx says, is his self-consciousness, his I, his self, his consciousness. As a result, "all alienation of the human essence is . . . nothing but ⟨alienation⟩ *of self-consciousness. The* ⟨alienation⟩ *of self-consciousness is not regarded as an expression* [*Ausdruck*] *of the real* ⟨alienation⟩ *of the human being—its expression reflected in the realm of knowledge and thought.* Instead, *the real* ⟨alienation⟩—that which appears real—is according to its *innermost*, hidden nature (a nature first brought to light by philosophy) nothing but the *manifestation* [*Erscheinung*] of the ⟨alienation⟩ of the real essence of man, of *self-consciousness. The science which comprehends this is therefore called Phenomenology.*"[20] Hegel is accused of conceiving man as egoistic subjectivity, on the one hand, and as a spiritual being, on the other. In consequence, he does not see the social nature of man or his sensuous reality, and he interprets real alienation as the alienation of self-consciousness. Concrete, total man does not exist for him, since he sees only abstract man, as part of a spiritual totality. Marx understands Hegel in his own way; he does violence to him, and his interpretation is a counter-interpretation. Yet, in doing that, Marx remains bound to Hegel, and he drives the Hegelian metaphysics in the direction of one set of consequences. Marx *will* dealienate man, who has up till now appeared only in the guise of his alienations.

The Marxian man is natural "and" social, individual "and" communal, and he *manifests* himself—alienatedly—in and by the sensuous activity of social practice. Without worrying about his own metaphysical presuppositions, since no one has yet seen this human person whose essence is alienating itself, Marx writes: "The individual *is the social being. His life, even if it may not appear in the direct form of a communal life in association with others—is therefore an expression and confirmation of social life.* Man's individual and species life are not different . . ."[21] What constitutes individual life "and" species life in man—his real, acting (*wirklich, wirkend*) nature, his actual, effective essence, his objectness (or objectivity)—is precisely sensuous activity. When he speaks of reality and objectivity and sensuousness and materiality, Marx always has in mind real, objective, sense-perceptible, material *activity*, productive making, *praxis*. In Hegel it is basically the speculative and "inactive" metaphysics that he denounces. Among those who are at war with Hegel, the Left Hegelians, what Marx finds blameworthy is a materialist metaphysics that is equally "inactive." What Marx himself is proposing is a "metaphysics" that renounces itself by accomplishing itself in (material) activity. The activity that idealism knows and recognizes is not activity, because it is mechanical and not human. The first thesis on Feuerbach puts back to back Hegel's idealism of the mind and the idealism of the thing propounded by the Left Hegelians: "The chief defect of all hitherto existing materialism (that of Feuerbach included) is that the thing, reality, sensuousness, is conceived only in the form of the *object or of contemplation*, but not as *sensuous human activity, practise*, not subjectively. Hence, in contradistinction to materialism, the *active side* was developed abstractly by idealism—which, of course, does not know real, sensuous activity as such."[22]

Marx wants practical labor in its transforming and producing action, the totality of social praxis as providing satisfaction for the totality of natural needs, to be recognized in both its real materiality and its alienation. This labor is essentially practical and objective, for theoretical labor derives from it and can at the most understand it. Yet *practice* and *reality* are not *true* as such; they are the sources of truth, but in alienation they have become alienating and alienated, putting a brake on the energy of true activity. No reality, no practice is truly real. True and real *praxis* does not consist of a cramped, limited, egoistic and commerce-minded, fetishist and reifying activity, what Marx calls praxis "in its dirty-judaical manifestation"; it consists rather of a practice that will have radically superseded alienation, a total, open, communal, perpetually revolutionary action that liberates men and things at one and the same time and that appropriates nonpossessively the whole world, that is, the world of Man and Nature. Only this praxis is truly revelatory and only the reality that corresponds to it is actually true. For the rest, every externalization of man remains a practice of alienation. The first thesis on Feuerbach, its first lines already quoted, continues as follows: "Feuerbach wants sensuous objects, really distinct from the thought objects, but he does not conceive human activity itself as *objective* activity. Hence, in *Das Wesen des Christenthums*, he regards the theoretical attitude, while practice is conceived and fixed only in its dirty-judaical manifestation. Hence he does not grasp the significance of 'revolutionary,' of 'practical-critical,' activity."[23]

It is just this permanent revolutionary activity, at once critical and practical and antispeculative, that Marx sets up against Hegel and Feuerbach. His task is to dealienate man practically and in actuality by dealienated praxis. Turning his back upon being as being-thought and upon self-consciousness, upon absolute knowledge and the dialectic of essences in thought, he sets his sights on the carrier of all these other things, namely, real but alienated man. As Marx sees it, Hegel "posits" man not as a concrete being who deploys a real activity of realization but as a self-consciousness; and he goes so far as to have thingness (*Dingheit*) be in general posited by self-consciousness. As a result, man is not a natural being; he is not endowed with objective forces of substance that are sensuous and does not have real, sensuous objects as the objects of his need and the objective of his action. He is only a subject that is spirit, a portion of absolute mind, a self-consciousness that encounters, through its manifestation, not the reality of things but the thingness posited by self-consciousness itself. The products of human activity appear in this perspective as products of absolute spirit, as spiritual elements, as ideal beings, as moments of ideal totality. Hegel is thus accused of not having the eyes to see the tragedy played out upon earth, which is the only tragedy, of considering it only the repetition of the play of heavenly ideas. Marx recognizes that Hegel's vision of the "unhappy consciousness" contains dramatic and critical elements, but it does so in alienated form, because the subject and the objects of historical becoming remain self-consciousnesses and objects posited by mind. The real alienation of man keeps appearing as an alienation of self-consciousness, and objectness dissolves into spiritness. In this way Hegel superseded the object only as object of consciousness, while maintaining both the alienation of real human being and the alienation of real things. He does all this, that is, he does all this *in thought*, because he has not seen that man is a natural, sensuous being, having his nature outside himself, participating in the movement of Nature, and constituting the subjective locus of objective forces, a being whose action of making has only alienated itself, a being dispossessed of its own products.

The Marxian interpretation of Hegel, as any creative interpretation will, distorts as it interprets. With Hegel, Western metaphysics reaches its culmination. It is a philosophy that culminates in absolute spirit and absolute knowing, in the grasp of the absolute as Subject and by Subject as historical. Marx takes the other leg of this idea and continues it, positing the human subject as the objectness of the essential, material forces in man, forces that are manifested in action. All action has up till now been alienated, yet it is practice alone that can dealienate man's being and objectively fill his needs. Marx reproaches Hegel with having confused self-consciousness in man (spiritual being) with the material reality of his alienated life, with having failed to grasp real alienation and instead taken the alienation of self-consciousness in an alienated way, that is, in an abstract and speculative way. Human life thus becomes philosophic life, and actual existence becomes thought existence. Thought and philosophy consequently keep human alienation in its material tragedy going, give a justification for the existing state of things, and move beyond it only in some metaphysical sense. The moral teaching that flows from this conception is conformist and hypocritical, compelling man to accept what is, to accept his alienated being, and to accept making action as alienating action. Moral conscience "overcomes" externalization in thought in order better to maintain it in reality: ". . . the supersession of the alienation becomes a confirmation of the alienation; or again, for Hegel this movement of *self-genesis* and *self-objectification* in the form of *self-alienation* [*self-externalization*] and *self-estrangement* [*self-alienation*] is the *absolute*, and hence final, *expression of human life*—of life with itself as its aim, of life at peace with itself, and in unity with its essence."[24]

In his attack on false interpretations of human activity and on the illusions of self-consciousness, Marx is particularly concerned with their practical presuppositions and their "ethical" consequences. For him the basis of all alienation is and will remain labor in the regime of private property; it is economic movement on the real, material level that conditions movements of alienation in ideology and ethics. *Morality*, an autonomy-asserting realm within which man pursues his alienated life, is not an autonomous tribunal. It is conditioned by the processes of production and serves to provide a mask for their meaning, that is, for their meaninglessness. Morality is ruled by the powers of having, being the expression in disguised fashion of the commandments of political economy. Moral conscience, far from manifesting and guiding man's being as it unveils itself in making, is one of the forms and forces of alienation. It is rooted in real alienation, and it serves the cause of the repression of natural, social, objective, and material needs. Morality is one of the key pieces of the (superstructural) edifice that maintains human alienation and its real structures. It has neither independence nor a sphere of becoming all its own; it is rather the expression of the vital, material process of human activity in a deformation of its meaning. It has no history of its own, being bound to the history of the development of material production, that is, to the history of the alienation-externalization of man. When a system of morality comes into contradiction with existing social conditions, this happens not for moral reasons but because existing social conditions have come into contradiction with existing productive forces and impede their development. In offering an alienated ideal for alienated human life, morality stands opposed to the full unfolding of human nature; it keeps labor safe in its alienating role and prevents needs from progressing toward their full satisfaction through a dealienated activity of making that nothing would hamper or halt.

Neither spiritualistic morality nor that morality which is passively materialistic succeeded in being truly educating powers. Spiritualistic morality allows no place for man's revolutionary activity, materialistic morality does not allow enough place for *activity*;

and it is only activity of this revolutionary kind which is capable of breaking the chains of alienation that make man an isolated subject and things reified realities. "The materialist doctrine concerning the changing of circumstances and upbringing forgets that circumstances are changed by men and that it is essential to educate the educator himself. This doctrine must, therefore, divide society into two parts, one of which is superior to society. The coincidence of the changing of circumstances and of human activity or self-changing can be conceived and rationally understood only as *revolutionary practise*."[25] Since human being manifests itself in making and since every manifestation has been nonetheless an alienation and making itself an alienating process, only an action of making that is dealienating can radically break the chains of alienation. The making that leads to having is the root of alienation; the making that abolishes having is the antidote for it. The morality that is seated in real economic movement and the moral conscience that lies in political economy belong to the sphere of what is to be superseded, since they watch over the preservation of alienated labor and having, at the expense of man's being and the true powers of his activity. For morality consummates human alienation as much as the alienation of self-consciousness; it opposes political economy only *in appearance*. True, one abstracts from morality to the degree that one does political economy, and vice versa: one abstracts from political economy when one undertakes the consideration of moral principles. But *in reality* morality serves and ministers to economy, putting a brake on production and consumption, that is, on the satisfaction of needs, not in opposition to economy but in league with it.

> The ethics of political economy [and of its real basis] is *acquisition*, work, thrift, sobriety—but political economy promises to satisfy my needs. The political economy of ethics is the opulence of a good conscience, of virtue, etc.; but how can I live virtuously if I do not live? And how can I have a good conscience if I am not conscious of anything? It stems from the very nature of ⟨alienation⟩ that each sphere applies to me a different and opposite yardstick— ethics one and political economy another; for each is a specific ⟨alienation⟩ of man and focuses attention on a particular round of ⟨alienated⟩ essential activity, and each stands in an ⟨alienated⟩ relation to the other.[26]

141 Being, Making, and Having

At the heart of all these alienations, then, lies human alienation, the alienation of the essential forces of man's objective subjectivity. In the course of the development of productive technique, man only alienated himself from his being, from his activity, and from the products of that activity. Man feels more deeply the increasing unsatisfaction of his needs. His making is hobbled, and the totality of what is, is refused him. In the propagation of the species, in (bourgeois) love, and in the family he is alienated as well, and in the hold that having has upon things his being is stifled. The awareness he has of himself is inadequate, and his self-consciousness is not truthful. Morality, lastly, contributes to the preservation of alienation, giving him but one way out: revolutionary practice.

The adjectives that Marx uses to characterize man in the alienation of his being are offered, of course, without much foundation. Everything is deemed to be *sensuous, material, objective,* and *real* in its essence, which is discovered so far only negatively. Onto this whole historical and anthropological materialism, inspired by metaphysics but without an explicit ontology, is grafted a spiritual component (spiritual needs, spiritual senses, etc.), which Marx accords limited importance. In this way, he extends metaphysics, through an operation of negating reversal, intending that metaphysics become realized on the plane of action. Action aiming for what? For the satisfaction of needs. The Marxian view of morality is a bit short-sighted, albeit radical. Marx has given little serious thought to ethics, preoccupied as he is with the definition of the *ethos* of man as he understands him, as Modernity seems to demand. Unconcerned about Antiquity, Marxian humanism wants to dealienate man not in order to *recover* his essence but in order to find it for the first time, in bringing it to reality. The Promethean dream has never yet been realized. The whole progress of technique coincided with the progress of alienation. Man has no place worthy of him, every habitat fails him. Alienated man, become worker, "dwells" in a strange and foreign world, a hostile and alienated world; and all men have become workers without habitation and without abode. Man has lodgings but nowhere any home. "For this mortuary he has to *pay*. A dwelling in the *light*, which Prometheus in Aeschylus designated as one of the greatest boons, by means of which he made the savage into a human being, ceases to exist for the worker."[27] Is the savage then less alienated? Does he have a place that is his own? "The savage in his cave—a natural element which freely offers itself for his use and protection—feels himself no more a stranger, or rather feels himself to be just as much at home as a *fish* in water. But the cellar dwelling of the poor man is a hostile dwelling, 'an alien, restraining power which only gives itself up to him in so far as he gives up to it his blood and sweat'—a dwelling which he cannot regard as his own home where he might at last exclaim, 'Here I am at home,' but where instead he finds himself in *someone else's* house, in the house of a stranger . . ."[28] Is there any need to point out that this situation goes far beyond the problem of worker dwellings and the housing crisis? It is the being of modern man, of man become laborer, that is here at issue; it is the essence of man as turned into worker, and paying mortally for whatever he has, that is being taken up here. And this *being* (*Sein*), this *essence* (*Wesen*), and this *existence* (*Dasein*) of man have no place and dwelling. Men have become men without a country, they no longer have an "at home" (*ils n'ont plus de "chez-soi"*), they are rootless. Caught in the engagements of property as if in a great machine, man loses his own being, finding himself deprived of all enjoyment and naked of any shelter. In vain he tries to take refuge in what is a tomb, a hostile lodging, and an alien house.

Part V.

Ideological Alienation

The technological development of mankind is the real base that determines mankind's ideological development, the superstructure, that is, law and morality, religion and art, philosophy and science. Technology takes priority over ideology, and we have already quoted the famous passage in the preface to the *Contribution to the Critique of Political Economy* that states and summarizes the essential theses of Marx's historical materialism. What Marx calls ideology, however, has to be well understood, even though Marx, in giving it a categoric definition, leaves it open to a great many possible meanings. To the instrumental development of technique correspond ideas, themselves instrumental things; but this instrumentality poses some problems. Human consciousness expresses, *translates*, faithfully or unfaithfully, and *conducts* the transforming activity of practice. This consciousness, this theory that arises from praxis, can be real and truthful, even if it has never yet been so entirely, or abstract (from reality) and alienated. For to the process of real alienation corresponds the movement of ideological alienation.

THOUGHT AND CONSCIOUSNESS:

TRUE AND REAL?

Men produce their material (and total) life in a specific way as determined by the nature of their surroundings, their bodily organization, and the degree of development of productive forces; and in this way they evolve in a history. Men also *know* a historical becoming. From the outset, their thought takes part in this movement of becoming history, and it is thought as well that can afterward return upon that becoming in order to try to understand it. What is *thought* or *consciousness* or *knowledge*? Marx does not distinguish these three terms and employs them interchangeably. The very first kind of consciousness, real because actual and active, is a practical consciousness: *language*. Language, the absolutely constitutive element of thought, is material in nature, according to Marx. Language and thought (consciousness and knowledge) arise only from practical need, from the necessity of intercourse among men. The animal, not having a relationship with other beings precisely as relationship, does not have language and thought. Consciousness is a social *product* and can only remain such. It has its source in material nature and manifests itself in the historical development of human society.

We have already had occasion to speak of the "three sides," the three "moments" that constitute human history: (1) *the production of goods* that allow the satisfaction of natural needs, that is, the production of life by labor; (2) *the movement* that leads the action of satisfaction and the already acquired instrument of satisfaction on to *the production of new needs*; and (3) the production of another's life through *reproduction*. These "three factors" coexist, presiding over what men do in order to be able to "make history." Another power has now been added: "We find that man also possesses 'consciousness,' but, even so, not at the outset, 'pure' consciousness."[1] Is *consciousness* itself, then, a *supplementary* power coming after the constitutive powers, material production and reproduction, as a spiritual production? So Marx seems to think, since he gives consciousness the very last place here. Nevertheless, in a marginal note to the text just quoted, Marx adds: "Men have history because they must *produce* their life, and because they must produce it moreover in a *certain* . . . way: this is given by their physical organization, as also their consciousness."[2] Now one can read this hastily written marginal comment as meaning (1) that *"the physical organization of men," "just as their consciousness,"* makes production and historical life possible and necessary or (2) that the *production of life*, just like *consciousness*, is made possible and necessary by the physical organization of men. But, whichever way Marx intended it, he continues to move within the dimension of thought in which there is a "separation" between the *physical* and the *metaphysical*, *reality* and *idea*, *matter* and *mind* (*spirit*), *practice* and *theory*. Marx's thought is metaphysical, even though he wishes to pass beyond opposition and dualism; it is metaphysical, as all Western European thought since Descartes is, in the very fact that it gives a privileged position to the physical and the sensuous in relation to the "metaphysical." Marx thinks and operates with the *distinction* and *difference* that exist in a metaphysical way between *what truly is* and *what only secondarily is*. For him, that truly and fundamentally is which is not suprasensible but sensible; and yet this tenet is stated within a *metaphysical oppositeness*. The World, the totality of being, remains double: material *and* spiritual.

Accordingly, "after," yet bound to, the instruments of economic production and the organs of human reproduction is the tongue as instrument of communication among men, the basis of thought, of consciousness, and of knowledge. The tongue is, consequently, a tool. It does not belong to the realm of ideological superstructure. It constitutes the immediate reality of thought, which is practical consciousness. Thoughts do not exist outside language. Language is taken in a twofold—metaphysical—manner by Marx. "From the start 'spirit' is afflicted with the curse of being 'burdened' with matter, which here makes its appearance in the form of agitated layers of air, sounds, in short, of language. Language is as old as consciousness, language *is* practical consciousness that exists also for other men, and for that reason alone it really exists for me personally as well; language, like consciousness, only arises from the need, the necessity, of intercourse with other men."[3]

At the beginning of the historical becoming of mankind, consciousness is only a sense consciousness in active relationship to the immediate sense surroundings, people and things in the environing world. Being a social product, it is engendered by the needs of interhuman communication. It is through these still narrow, limited, and confined connections binding men to nature and with themselves that the individual, in society, begins to *become* conscious, to acquire thought and consciousness little by little. With regard to the terrible, terrifying thing nature is, men still act in an animal fashion; they are subject to it, having not yet become themselves the subjects of the historical transformation of nature. Still, consciousness is beginning to break through and men are beginning to take notice of their social life, through their restricted contacts and limited relations. "This beginning is as animal as social life itself at this stage. It is mere herd-consciousness, and at this point man is only distinguished from sheep by the fact that with him consciousness takes the place of instinct or that his instinct is a conscious one."[4] The degree of social productive development, that is, of nondevelopment or underdevelopment, conditions this phase. The "identity of man and nature," this nonrise of human history from nature, "appears in such a way that the restricted relation of men to nature determines their restricted relation to one another, and their restricted relation to one another determines men's restricted relation with nature,"[5] as Marx dialectically sees the situation. Equally "dialectical" is the original relationship between consciousness and history, since consciousness develops only in and by history, even while being one of the factors and preconditions of historical becoming.

Marx seems to be less committed to the producer role of thought (and consciousness) than to its derived role, since he considers it as a product of the development and expansion of productive forces, the multiplying of needs, and the increase of population and social relations. Nevertheless, consciousness has a part in all that and, in a certain way, coproduces that movement.

Do thought, consciousness, and knowledge really and truly take hold of being before alienation occurs, before that moment in social development when the *division of labor* is instituted as such? Before labor is divided into *material labor* and *intellectual labor* (*geistige Arbeit*)—a division in which labor really does acquire an alienated form—does "theory" actually express "practice"? Are there forms of thought that occur prior to alienation? The social

148 *Ideological Alienation*

process, product of human activity and producer of mankind, engenders and imposes the necessity of language, thought, consciousness, knowledge—*social mind*; yet does what is brought into language, what is thought and made conscious and known, correspond to true reality? We have already said that Marx does not define fully what he calls *reality*, and the same can be said of *truth*. Real, actual (*wirklich*) means above all "active," "efficacious" (*wirkend*), and what is true, really true, is that which sets action going. True thought would consequently be that theory which serves a practice, a nonalienated, "true" practice. General thought—real and true knowledge and consciousness, recognizing reality and contributing to action—"is only the *theoretical* shape of that of which the *living* shape is the *real* community . . . The *activity* of my general consciousness, as an activity, is therefore also my *theoretical* existence as a social being."[6] True consciousness would then be this general, theoretical consciousness by which man the individual would express his social life, reflecting and repeating his real existence in thought. The social and common species being—the practical activity of men—would thus be manifested in a general, theoretical consciousness. Nevertheless, this distinction between *theory* and *practice* already implies alienation, since it is a work of the alienating division of labor. Marx works with this distinction, even when he takes the two terms together in their dialectical unity, without having provided their grounds. Although he considers the opposition between the *material* and *practical*, on the one hand, and the *intellectual* and *theoretical*, on the other, or between *real structure* and *idealistic, ideological superstructure* as coextensive with alienation, he in no way succeeds in locating himself at some point prior to or beyond that division. Since every "theory" expresses a "practice" and since all human activity has been alienated, we have the right to suppose that Marx does not allow that a truly true or really real knowledge existed in the past, even before the division of labor. All thought, all consciousness, all knowledge were "ideological," alienated, restricted, and determined by the basic alienation: economic alienation.

It is certainly true, of course, that to the real need of men seeking to satisfy their drives by production and reproduction corresponded a conductor consciousness and to real practice corresponded a "true" theory, for consciousness contributed to the building of social history. It is true further that men have known a progressing historical evolution; but, because all history has remained a history of alienation and because history further creates the conditions for the superseding of alienation, all history of consciousness is just as much a progressive becoming of alienated consciousness. For the intellectual (*spirituel*)—ideas, thoughts, and representations—possesses no autonomy, being in constant opposition, for Marx, to the practical, the material—activity and real presences. Real ideas, because they were active, have been and are, of course, actual conductor agencies within global alienation; but the best thing they do is, at the very most, to allow a reflective grasp of alienation. Until the radical surpassing of alienation, all "reality" remains not real and no human activity has ever been true. The criterion of theoretical truth belongs to the practical order, it is active; and yet, despite all that, there has not yet been any true, dealienated practice. It is therefore *true* practice, real reality, that is the criterion of truth, not simple activity (i.e., alienated activity). We shall be able to see what true praxis is only when we sketch out Marx's vision and program for superseding alienation in the fully achieved naturalism, the communist humanism, the activism that wipes away all ideologies. For the time being, we are limiting our efforts to understanding how Marx brings all thought back to praxis, the praxis that satisfies real needs and that is not some hunt after chimeras. The second thesis on Feuerbach states: "The question whether objective truth can be attributed to human thinking is not a question of theory but is a *practical question*. Man must prove the truth, i.e., the reality and power, the this-sidedness of his thinking in practise. The dispute over the reality or nonreality of thinking that is isolated from practise is a purely *scholastic* question."[7] Since all practice has only been alienated activity, it necessarily follows *a fortiori* that all thought has existed only in alienation.

Truth is something not to find but to do and produce, something to make through true dealienated praxis. But, in saying this, is Marx not doing theory? On the other hand, in making *reality* and *power*, *"this-sidedness"* (*Diesseitigkeit*) [i.e., earthliness], the criterion of (practical) truth, does he not relinquish all theory?

It is really hard to grasp the genesis, the role, the function, and the mission of thought, that is, of consciousness and knowledge, of theory and idealistic "superstructure," of the sphere of intellect, of what Marx calls spiritual need. Before the division of labor, men *speak* and *say not what is* but what *seems* to them, because of their limitations, their nondevelopment of technique, and the restricted character of their relationship with nature and other men. Their language is not the saying of truth, since their practice is so little true, so restricted, so little an action of transformation. At this level, at the beginning of the process by which nature becomes history, men have only "a herd-consciousness."[8] Their very language (which is not something belonging to the realm of superstructure) and consciousness merely stammer along, so little true they are, that is, having so little reality and power. Yet, it is by this very consciousness that the history of mankind is built up. And man, the animal that makes tools, that is just barely rational, thus enters into the history of alienation. One can say practically nothing about this phase prior to the division of labor; to apply categories such as *practice* and *theory* is inadequate. For this "herd-consciousness" acquires its development from the development of the productive forces that condition the division of labor. The division of labor is instituted by, and itself institutes, the division between *real, practical, material labor* and *derived, theoretical, intellectual labor*. Before the division, there was almost no "theory," just as there was very little practice; this very distinction did not even exist. "Herd-consciousness" had neither truth in any lofty sense nor reality and power to any great degree before labor progressed and became alienated in being divided. "From this moment onwards consciousness *can* really flatter itself that it is something other than consciousness of existing practice, that it *really* represents something without representing something real; from now on consciousness is in a position to emancipate itself from the world and to proceed to the formation of 'pure' theory, theology, philosophy, ethics, etc."[9]

Men produce material goods, really fabricate the instruments of production, reproduce, and *also* produce ideas, thoughts, concepts, intellectual (*spirituel*) instruments. What men produce with their hands is most real and most powerful, even if inscribed in an alienated practice (while with their genital organs men actually reproduce other men). The products of their heads, on the contrary, in a certain way express what they make, but it is by passing beyond material reality, by alienating themselves. Consciousness is a product with the function of translation that at the same time is a betrayal; it is a knowing that derives from a making, but as a deceitful accompaniment. The first twenty-odd pages of part one of *The German Ideology* try to make explicit the Marxian theory of ideology, and it is their naturalistic, humanistic, and activistic radicalism that must be radically understood. One must render to Marx what is Marx's, namely, his genius and his narrowness, and not try to dissolve his language into a language without contours. Here is what he says, which we must try to understand: "The ideas [*Vorstellungen*] which these individuals [i.e., productively active individuals] form are ideas either about their relation to nature or about their mutual relations or about their own nature. It is evident that in all these cases their ideas are the conscious expression—real or illusory—of their real relationships and activities, of their production and intercourse and of their social and political ⟨organisation⟩.[10] The opposite assumption is only possible if in addition to the spirit of the real, materially evolved individuals a separate spirit is presupposed."[11] Marx seems here to admit the possibility, the reality even, of a real conscious expression. Yet, he denies it in fact, since every "conscious" expression, even one that is somewhat real because it is an active agency, is only the alienated theoretical expression of real practical alienation.

We shall have the occasion yet, in the pages that will follow, to show what Marx actually thinks of human thought, namely, that it has always hitherto been false. But right now, let us continue reading our present text:

> If the conscious expression [*Ausdruck*] of the real relations of these individuals is illusory, if in their ⟨ideas⟩ [*Vorstellungen*] they turn their reality ⟨on its head⟩, then this in turn is the result of their limited material mode of activity and their limited social relations arising from it.[12] The production of ideas, of conceptions, of consciousness, is at first directly interwoven with the material activity and the material intercourse of men, the language of real life. Conceiving, thinking, the mental intercourse of men, appear at this stage as the direct efflux of their material behavior. The same applies to mental production as expressed in the language of politics, laws, morality, religion, metaphysics, etc., of a people.[13]

Corresponding to the gestures and relationships of *material* and *practical* production, there is language as resulting from *intellectual* and *theoretical* production. Production is the common denominator; yet this production is a *double*, even *dualistic*, phenomenon, and Marx hardly overcomes a two-sided conception of man: physical and metaphysical. To the alienation of activity corresponds an ideological alienation, the illusory expression of a real alienation. And here we continue our reading of this rather forcefully expressed text:

> Men are the producers of their conceptions, ideas, etc.—real, active men, as they are conditioned by a definite development of their productive forces and of the intercourse corresponding to these, up to its furthest forms. Consciousness [*Bewusstsein*] can never be anything else than conscious existence [*das bewusste Sein*], and the existence of men is their actual life-process. If in all ideology men and their circumstances appear upside-down as in a *camera obscura*, this phenomenon arises just as much from their historical life-process as the inversion of objects on the retina does from their physical life-process.[14]

Sketching the genesis of ideology with broad strokes, Marx wants to bring everything that gives the appearance of the *ethereal* and the *celestial* back to the *earth* and the *soil* from which it issues. It is on the earth that real, concrete, active, productive men act, determined by the degree of development of productive forces. These same historical men thus produce the products that mount from the earth toward heaven—ideas and representations, thought and consciousness. All these elements of idealistic and intellectualistic superstructure emanate from being—not *Being* but the *being* of *man*—express being, reflect and invert being; they constitute the material of ideological alienation, of the concealing of truth; they are deceitful, illusory mirages that express through consciousness the alienation of that of which consciousness is the alienated awareness. The thinking that truthfully expresses (human) being, real consciousness and thought as guide to practice—not a simple producer practice but a total praxis whose goal is the total satisfaction of needs—this thinking has never yet existed within alienation. At best, a half-true thinking, a partially discovering consciousness, has been able to show through, but that does not make Marx feel any particular attachment to intellectual productions; they have but minor significance and importance. The *camera obscura* that inverts the position and sense of phenomena in the world of ideology is the pendant of the real inverted world; within it no clarity is to be found, and all meaning fails.

Intellectual production is in no way *identified* with material production; it is reductively *brought back down to it* (*ramené à elle*). In opposition to Hegel, who saw in history the development of spirit, Marx offers his own design, an inversion of Hegel's. Marx inverts *metaphysics* by no longer seeking in the invisible and suprasensible the foundation for the *physical* (the visible, the sensible, or sensuous) and by making the material, the practical, the empirical the foundation for the spiritual. He moves here completely along the axis of the distinction, the difference, the cut, the fault, between what *truly is* and what by that standard *is* only in a *derived* way; and so he remains still a metaphysician. He is a metaphysician of the inversion of the order of the "two worlds," for he speaks of the *base*, the *real structure*, which has now become what truly is, and of *superstructure*, of *idealistic, spiritualistic emanation*, which, measured against the true, is now only derived. Marx's materialism is not concerned with *cosmic matter*, first and basic being, ontological *archē*. Marx, having little "talent" for "ontology" and "cosmology," turns his penetrating gaze toward *the matter that matters for human being*, the matter that makes everything in existence come to appear as *material for* productive, transformative *labor*. This material, practical, sensuous labor, this activity of making real, is really alienated; and as a result the spiritual production that emanates from it is additionally alienated. Before all else, the issue for revolutionary praxis is to dealienate objective human labor so that it can unfold to its full perfection, so that it can take place in *real* conditions that would prevent the clouds of ideology from obscuring the horizon of human social activity. In the transformation of Nature into History by Technique, man will annihilate all ghosts and fantasies.

Nevertheless, if he were reduced to practical activity alone, man, being defined by his social, productive nature, would not be *human being* but an animal of some higher species. So Marx is obliged to introduce metaphysics (as a derived world)—that is, concepts and ideas, consciousness and thought—making this world a world *complementary* and *supplementary* to the world of the physico-historical and the animal-human. In the manner of the dialectician he wants to be, Marx sees the human world, whose objectness is defined in terms of the object of human labor and whose reality is that it be the object of real action, as the base, as the *first* and fundamental reality. Upon this base a supplementary structure is erected, a *second* world, the world of subjective conceptions (*représentations*). Despite what Marxists or anti-Marxists might wish, Marx does not succeed in reducing the duality, even though he struggles to do so. He begins from it in what is a continuation of the metaphysics of Subjectivity, the metaphysics of Man—man the subject or objective subject who by his making activity, by his will and conception, takes hold of a world outside himself and "grasps" it in a representation of it. We shall see that even the superseding of alienation will not open up into the identity of the two worlds; the "spiritual" world will continue to lead its own existence. This spiritual world, however, which Marx does not succeed in making into nothing, is violently reduced to the world of material production, of which it is a kind of pathological excrescence. But would man be human without this excrescence? We shall try once more to listen to Marx's words, as he struggles against ideological alienation, the *camera obscura*:

> We do not set out from what men say, imagine, conceive, nor from men as narrated, thought of, imagined, conceived, in order to arrive at men in the flesh. We set out from real, active men, and on the basis of their real life-process we demonstrate the development of the ideological reflexes and echoes of this life-process. The phantoms formed in the human brain are also, necessarily, ⟨supplements⟩ [*Supplemente*][15] of their material life-process, which is empirically verifiable and bound to material premises. [Is he not here working his metaphysically anti-metaphysical inversion of the "spiritual" in favor of what is *different* from it, namely, the material?] Morality, religion, metaphysics, all the rest of ideology and their corresponding forms of consciousness, thus no longer retain the semblance of independence. They have no history, no development; but men, developing their material production and their material intercourse, alter, along with this their real existence, their thinking and the products of their thinking. Life is not determined by consciousness, but consciousness by life.[16]

The social nature of men, whose humanity has its historical foundation in labor, is always expressed by an ideology. According to Marx, men are such entities as manifest themselves. And, if their manifestations are alienated, it is their being that is alienated. Alienated being (*Sein*) in turn determines the alienation of consciousness, of conscious being (*Bewusstsein*); and this alienation intensifies still more the errance of mankind, which is all "prehistory" prior to Marx. Theory that derives from alienated practice remains restricted, limited, deceitful. Consciousness of every kind—"herd-consciousness" as well as immediate sense-consciousness, elementary conceptions about nature and interhuman relationships as well as more elaborate knowledge, and, lastly, systematic, apparently independent ideologies—have provided spiritual, ideal, expression for the material, real alienation of alienated men, without recognizing the truth of what they were expressing. When a certain idea arose in a quasi-real way, when some real awareness came to light, this as well took place within alienation as a total situation and in particular within the alienation of ideology.[17] Yet, men *also* have consciousness, as Marx puts it, and without thought and consciousness human history could not have been built up. Taken literally and pushed to its ultimate consequences, the Marxian theory of ideology, even while recognizing an instrumental character for ideas and not ignoring consciousness as active and "real"—though never really and truly real—would make all language, all thought, all art, all ethics, all law, all religion, all conceptuality, and all philosophy a shadowy cloud that covered the earth with darkness.

Taken in all its radicality, Marx's understanding of the spiritual "productions" of mankind would exclude the very idea of some theory as something worthy of attention. Nevertheless, thinkers do not readily push their ideas as far as they might go, and the Marxian theory of ideological superstructure remains ambiguous. When Marx brings ideology back to production in its sacrosanct form and makes it a second, secondary, and derived species of that production, he does not annihilate it. On the contrary, he reassumes in inverted position the tradition of ("dualistic") metaphysics with its positing of an "opposition" between the level of the senses and the level of ideas, between nature and spirit, between thought and reality, a tradition that reaches from Plato to Hegel. Marx does not succeed in grasping the single foundation of being, neither for human being nor for all that is. Practice *and* theory, reality *and* ideas, materiality *and* spirit exist, despite all interactions, interdependencies, interpenetrations, and other dialectical bonds. Is this duality consequent solely upon the alienation and division of labor into material labor and spiritual labor? It does not seem so. But let us try to foresee what the "spiritual" might become after the supersession of private ownership of the means of production. For the moment let us keep in mind the twofold movement by which Marx encompasses the double movement of production. Ever since Plato, who stands as the true origin of Western European metaphysics (i.e., philosophy), ever since Judeo-Christianity and its theology and on up to modern European thought (Descartes, Kant, Hegel), the *idea* separates from the *entity* in dominion over it; *thought* stands opposed to *nature* in transcendence beyond it; and idea, God, spirit, or thought is being par excellence, *true Being*, in relation to which the *sensuous entity* is different and derivative. From Descartes on, thought, conception, ideas, and consciousness become the measure of the reality of anything that is, and the *ego* of man is their locus. Hegel brings this metaphysics, that is, metaphysics simply put, to its consummation, and Spirit, the foundation of the totality of being, recognizes itself in universal history and the fully achieved forms of self-consciousness. Truth according to Hegel is, of course, "the movement of its becoming"; but truth is idea, for "the absolute idea alone is being." From the preface to the *Phenomenology of Spirit* to the *Science of Logic* the being-becoming of totality simply is, for Hegel, *idea*. Yet even though after Hegel the reaction is generally one of opposition to his metaphysics, this opposition remains within metaphysics. Marx is the first one to work its inversion. For him, sense nature, the essentially active forces of man as social, the development of productive forces, and the material economic and social conditions are what contain truth and reality and constitute the *material base*, in relation to which *idealistic superstructure*—thought, ideas, conceptions, consciousness—is derivative and different and constitutes the domain of spiritualistic ideology. Nonetheless, this reversing of metaphysics remains metaphysical, and we shall see that Marx does not end up abolishing all "superstructure."

The basic unity of thought "and" being, of saying "and" making (of λέγειν, ποιεῖν, and πράττειν), as it was grasped by the Pre-Socratics, before any philosophical, metaphysical, and dialectical systematization, is something Marx does not reach, either to find it or to reconstitute it, despite his dialectic and its aspirations of unification. His enterprise of unification starts off from opposition and difference (that of theory *and* practice, for example), and any unity he achieves has at its heart one of the two elements in a position of determination, domination, and foundation with respect to the other. In this sense the Marxian course follows the path of Platonic and post-Platonic—indeed, of Christian—metaphysics and listens much less to the thinking word of the Pre-Socratics. Its grasp does not take in the unity of the being of totality. Its axis lies in the fissure that opposes thought to reality and theory to practice. True, alienation is the "cause" of this fissure; yet Marx does not get out of it and perhaps even deepens it. Neither Heraclitus's saying, "When you have listened, not to me but to the ⟨Logos⟩, it is wise to agree that all things are one,"[18] nor Parmenides's thought, "For it is the same thing to think and to be,"[19] any longer stakes out a path of approach to that unity of being.

According to Marx there is no Logos of the *one* being of Totality, and thought *only repeats* the activity of being—not of Being but of real human being. Logos and truth, meaning and thought are here considerably narrowed, becoming simply consciousness. "In his *consciousness of species* man confirms his real *social* life and simply repeats his real existence in thought.... Thinking and being are thus no doubt *distinct*, but at the same time they are in *unity* with each other."[20] Marx is all the same concerned with demonstrating the *difference* and not the unity, the difference, that is, that characterizes both kinds of alienations, the practical *and* the theoretical. As we pointed out earlier, according to Marx there seems never to have been one real thinking awareness that adequately expressed true reality instead of alienated reality. Even when thought "correctly" expressed what was objectively the case, it continued to express a reality that was operative but alienated. The *adequatio intellectus et rei* was adequate to the double alienation. Alienation as such was simply not yet grasped by thought.

Marx reduces ideological instruments to economic instruments. Yet ideas remain different from realities. This duality between base and superstructure—which, by the way, is not a duality between equal powers—permits ideas and thought to exercise an influence on the economic base and relations of productions in a return action. Deriving from practice by way of being a translation of it, forms of consciousness can act upon action to direct it. Nevertheless, any such action on the part of theory remains secondary. It will be Engels in particular who, in a series of letters, will insist upon the thesis of a reciprocal action binding together base elements and superstructural formations.[21] The relation of cause and effect remains what it is, but it is further "dialecticized." Even up to our own day, Marxists official or unofficial, noteworthy or routinely academic, repeat the thesis that gives ideology a kind of importance. Lenin, Stalin, Mao Tse-tung, and all the rest continue to emphasize that theoretical formations re-act upon the reality of practice. Erected on the foundation of economy, technical development, and social practice, the "spiritual," ideological forms of the superstructure can enter into action and play a role in historical development. Thought and consciousness, while never escaping the main determination of economic evolution, are thus viewed as exercising a re-active influence (positively or negatively, as conducive or inhibitive) upon that from which they emanate. Thought and consciousness can therefore be "real," for they are thus really active; however, this does not prevent them from being alienated for the whole of human history, which is the history of the alienation/manifestation of men. Only radical superseding, the actual abolition of real alienation, will dealienate at once language, thought, and consciousness, all of which will remain, however, different from practice. Thought and consciousness will become real and true, will become adequate expressions of what making discovers, when they finally *unite* with the historical action of men, after the superseding of alienation. Yet even then, as we shall shortly see, they will not be identified with praxis. Theory and spiritual manifestations, thought, consciousness, and knowledge will always *soar over* the reality of practice.

9.

RELIGION AND IDEAS

Religion is for Marx the first and primordial form of ideological alienation, and the most tenacious; and it is in religion that ideology is most active as mystification. Religion is first of all a product of that animal awareness of nature that has just been discussed. When it first arises, at the beginning of the historical process of mankind, it is "natural religion" and expresses the impotence of men face to face with an all-powerful nature whose strange and alien force has been scarcely touched by the transforming action of labor. The first forms of a wholly sensuous consciousness—a "herd-consciousness," this "animal" grasp of nature in a society still half-animal, half-human (because still deprived of technique)— are reflected in primitive natural religion. The active relationship that binds men to nature and among themselves is still extremely limited and narrow, and the power of productive forces is not yet truly manifest; so men produce religion, whose superior and transcendent powers embody on a spiritual level their own practical powerlessness before nature on a material level. Religion is, therefore, basically but one particular mode of production, falling under its general laws and having no historical development all its own: it is always a reflection of material historical development.

In religion man both experiences and creates a representation of those things that constitute his present vital weakness. Religion is generated on the base of undeveloped productivity and thereafter continues to be the sublimated, unreal reflection of the meaning and meaninglessness of the social economic process. For, with the increase in population, the growth of productivity, the evolution of productive forces, and the multiplication of needs, with the division of labor into material and spiritual work, with the institution and consolidation of the tyranny of private ownership, religion begins to express the alienation of man in relation to the products of his labor as the imaginary satisfaction of unsatisfied real drives. The nondevelopment of productive forces determines the genesis of religion, and this later development determines its subsequent "evolution." Being the expression of impotence and alienation, religion in turn, in its own modality, alienates man from his life and his essential forces. Far from being some kind of index of the strength of human being, religion comes about only owing to man's weakness, his frustrations, his dissatisfactions, his alienation. An abstraction from concrete conditions, religion is a product of the alienation of man on the level of both practice and theory. Mystery, far from implying a truth of its own, veils the truth of reality and masks its own mystification. It is with an extraordinary fanaticism that Marx wages war against religion *of any kind*. His critical analysis of practical alienation (alienation on the level of the material base) and of ideological alienation (alienation in the spiritualistic superstructure) reaches its high point in the criticism of religiosity and religion.

The first pages of Marx's first "Marxist" writing, the introduction to his "Contribution to the Critique of Hegel's 'Philosophy of Right,' "[1] are filled with this extraordinary antireligious zeal.

> The critique of religion is the prerequisite of every critique. ... The foundation of irreligious criticism is this: man makes religion; religion does not make man. Religion is, in fact, the self-consciousness and self-esteem of man who has either not yet gained himself or has lost himself again. ... Man is the world of man, the state, society. This state, this society, produce religion, which is an inverted world-consciousness, because they are an inverted world. Religion is the general theory of this world, its encyclopedic compendium, its logic in popular form, its spiritualistic *point d'honneur*, its enthusiasm, its moral sanction, its solemn complement, its universal basis of consolation and justification. It is the fantastic realization of the human being because the human being has attained no true reality. ...
>
> The wretchedness of religion is at once an expression of and a protest against real wretchedness. Religion is the sigh of the oppressed creature, the heart of a heartless world and the soul of soulless conditions. It is the opium of the people.[2]

What Marx criticizes before all else is that very thing itself which is the source of all true reality: the practical activity of men, that is, their alienated life. It is because religion is the solemnized complement of this life made empty of its essence that he subjects it to a destructive criticism. Criticism is not itself his own goal. On the contrary, Marx assails those who *stay* at the level of criticism or of critical criticism, in particular the Left Hegelians: D. F. Strauss, Bruno Bauer, Max Stirner, and Ludwig Feuerbach—the Young Hegelians. Unlike Hegel and the Old Hegelians, who *understood everything* by way of spirit and idea, the Young Hegelians *criticized everything*, that is, criticized everything by making everything a matter of religion and religious representation. The total domination of religion was presupposed, and religious concepts dominated all realities and all ideas; so that, after first interpreting everything in a religious and theological way, these critical critics would attack that very domination as a usurpation of the true and natural life of man. They wanted to free men from their religious bonds. And yet, since they are the ones who viewed everything through religion, their negation of what held men in chains remained ideologically critical, abstract, theological in an antitheological form, and simply long-winded. Marx's philosophical writings (the introduction to "A Contribution to the Critique of Hegel's 'Philosophy of Right,'" the *Manuscripts of 1844*, *The Holy Family, or the Critique of Critical Critique*, the *Theses on Feuerbach*, and *The German Ideology*) are all inspired by his twofold philosophical struggle: on the one hand against the "idealism" and "spiritualism" of Hegel and on the other against the pseudoradicalism, the pseudohumanism, and the pseudomaterialism of the left-wing Hegelians. Marx undertakes a critique of reality as it is and of the ideology that corresponds to it, a critique that would end by compelling the practical and revolutionary transformation of everything in existence. The battle is engaged not in the name of "philosophic truth" but in order to supersede alienation on a practical level and free both productive forces and men.

The critique of religion focuses, then, on the world from which it issues, and it is this root of alienation that must be extirpated. "The struggle against religion is indirectly the struggle against that world of which religion is the spiritual aroma.... The abolition of religion as the illusory happiness of the people is a demand for their true happiness. The call to abandon illusions about their condition is the call to abandon a condition which requires illusions. Thus, the critique of religion is the critique in embryo of the vale of tears of which religion is the halo."[3] All religious transcendence is illusory and alienating and must consequently be *annihilated*.

162   Ideological Alienation

Religion expresses in an ideological way the weakness of the technique and practical organization of men; only an all-powerful technique and practice will return man to himself and to his real works. Marx's intent is to make dealienated practice and theoretical understanding "coincide," without inquiring too far into the meaning and content of this kind of "demystification" enterprise; and this is the point of his fourth and eighth theses on Feuerbach:

> IV. Feuerbach starts out from the fact of religious self-alienation, of the duplication of the world into a religious world and a secular one. His work consists in resolving the religious world into its secular basis. But that the secular basis detaches itself from itself and establishes itself as an independent realm in the clouds can only be explained by the cleavages and self-contradictions within this secular basis. The latter must, therefore, in itself be both understood in its contradiction and revolutionised in practice. Thus, for instance, after the earthly family is discovered to be the secret of the holy family, the former must then itself be destroyed in theory and in practice.
>
> VIII. All social life is essentially *practical*. All mysteries which lead theory to mysticism find their rational solution in human practice and in the comprehension of this practice.[4]

It is not only *Christianity* that Marx is fighting: his aim is the *annihilation of all religiosity* and all religion. The fulminations he levies against *Judaism*, the religion of his forebears, are still more violent.[5]

Marx confronts the Jewish question in a manner that is neither theological or religious nor political, but rather social, that is, within an economic perspective. He does not look for the secret of the Jew in his religion, since, inversely, it is in the Jew as an actual reality that the secret of this religion is found. The mundane and profane foundation, the true foundation, of Judaism he finds in *practical need*, in *interest*. *Huckstering* and *money* are, according to Marx, respectively the worship and the God of Jews as they are in reality and not as idealized Sabbath Jews. The pages he has devoted to the Jewish question give evidence of an extraordinary vehemence on his part, for the objective expressed in them is the abolition of Judaism and its practical, utilitarian, and commercial spirit.

> Very well: then in emancipating itself from *huckstering* and *money*, and thus from real and practical Judaism, our age would emancipate itself.
>
> An organization of society which would abolish the preconditions and thus the very possibility of huckstering, would make the Jew impossible. His religious consciousness would evaporate like some insipid vapour in the real, life-giving air of society....
>
> We discern in Judaism, therefore, a universal *antisocial* element of the *present time*, whose historical development, zealously aided in its harmful aspects by the Jews, has now attained its culminating point, a point at which it must necessarily begin to disintegrate.
>
> In the final analysis, the *emancipation of* the Jews is the emancipation of mankind from *Judaism*.[6]

Jews have already been emancipated in the Jewish fashion, namely, by turning everything into *commodity*, by making the practical mind of the Jew the spirit of Christian peoples, by transforming Christians into Jews. And in this way they have fervently cooperated in the building up of bourgeois, capitalist society, which produces the Jews out of its own bowels. Jews are accused of having stripped all value from the entire world both of man and of nature, in that they have made everything a matter of alienating commercial value. This depreciation of man and nature, which exists ideologically in the Jewish religion, is what the real Jews bring about on the level of practice. "That which is contained in an abstract form in the Jewish religion—contempt for theory, for art, for history, and for man as an end in himself—is the *real, conscious* standpoint and the virtue of the man of money. Even the species-relation itself, the relation between man and woman, becomes an object of commerce. Woman is bartered away. The *chimerical* nationality of the Jew is the nationality of the trader..."[7] The Jewish law is a merely formal and lying caricature, a law that implies the taking of revenge, a law that is constantly abrogated in scheming practice. It is this practice, narrow, limited, utilitarian, egoistic, commerce-minded, vulgar, and atheoretical, that Marx inveighs against in the name of true practice, namely, dealienated, revolutionary communal practice that implies its own theoretical understanding, a practice open to the future.[8] "*Judaism* could not develop further as a *religion*, in a theoretical form, because the world view of practical need is, by its very nature, circumscribed, and the delineation of its characteristics soon completed. The religion of practical need could not, by its very nature, find its consummation in theory, but only in *practice*, just because practice is its truth. Judaism could not create a new world."[9]

Christianity, issuing from Judaism, ends up by rejoining Judaism. Christianity has not won over real Judaism; it has sublimated the practical need and the vulgar practice of Judaism without eliminating them. Judaism in its actual essence, rather than the Judaism of the Pentateuch and the Talmud, is what has been sublimated into Christianity; and the latter, becoming more and more practical and vulgar, relapses into Judaism. "Christianity is the sublime thought of Judaism; Judaism is the vulgar practical application of Christianity. But this practical application could only become universal when Christianity as perfected religion had accomplished, in a *theoretical* fashion, the alienation of man from himself and from nature."[10] Christianity thus made possible the unfolding of *bourgeois society* in its stifling of species man and its breaking down of the world of men into a world of alienated "atomistic" individuals (*atomistischer Individuen*). Judaism, origin of Christianity and bourgeois society, reaches its apogee with the achievement of bourgeois society, which rests upon egoistic, commerce-minded interest. Unable to create a new world, it attracted into its sphere of action all other creations and all other conceptions, while itself remaining the passive element; then it turned them all into commodities through its practical spirit. The Jewish spirit became as a result the practical, egoistic spirit of bourgeois society, within which the world of egoistic needs and practical objects remains subjected to the secularized Jewish god, money. Far from wishing to contribute to the cause of the emancipation of the Jews, Marx wants human society as dealienated to abolish the empirical essence of Judaism, hucksterism and its conditions, by making Jews impossible as Jews. Far from denouncing the social limitation of the Jew, he wants to abolish and surpass the Jewish limitation of society. Marx's writing, "On the Jewish Question," ends with this statement: "The *social* emancipation of the Jew is the *emancipation of society from Judaism*."[11]

According to the Marxian interpretation of religion, this special but determinative sphere of ideological alienation and spiritualistic superstructure, *all* religion reduces to practice. The undeveloped state of practice and technique conditions the birth of primitive, "natural" religion; the underdevelopment of productive forces and the barriers inherent to the relations of production in an era of private ownership determine the character of religion as idealistic justification of the existing state of affairs and as pseudoprotest against real misery. The full development of practice and technique will abolish all religion simply by making it completely *useless*. Religion rests on a want, a defect, a limitation. Its truth resides in practice, though religion itself, as religion, possesses no practice, just as it does not have a history of its own. Since practice, of which religion is always the sublimation, did not contain real truth, religion has been only the alienated expression of a real alienation and, of course, has contributed to the continuance of that alienation. Marx does not recognize any formative and basic role for religion. Neither *politics* nor *religion*, neither *art* nor *philosophy* is for him a power[12] constitutive of human history, even in an alienated state, or a mode of union between man and world. Practice alone is the source of truth and reality. Practice alone binds man to the world, and whatever intrudes additionally is the expression, reflection, and reflex of alienation. To the question, Where does religion get the power to do this? Marx accords no religious importance. His view of religion is short-sighted. In addition, he carries on a fight not only against historical religions but also against all religiousness, against all discovery of the divine itself or of the holy. What alone is "sanctified" is the practice of men *and* its corresponding understanding. There is not even any question of the "divine" or the "sacred"; these are but products of the alienation of religious imagination, which is itself a by-product of alienated material production. Alienated human life has given birth to religion; the death of religion, its abolition, will coincide with the true birth of man, with man himself as his own foundation. In effect, then, the thought and project of Marx and the world of which he speaks are placed under the sign of the *death of god*, the murder of God by men.

Marx quickly disposes of primitive, natural religions; he says nothing of oriental, Asian religions, and he finds little interest in pagan polytheism. He keeps to the movement that has led to the modern West, to bourgeois capitalist Europe, which, once superseded, will make possible the founding of universal human society. Judaism with its monotheism prepares the way for Christianity, and this latter is consummate religion, religion κατ' ἐξοχήν ["par excellence"], the essence of religion, the religion of *man deified*.[13] Without spending much time on this *essence of religion* and the serious problems that religion in general and Christianity in particular raise, Marx only awaits the superseding of monotheism and of man made into God that will come with *man's* becoming truly man and living on this earth as that natural social being that he is, with no need for any kind of religious heaven. As for *theology*, Marx raises hardly any questions about it, since he considers it only the "corrupt apex" of this other alienated theoretical activity which is philosophy.

Nevertheless, religious alienation is far from being the only alienation that expresses alienated practice in the form of theory. Other ideological powers as well reflect human impotence, without recognizing their character as reflection. Ideologies, forms of superstructure, and (alienated) theory, though in fact only reflected forms, do not recognize their role, are not aware of what they are. As forms of social consciousness, as forms and ideas of a certain illuminating *Bewusstsein*, as representations, they translate in alienated fashion that which in its own alienation conditions them: the things with which human being has dealings, man's actual presence in the world. They give some illumination, of course, but what they shine upon is dark and full of shadows. They are abstracted from the concrete becoming of men, they abstractedly immobilize both it and themselves, they constitute abstractions. Their "truth" is untrue, for they conceal in illusion a reality that is not actually in things (*une realité non-effectivement réelle*). We return here once again, for purposes of really understanding ideological alienation, to Marx's basic schema, to the path of his thinking, to the heart of the movement which does not divide into method and doctrine.

Fundamental reality, active and real (*wirkend* and *wirklich*), truly first reality, the source of all truth, is *the development of material productive forces*—tools and human labor—which is what allows man's being to posit itself in opposing nature; nature is of concern to man only through natural human history and not because it "preexists" man. Sensuous labor is the foundation of the sensuous world, of the totality of what has being by and for man. The development of productive forces generates *relations of production*, relations which men entertain between themselves and with the productive forces. This ensemble of productive forces and relations constitutes *the economic structure of the society of men*, the *material, real* base of the historical becoming of mankind. And it is the alienating externalization of productive labor which is the basis of general alienation.

Upon the material, real base is raised *the legal and political superstructure*: the juridical expression of actual relations of production, social organization, political life, and the State. This superstructure is *idealistic* and *spiritualistic*, for it is an *ideological* travesty of the meaning of the real, material state of things. To this ideological superstructure correspond, as part of it, conditioned and determined forms of social consciousness. *Forms* of consciousness are *theoretical*, in opposition to practical activity, and things which they represent to themselves in the form of concepts remain disguised to them; the true motive *forces* that animate them remain unknown to them. And so these forms of consciousness are, properly speaking, unconscious. Thought, founded by practice, ideologically believes itself to be founded on itself; what is accomplished by the intermediary of thinking appears to thought to be accomplished owing to and by thinking. Conceptions, ideas, thoughts, theories, forms of awareness, and so on, are abstract, an ideological reflection—that is, the alienated reflection—of a real, material alienation. *Religious, moral, artistic, philosophical*, and *scientific* creations form the main *spiritual* productions, and they are in truth the several sublime species of material production, the industrious and industrial activity of man. These reflexes are the illusory reflections of true reality, of the internal motor of development, namely, the development of the technique employed by men in the production of their life, the economic conditions of production. Already certain aspects of the *relations* of production themselves, juridically constituted or able to become so, mask or restrict the meaning of development and emerge thus into the superstructural. To law is found linked "morality"; as for politics, it is far from constituting a truly real activity: it "organizes," in alienation, the life of the social organism. Finally, ideological forms, which sometimes are partially "true," partially active and "real," constitute essentially the forces of that alienation that maintains the impotence of man, because they translate things into ideas and confuse ideas with things.

This is the "metaphysics," violently antimetaphysical, that must be understood strictly in all its *radical, reductionist*, and *one-sided* depth, in order for Marx's antiphilosophical philosophy to be philosophically understandable. Lazy, wide-ranging interpretations that dull and flatten Marxian thought by removing its teeth and bite never get to the center of his thinking. They are mere embellishings that try to make Marx's brutal thinking more "digestible." Marx himself wants his thought to be free of philosophical presuppositions, to begin from experience, from the real, material premises of the natural, social history of men. Yet the thinking he finds himself called upon to follow continues to move along the axis of metaphysics, even if by way of being its negation. The Marxian criticism of *ideas* in general, of ideas as the soul of ideology, is unrelenting.

It is because "ideas" are the sublimated expression in idealistic, ideological form of the vice of limitation, of the impediments inherent in real human history, the history of real alienation, that they are fought in the name of the practice that gives birth to practical truth and history. "We have to take up the history of men for this reason, that almost the whole of ideology reduces either to a distorted conception of this history or to a complete abstraction from it. Ideology itself is only one side of this history."[14] Marx actually takes up the struggle against *all* ideology, which is from the outset qualified as idealist. He fights against every species of Platonic idealism, Christian or modern (Christianity is popular Platonism, Nietzsche will say), against all autonomy for thought. Marx writes: "Moreover, it is quite immaterial what consciousness starts to do on its own . . . It is self-evident, moreover, that 'spectres,' 'bonds,' 'the higher being,' 'concept,' 'reflective scruple,' are merely the idealistic, spiritual expression, the conception apparently of the isolated individual, the image of very empirical fetters and limitations, within which the mode of production of life and the form of intercourse coupled with it move."[15] And here he is talking not only about the Left Hegelians but also about all "ideas" considered as idealistic expressions of real barriers. The secret of the formation of ideas lies in real praxis; their "wealth" corresponds to the production of social wealth, and their emptiness expresses the poverty within which mankind struggles despite the deployment of an immense productive labor. A *real relationship* between man and man and between men and things is sublimated by the mediation of ideology to become a thought, an idea.[16] As for *praxis*, which stands in the primary and fundamental position, Marx in the end does not make its essence very explicit; praxis, rather, itself serves to explicate both practical and theoretical activity. Here he does what thinkers generally do: he leaves unexplained that with which he explains being in its totality.

Human energy finds its path blocked by ideas, for ideas impede the unfolding of the essential forces of man. Ideas, like the instruments of production, are tied to the dominant class. Those people who hold dominion over the earth and exploit the productive labor of workers are the ones who produce as well the ideas that as "supernal" hold a position of dominance. Far from expressing something universal, ideas as reflex formations reflect a particular that takes itself for the universal. "The ideas of the ruling class are in every epoch the ruling ideas: i.e., the class, which is the ruling *material* force of society, is at the same time its ruling *intellectual* force. The class which has the means of material production at its disposal, has control at the same time over the means of mental production, so that thereby, generally speaking, the ideas of those who lack the means of mental production are subject to it."[17]

The thinking, the ideas, the categories, the principles that hold sway are, consequently, the ideal, idealist, and ideological expression of the dominant material conditions. Those who hold the means of production produce as well the ideas of their time, and they regulate the distribution and consumption of those ideas. The dominant class disposes of that which the others, the dominated, do not have, and ideas serve to compensate for and sublimate this *twofold* lack, material *and* spiritual. Ideas thus minister to those they subjugate. There are no eternal laws and eternal ideas: there are social and historical laws with a base in the economic order and ruling ideas. Ideas and thinking have no history and development which belong to them on their own, since they constitute only the spiritual, spiritualistic sublimation of material history and material development. Their premises are real and material; if their status becomes abstract, ideological, and unreal, that is the result of the double nature of alienation, which is economic on the basic level and ideological on the derived. Thinking and ideas, representations and forms of consciousness, theories and concepts must be analyzed and criticized according to their concrete social and historical character, their empirical determinations, the phase of the development of production, and the mode of corresponding relations of production. Ideas conceive themselves as autonomous productions, but in truth, that is, in reality, their autonomy is only something they have set up for themselves, and they are not for all that actually independent. Their own logic comes to them from elsewhere.

The regime of private property and the division of social labor alienate thinking and ideas. Just as within society as a whole productive labor is divided, so intellectual labor is divided within the dominant social class. Some people are the producers of ideas, and these are thinkers and active ideologists; others are the passive receivers-consumers of produced ideas. Inversely, those among the passive consumers of ideological products who belong to the dominant class are the socially active members of that dominant class. The ideas that hold sway, dominant as they may be, are nonetheless mortally bound to their social support, able only to live its life and know its death. Their only act is to translate unreal realities theoretically into an illusory language, going so far as to entertain the illusion that they are independent of concrete individuals and social conditions, if they do not take themselves as the motive agency of historical becoming. Their "truth" resides in their lying, in their ingrained erring.

The character of ideological alienation is, so to speak, of a double sort: ideas express in deformed fashion the real relationship of material powers, constituting an unfaithful and inadequate reflection, and, on the other hand, they create the illusion of their own power, transforming the real life of social individuals into an ideological life. Men thus live their "moral," "religious," "theoretical," "artistic" life in an abstract and alienated way, in complete and unacknowledged contradiction to actual, concrete reality and practice. For man is not alienated in respect of his being theoretical; it is not the alienation of his consciousness that is the basic alienation. His real and material alienation, the alienation that concerns the material of his real labor, is the root of all alienation; this is the radical alienation that is reflected in veiled fashion in whatever is ideology. Men have even come to think that ideas and concepts are what rule history (Hegel) or have sought liberation by means of further ideas (Left Hegelians); but that is all false and falsifies the true meaning of becoming. One must look to the empirical base, the infrastructure, for the secret of the genesis, the order, the connection, and the function of the ideas that come and go in social space and time; one must radically demystify any idea of a mysterious relationship endowed with its own specific dialectic which would constitute the essence of the world of ideas. By the substitution of the mental idea for the real, material relationship, all is changed into ideas, and beings and things lose their truth to become *ideals*. By way of these ideological constructions, human relationships are hidden away. The totality of relationships that unite men to nature and among themselves turns into something autonomous, becoming a power incomprehensible to men, a strange, alien, and hostile force. All that detaches itself from the relationships established in the course of the life-process goes on now over men's heads and beyond humanity, preventing men both from living through the production of their life and from understanding this practice through theory.

Marx seems, therefore, to commit to the flames of hell theory, thought, consciousness, conception, ideas, and spiritual productions, all of these being forms and forces of ideological alienation that lack real truth and mask practice. His critique aims for the abolition of alienation in its twofold aspect. We have already pointed out how Marx, *thinker* of economic alienation and its supplement alienation, ideology, does not manage to transcend the conception that opposes reality to thought and (man's) being to (self-)consciousness. He works in a double manner, even while referring what is second and secondary to that which is first and primordial. Ideas and things remain separated. Thus, though every theory, form of consciousness, or systematic idea has hitherto been alienated and ideological, it is theory, consciousness, and the power of ideas that will lead to the abolishing of alienation, in being, of course, the expression of the development of productive forces. The moment alienation is recognized, at some level of development of technique and industry in its engendering of the proletarian class, it seems that thoughts can become true, real, and active. It is indeed the underdevelopment of technique that determines the ideological lie, while the current degree of maturity in technique may permit, for the first time in the history of mankind, the unfolding of thoughts and ideas as true and real.

The proletariat that undertakes to abolish alienation cannot accomplish that task without being guided by revolutionary ideas that illuminate its struggle. Where does it get these ideas? From bourgeois intellectuals who reject their class, the dominating class, and rally to the proletariat, the oppressed class. It is the *awareness* of the historical situations that determines this change of sides on the part of renegade intellectuals who quit their class; it is *consciousness* which, *from outside*, they bring to the proletariat. Consciousness, theory, thoughts, and ideas are no longer here a simple ideological epiphenomenon; they *transcend* particular empirical conditions to rise to a level on which the whole is seen. The second aspect of the world here, in some way, takes over; and the theoretical weapons which bourgeois intellectuals *bring* to the proletariat (for it is not the practice of the proletarians that generates revolutionary theory in a direct way) are fundamentally necessary to it for purposes of its practical emancipation. And it is Marx, the bourgeois intellectual and author of the *Communist Manifesto*, who, without concerning himself too much with the dualism apparent here, writes the following:

> Finally, in times when the class struggle nears the decisive hour, the process of dissolution going on within the ruling

> class, in fact within the whole range of old society, assumes such a violent, glaring character that a small section of the ruling class cuts itself adrift and joins the revolutionary class, the class that holds the future in its hands. Therefore, just as, at an earlier period a section of the nobility went over to the bourgeoisie, so now a portion of the bourgeoisie goes over to the proletariat, and in particular a portion of the bourgeois ideologists who have raised themselves to the level of comprehending theoretically the historical movement as a whole.[18]

This *theoretical awareness* of the whole movement of history, these new and revolutionary ideas and thinking, come to the vanguard of the new world by the mediation of the intellectuals of the world of alienation, and they give the struggle a fruitfulness in two orders: in its *practical*, economic, and political aspect and in its *theoretical* (ideological or "ideological"?) aspect.

The revolutionary class, therefore, *receives* revolutionary ideas, and this theory guides practice; what takes the lead now is being precisely as being-conscious rather than as being-man (*l'être conscient prend le pas sur l'être des hommes*). Even militant, victorious Marxism did not manage to surmount this conception of the revolutionary consciousness that, although expressing the material, really revolutionary situation of the proletariat, nevertheless came to it *from the outside*, thanks to revolutionary intellectuals. Lenin—turning to "the burning questions of our movement" in *What Is To Be Done?*—treats the problem of the revolutionary consciousness and the theoretical struggle as a vital, burning question. What is continually before his mind is the thought that, "without revolutionary theory, there is no revolutionary movement"; and, while recognizing three forms in the great struggle—the economic, the political, and the theoretical—he places the third, theoretical struggle, on the same level as the other two.

> We have said that *there could not have been* Social-Democratic consciousness among the workers. It would have to be brought to them from without. The history of all countries shows that the working class, exclusively by its own effort, is able to develop only trade-union consciousness, i.e., the conviction that it is necessary to combine in unions, fight the employers, and strive to compel the government to pass necessary labour legislation, etc. The theory of socialism, however, grew out of the philosophic, historical,

and economic theories elaborated by educated representatives of the propertied classes, by intellectuals. By their social status, the founders of modern scientific socialism, Marx and Engels, themselves belonged to the bourgeois intelligentsia. In the very same way, in Russia, the theoretical doctrine of Social Democracy arose altogether independently of the spontaneous growth of the working-class movement; it arose as a natural and inevitable outcome of the development of thought among the revolutionary socialist intelligentsia.[19]

Primitivism, economism, practicism, populism, blind confidence in the spontaneity of the exploited masses—these doctrines have never been preached by Marx. Revolutionary *action* cannot get along without revolutionary *theory*—this is his constant position. Nevertheless, he has not examined in depth the problem that this relation posed for him then and still poses now.

10.

ART AND POETRY

Your favourite virtue	Simplicity
Your favourite virtue in man	Strength
Your favourite virtue in woman	Weakness
Your chief characteristic	Singleness of purpose
Your idea of happiness	To fight
The vice you detest most	Servility
Favourite occupation	Bookworming
Poet	Shakespeare, Aischylos, Goethe
Prosewriter	Diderot
Hero	Spartacus, Kepler
Heroine	Gretchen
Favourite maxim	Nihil humani a me alienum puto
Favourite motto	De omnibus dubitandum[1]

Art, too, is one form of superstructure, one ideological sublimation, a reflection and complement of economic life, so real and so alienated. Art, taken as the whole domain of all the art forms of poetry, literature, the theater, the plastic arts, and music, has no history of its own in its internal development; it is simply a particular species on the level of mind of the general movement of material production. Marx is very positive in his negation of autonomous historical becoming for the forms of alienated superstructure: "There is no history of politics, law, science, etc., of art, religion, etc."[2] It is the ideologists who invert the lines of force of history and make it walk on its head.

Τέχνη as *art* is one of the creations of τέχνη as *technique*. Technique in economic production and social conditions determines all human creations, including, therefore, works of art. One should try to understand well what Marx has in mind in what he says: Economic determination prevails, it is true, only *as what is ultimately decisive*—as a matter of the last analysis—and by way of a passage through a whole series of mediations. Yet, productive forces and the corresponding relationships are what constitute the constant, basic cause of the development of all human productions. Marx is quite explicit here, and this idea is the pivot of his thought.[3] Since production and economic and social life are radically alienated, it follows as a consequence that artistic production also must necessarily be radically alienated.

It is *above all* bourgeois, capitalist society of the last centuries that has consummated the alienation of art. In making the work of art an object of exchange, so that a quantity of crystallized labor is exchanged for another quantity, in making the work of art a commercial affair, modern European society took a decisive step; it has commercialized and reified artistic production and made it depend on the law of supply and demand. In order for this to be done, however, art had to be, by its very nature, the object of a production, the realization of a labor, a material (and spiritual) work. The present situation of art aggravates this very characteristic of art. Alienation seems "intrinsic" to it. The regime of private ownership is the cause of all economic and social, as well as ideological, alienation; and art is thereby included. Here we quote for the second time a most important passage from the *Manuscripts of 1844*: "This *material*, immediately perceptible private property is the material perceptible expression of ⟨*alienated*⟩ *human* life. Its movement—production and consumption—is the *perceptible* revelation of the movement of all production until now, i.e., the realization or the reality of man. Religion, family, state, law, morality, science, art, etc., are only *particular* modes of production, and fall under its general law."[4] The realization of man has not yet taken place, however, except in alienation, and man's spiritual production alienates him further. Marx, aiming for the abolition of the root of all alienation and the superseding of the *double* aspect of alienation (material and spiritual), wants man to quit these forms of an alienated superstructure—politics, the family, religion, science, and art—in order to reintegrate his human existence, that is, his social existence; for man is himself his own proper foundation and root.

Marx's position on art remains, however, extremely ambiguous and ambivalent. As a form of alienated ideological superstructure, as a world that is ideal and not at all real, as a particular spiritual species of material production that is fundamentally alienated, ought art to be abolished and surpassed, like all the "powers" of the superstructure, so that real, material, productive technique might alone deploy all its forces in the task of practically transforming everything that is? Is not art one of these forms? The few lines of Marx's writing just quoted speak of religion, the family, the State, law, morality, science, and art as but particular modes of material private property (i.e., the sensuous expression of alienated human life and of alienated productive movement). And then we read the following: "The positive transcendence of *private property*, as the appropriation of *human* life, is therefore the positive transcendence of all alienation—that is to say, the return of man from religion, family, state, etc. [art is not mentioned here, as it was previously], to his *human*, i.e., *social* existence."[5] The forms of human alienation, the forms wherein human life is externalized—*religion, the family, the State*, and *all that is included in Marx's "etc."*— will therefore be abolished and superseded; but will not *art* also be abolished and superseded, being precisely one of those things? The question we are raising here is not the essential, decisive issue of what results from this return of man to his human existence after the annihilation of the family, morality, religion (science also), and the State. We are leaving open the question of the meaning and content of this human life that would rest on its own roots, pruned of any alienating branches, namely, the family, morality, law, politics, the State, religion, science, philosophy. We are trying to shed a little light on the problematic of the essence of art.

In enumerating the ideological forces of alienated superstructure, the forms and forces that have no history of their own, in defining the forms of spiritual production that sublimate deficiencies in material production, Marx makes mention of art with the idea that it is a part of this second world, the ideological, idealistic, ideal world, the theoretical world. In wanting man to reintegrate his human existence (the only one there is), to abandon the powers that complement his practical impotence, Marx does not speak of art (see the last quotation given above). Does alienation, then, not affect art *mortally*, and does the abolishing of all alienation not abolish art also, as art?

To repeat: Marx's position regarding art remains extremely ambiguous and ambivalent. What seems to happen is that Marx, while wishing to annihilate radically the family and religion and politics and theoretical knowledge so that man can give himself totally to productive praxis, hesitates with art. Art is taken in a twofold way. On the one hand, it is only a species of spiritual production, falling under the laws of (material) production and consequently of alienation. Seen in this perspective, art ought certainly to be abolished. On the other hand, Marx, as a concrete individual and a (great) thinker, was continually preoccupied with art, and it would be really superficial to understand this concern as merely a manifestation of "general culture" or "aesthetic interest."

In his adolescence, Marx wrote lyric, romantic poems, love poems. In 1836 he sent to Jenny von Westphalen, his fiancée, whom he loved passionately, three collections of poems that he had authored: the *Book of Songs* (*Buch der Lieder*) and two volumes entitled the *Book of Love* (*Buch der Liebe*). The poems were lacking in great *poetic* intensity, to be sure, and Marx himself a little later judged them to be mediocre, though full of a human warmth. In them nostalgia, melancholy, revery, idealism, and despair raise their voices. Yet, even in these poems the Marxian voice of Marx is already evident, as he writes:

> I would to win all things for mine,
> Attain each fairest godly grace,
> Press boldly on in utter knowledge,
> Lay hold too of song and art. . . .
>
> Let not a brooding course be ours
> In anguish 'neath the lowly yoke,
> For still to us there these remain,
> Desire, longing, action.[6]

Marx also composed part of a gloomy verse play, *Oulanem*; a satiric novel, *Scorpion and Felix*, in which he bitterly recognized that "our time can no longer create an epic";[7] a dialog, "Cleanthes, or the Starting Point and the Necessary Progress of Philosophy";[8] and he proposed, much later, to write for his daughters a play about the Greeks, which he never did. The strain of his philosophic thinking, the scientific and economic work, and the directing of political and practical action drew Marx away from poetry; but art did not lose its fascination for him. Homer, Aeschylus, the poet of the tragic revolt of Prometheus, Dante, Shakespeare, and Goethe, author of *Faust*, in which a man wants to conquer the world through action—these remained his poets; and he regarded them, especially Aeschylus, Shakespeare, and Goethe, as the most powerful poetic geniuses of mankind. Every year he reread Aeschylus in the original. Among prose writers and novelists, Cervantes, Diderot, and Balzac impressed him. He himself, though against all epigonous formalism, always wished to give as much care as possible to the style of his writing. Indeed, he wrote to a friend: "If you are a *poet*, I am a *critic*";[9] but it is completely superficial to make Marx a vulgar sociologist of art. Did he not think of these great poets with whom he communicated, these poets of divine and human tragedy and comedy, as workers of the spirit who expressed through ideological, poetic alienation the real alienation of human productive activity—though without recognizing it? But then what is the meaning of this double attitude towards art: art as a human creation resting on the alienating division of alienated labor and art as a creation that reflects a world cut in two?

The answer to this question does not stand out clearly. Let us try, nevertheless, to circumscribe a little more closely the problematic of Marx's thinking on art. Marx's ideas about art assert themselves as a force to counteract bloodless philistines, the pure consumers of art, and to work against all autonomy of the realm of poetry and the plastic arts, against all aestheticism whether idealistic or realistic, and against all petty bourgeois artistic production. This kind of flat and nonessential production in particular draws Marx's fulminations. Petty bourgeois artistic creations pass off as universal something that is quite particular, limited, and restricted, something that represents its own narrow class interest, its own hesitation between *pro* and *con*, its own practice of seeing two sides to everything in a go-between action that works unconsciously. Yes, Marx came out against all that; yet he also came out for certain things.

We should not expect to find in Marx a history of art, a philosophy of art, or an aesthetics as a philosophic or scientific discipline. The imaginary, or real, museum of world art hardly interested him. Nevertheless there is a typical form to the way his thought marks out three stages of progress for human history on its way to becoming universal: *Greece*, the *Christian Middle Ages*, and *capitalist Modern Times*. Thus, in beginning his philosophic work Marx turned to the genesis of the Western world; his doctoral thesis, "The Difference between the Democritean and Epicurean Philosophies of Nature," is a study of the origin of a certain kind of naturalism, materialism, and atheism. In the same way, he turns to Greek art as the origin and first stage of Western, then European, art. And what does he see there? The freedom of mortals who have conquered themselves in their Promethean struggle against the "immortals." "The struggle of the ancients could only end in the destruction of the visible heaven, of the substantial bond of life, of the weight of political and religious existence; for nature must be shattered for spirit to find its oneness in itself. The Greeks destroyed it with the art-laden hammer of Hephaestos, they smashed it into statues."[10] So Marx puts it in the preparatory notes to his doctoral thesis. Marx, who could sometimes see, perhaps only fleetingly, that even the praxis of thought is theoretical and that nonpractical contemplation had its own flash of illumination, looks at these statues, broken as they are. "And yet these gods are not a fiction of Epicurus. They existed. *They are the plastic gods of Greek art.* ... The theoretical calm is a principal element of the character of the Greek gods, as Aristotle says: 'What is the best has no need of action, for it is itself the end.'"[11]

What Marx admired in the plastic art of the Greeks was the luminous humanity in it; it was an art that crystallized into individual forms the life of the ancient city, though that city rested on the distinction between men who were free citizens and those who were producer slaves. Does this same luminous humanity characterize as well the era of tragedy? Marx does not have a unitary outlook that encompasses both the calm beauty of plastic art and the holy dread of tragic poetry. What he had to admire in poetry was particularly the individual freedom of *human being*, of the heroes of the Trojan War, of Ulysses, of Prometheus, of Antigone, though this admiration may well have led him to risk interpreting Antiquity in too modern and humanistic a fashion. As for oriental and Asian art, he simply turned away from it; it was an art that did not interest him, any more than oriental and Asian history did. The world really begins to become a *world* with the Greeks. It is in the West that light dawns. Oriental art is gross, extravagant, tenebrous, and inhuman, for it expresses tyranny and despotic superstition.

Yet Greek art had to die. And this, it seems, was because "abstract individuality is freedom regarding existence and not freedom in existence. It does not have the capability of shining out in the light of existence."[12] Greek art was to die, it seems, just as Greek philosophy was to die, not only that of Aristotle, the Alexander of Macedonia in Greek philosophy, but also the three "isms" that came after him. Stoicism, Epicureanism, and Scepticism are the fully achieved forms of subjective, abstract, and formal self-consciousness that, unrealized *in* existence, remain as a result instances of individual, alienated consciousness. "An immortal death carries off the mortal life"[13] of thought and art, of the individuals and the cities of Greece; self-consciousness cannot succeed in expressing the world-in-becoming as a whole, since its freedom remains "negative," a freedom in regard to existence that is not realized in the being-becoming of the whole of things. The forms of art and the forms of consciousness as natural, individual, and abstract succumb by reason of their own weaknesses rather than the weakness of the development of productive forces. At least this is what Marx thinks in his younger days.

But what do we, modern men, see in ancient art? How is our view defined by Marx in his maturity? Finding ourselves at another level of the development of material (and spiritual) production, what might we still look for in an era already past forever? The problem is a bothersome one, and so it is and remains for the founder of Marxism. He tries to pose it, if not resolve it, in an uncompleted study written while he was working on the *Contribution to the Critique of Political Economy*. In this piece, an "introduction" to the critique of political economy,[14] Marx notes "the unequal development of material production and, e.g., that of art"; and Marx himself puts us on our guard against any kind of formal progression in these words: "The concept of progress is on the whole not to be understood in the usual abstract form."[15] The particular sector of production which is art would not follow, then, the progressive movement of technique.

> As regards art, it is well known that some of its peaks by no means correspond to the general development of society; nor do they therefore to the material substructure, the skeleton as it were of its organisation. For example the Greeks compared with modern [nations], or else Shakespeare. It is even acknowledged that certain branches of art, e.g., the *epos*, can no longer be produced in their epoch-making classic form after artistic production as such has begun; in other words that certain important creations within the compass of art are only possible at an early stage in the development of art. If this is the case with regard to different branches of art within the sphere of art itself, it is not so remarkable that this should also be the case with regard to the entire sphere of art and its relation to the general development of society. The difficulty lies only in the general formulation of these contradictions. As soon as they are ⟨specified⟩ they are already explained.[16]

The often regressive development of art is not positively obedient to the development of productive forces. In fact, the opposite is often true. The development of some productive, material forces retards or even kills the development of artistic, spiritual forms. At

182 Ideological Alienation

the point where art is manifested as artistic *production*, some important formations become impossible. But is this contradiction explained as soon as it is specified? Does Marx solve the specific problem that bothers him, namely, ancient Greek art in its relationship to modern technique? The text we are concerned with continues:

> We know that Greek mythology is not only the arsenal of Greek art, but also its basis. Is the conception of nature and of social relations which underlies Greek imagination and therefore Greek [art] possible when there are self-acting mules, railways, locomotives and electric telegraphs? What is a Vulcan compared with Roberts and Co., Jupiter compared with the lightning conductor, and Hermes compared with the *Crédit mobilier*? All mythology subdues, controls and fashions the forces of nature in the imagination and through imagination; it disappears therefore when real control over these forces is established. What becomes of Fama side by side with Printing House Square?[17] Greek art presupposes Greek mythology, in other words that natural and social phenomena are already assimilated in an unintentionally artistic manner by the imagination of the people. This is the material of Greek art ... [18]

The economic and social development, that is, the *technical* underdevelopment of Greek antiquity, favored the *mythological* relation (one should say mythic); the very relation of men to nature was productive of mythology, and this productive mythology, itself having a genesis, engendered as well artistic production. But we are modern men and have replaced mythological imagination, which is a product of a determining and determined impotence, with technical power as a real (or nearly so) determinant and dominator of nature; what then do we do with artistic production?

> Is Achilles possible when powder and shot have been invented? And is the Iliad possible at all when the printing press and even printing machines exist? *Is it not inevitable that with the emergence of the press bar the singing and telling and the muse cease,*[19] that is the conditions necessary for epic poetry disappear?
>
> The difficulty we are confronted with is not, however, that of understanding how Greek art and epic poetry are associated with certain forms of social development. The

> difficulty is that they still give us aesthetic pleasure and are in certain respects regarded as a standard and unattainable ideal.[20]

Marx does not take up this difficulty inherent in (Greek) art and the problematic of our relationship to it from the angle of technology. He moves from social history to "individual" history and from the mature age of mankind to its childhood, in order to try to find points of illumination there. His thoughts go on, then, in this nostalgic vein:

> An adult cannot become a child again, or he becomes childish. But does the naïveté of the child not give him pleasure, and does not he himself endeavour to reproduce the child's veracity on a higher level? Does not the child in every epoch represent the character of the period in its natural veracity? Why should not the historical childhood of humanity, where it attained its most beautiful form, exert an eternal charm because it is a stage that will never recur? There are rude children and precocious children. Many of the ancient peoples belong to this category. The Greeks were normal children. The charm their art has for us does not conflict with the immature stage of the society in which it originated. On the contrary its charm is a consequence of this and is inseparably linked with the fact that the immature social conditions which gave rise, and which alone could give rise, to this art cannot recur.[21]

The manuscript stops there, uncompleted, and the answer as well remains incomplete and fragmentary. Greek art, the art of the past, remains "living," but it lives as remembrance; we, adults, *still* contemplate works done by children. One kind of artistic production has become impossible thanks to the development of modern productive technique. Yet we can still admire the works of art that a bygone production offers for our consumption. This *consumption* is, after all, rather harmless.

Christian art—monumental, "overwhelming," "venerated"—does not seem to have held Marx's attention; and he certainly does not consider it worthy enough to continue to "live" in the historical and "artistic" memory of mankind. His violent anti-Christian position keeps him from allowing any justification at all for the plastic art of Christendom. This art has a religious essence that is much more than "mythological," and, being inseparable from the religion of illusion that inspires it, it constitutes the height of religious alienation in artistic form. And yet—and yet Marx, dual thinker of a thought structured by dualism, had the greatest admiration for Dante, the poet of the Christian, feudal Middle Ages. Was not the poetic cathedral of the *Divina Commedia* just as alienated as the cathedrals of stone?

To Christian art and to those who turn to it, Marx opposes pagan, Greek art, an art that nevertheless was born of the (alienating) division of labor, that gave plastic form to the gods, and that rested on the foundations of a slave-holding society in which productivity was underdeveloped. Does the fact that Greek art expressed a society which gave a privileged position to use-value—that is, to the specific quality of things—and was not yet subservient to exchange-value—that is, under the rule of quantity as reified and commercialized—does this fact "justify" Greek society and the art that corresponds to it?[22]

As for the "modern feudal" art of "Christian chivalry," that is, *romantic* art, Marx regards it as quite frankly reactionary. The first manifestation of Romanticism is considered a reaction against the French Revolution and its liberating, desacralizing work; it is a romantic reaction that envisages the return of the Christian Middle Ages, with their feudalism and darkness. Marx is against all Romanticism, because he thinks Romanticism is turned toward the past and the superseded. The second manifestation of Romanticism constitutes a reaction against the process of socialist, communist revolution. It looks beyond the Middle Ages to the primitive and archaic epochs of different peoples. Romanticism hardly labors with the future in mind. "The impreciseness, the intimate niceness, and the subjective accent" of Romanticism, its cult of the hero, and the Christian, chivalrous, feudal principle that governs it, together with its nostalgia for a return to nature (as it is presented to be), all show the whole impotence of Romanticism for taking hold of the present and the movement of objective forces that spread out before its very eyes.

Let us not expect to find in Marx a "systematic" historical, aesthetic, or philosophic analysis of the various epochs of art and of its different genres and styles. His views on art and poetry are at the same time profound and summary, penetrating and fragmentary. Can one even state categorically that Marx is the apologist of *realistic* art—supposing that we know what *realism* is and, above all, what *poetic* and *artistic* realism is? The problem can hardly be posed in those terms. Of course, Marx wishes art to express "real relationships" truthfully and with a depth of understanding. This is the truthfulness and understanding that he appreciates so highly in Balzac, to whom he planned to devote a study after completing his own *Comédie humaine—Capital*. Marx did not dwell either on the possible contradictions between the (subjective) ideology of the creator of a work of art and the (objective) content of that work. Both passion and a sense of the scientific characterized his fondness for the author who nonetheless wrote the following line in the preface to his *Comédie humaine*: "I write in the light of two eternal Truths: Religion, and the Monarchy." And in a letter written April 19, 1859, to Ferdinand Lassalle, Marx had the following to say about the hero of Lassalle's tragedy, *Franz von Sickingen*:[23] "You should have indeed *Shakespearized* more, whereas I take your greatest fault to be your *Schillerizing*, the transformation of individuals into mere mouthpieces for the spirit of the age (*Zeitgeist*). . . . Hutten is to me too much the mere representative of 'enthusiasm,' which is boring. Wasn't he also very clever, a regular witty devil . . . ?"[24]

Besides, does not even "realistic" art remain *art*, that is, a form of externalization—in other words, alienation? Is not the historical epoch that gives birth to this realism the very one that consummates ideological alienation in general and artistic, aesthetic alienation in particular? For, by setting up the reign of a world market and divesting the spiritual activity of poetry and art of its aura, the bourgeois, capitalist society of modern Western Europe on the one hand showed a hostility toward some poetic and artistic production, while on the other hand it generalized the transformation of spiritual production into commodity. This same modern society "has stripped of its halo every occupation hitherto honored and looked up to with reverent awe. It has converted the physician, the lawyer, the priest, the poet, the man of science, into its paid wage-laborers."[25] And yet modern society has only generalized and consummated the alienation of which law and the jurist, religion and the priest, poetry and art and the poet and the artist, science and the scientist are the alienated expression. It has exposed and stripped of their halo the faces of alienation, while at the same time universalizing that alienation and creating the conditions for the superseding of it on a universal level. "In place of the old local and national seclusion and self-sufficiency, we have intercourse in every direction, universal interdependence of nations. And as in material, so also in intellectual production. The intellectual creations of individual nations become common property. National one-sidedness and narrow-mindedness become more and more impossible, and from the numerous national and local literatures there arises a world literature."[26]

It seems that Marx considers art, art in general and the art of the modern era, as a reflection of the technical impotence of man and as a special sphere of alienation. *Artistic alienation* is complementary to the *alienation of technique*. "The history of *industry* and the established *objective* existence of industry are the *open book of man's essential powers*, the exposure of the senses of human *psychology*. Hitherto this was not conceived in its inseparable connection with man's *essential being*, but only in an external relation to utility, because, moving in the realm of ⟨alienation⟩, people could only think of man's general mode of being—religion or history in its abstract-general character as politics, art, literature, etc. . . . ."[27] And, "since all human activity hitherto has been labor—that is, industry—activity ⟨alienated⟩ from itself,"[28] is it not "clear" that artistic productive activity is also alienated?

Art as such seems to imply alienation, which bourgeois, capitalist society can only aggravate and generalize. Through this particular kind of production which is art, just as through all modes of production, the movement of alienation is also achieved, the movement, that is, of the plundering, the dispossession, the loss for man of what constitutes his species being. Through art the subject is separated from the object, the producer from the product, the thing (content) from the idea (form), sensuous presence from imaginary representation, reality from ideology. Through art, that which *is*, is found overshadowed by that which is *thought*. And so "my true artistic existence [becomes] existence in the *philosophy of art*; . . . Likewise the true existence . . . of art is the *philosophy* . . . of art."[29]

Should not art, then, be superseded just as all the other forces and forms of material and spiritual alienation are to be, in order that man may be able to regain his species being? For to manifest himself he truly needs only a dealienated productive activity in the real order. Should art, for which Marx *at times* professes such an admiration, cease to be? Here is what we read from Marx's pen: "The poet falls from his sphere as soon as poetry becomes for him a means. The writer in no way takes his writings for a *means*. They are an *end in themselves*, being so little a means for himself and for others that he sacrifices *his* existence to *theirs* when necessary, just as, in another way, the preacher of religion makes this his principle:

'Obey God rather than men,' though it be among men that he is confined with his human needs and desires."[30] One has always to try to distinguish between what Marx attacks and what he defends— even though this distinction is difficult to get hold of—because he attacks and defends in a concrete perspective. Marx criticizes and rejects any conception of art as independent and evolving in its own proper world, as constituting an autonomous sphere; but he criticizes just as much what vilifies art. He is, of course, less interested in art *as art* than in the social state upon which it grows; he attacks and rejects bourgeois society which alienates art, yet art itself is deemed only one form of alienation. Marx defends art against capitalist society for the sake of the cause, but not unconditionally. Art is a species of production, and, when that production is capitalist, Marx defends art; his defense of art flows from his attack on bourgeois society. In the text just quoted, we can see this too: "*The first freedom of the press consists in not being an industry.* The writer who debases it into a material means deserves as punishment for that inner captivity the outer captivity of censorship, or rather his existence is already a punishment."[31] Marx writes these lines to attack bourgeois censorship, the bourgeois press, and capitalist industry, all the while keeping firmly in mind that *industry is the prototype of all human activity, art included*, and that art cannot and ought not be separated from the production of which it is only one branch. To defend the freedom of the press, he attacks its character as industry, even though he is the one who thinks that everything is industry. This man, whose thought centers on the total liberation of industry and on the transforming action by which all that is, is made into material for industry, is often touched by compassion for what is denied by technicist society: the family, morality, poetry, art. His feelings of commiseration for what is denied launch him all the more deeply into the negation polemic that aims for the radical abolition of all these alienations, so that man, abandoning them as false abodes, may regain his unique human existence—an existence reduced to the wealth of his needs. The terms *industry* and *art* can, then, have at least two meanings under Marx's pen: industry is "evil" as capitalistic and good as the type of human activity; art is "good" when bourgeois society does harm to it, but it does not thereby stop being "bad," since it is the expression of alienation.

Marx asserts that art is tied to all production, that it is an object produced by the producing subject and offered for the consumption of consuming subjects, production and consumption proceeding from the same source. Art holds no kind of "transcendence"; it is enclosed in the cycle that binds production to consumption. "Production therefore creates the consumer. Production not only provides the material to satisfy a need, but it also provides the need for the material. . . . An *objet d'art* creates a public that has artistic taste and is able to enjoy beauty—and the same can be said of any other product. Production accordingly produces not only an object for the subject, but also a subject for the object."[32]

We might, as a result, think that art, in the perspective of the transcending of *all* alienation, ought to be superseded *as art*. Art would be absorbed by and in human species activity as productive on the level of the real, for the total free movement of production would preclude any alienation, whether real and material or ideological (ideal) or spiritual. Human history itself would include "poetry" and "art"; but art and poetry would no longer constitute a world within, or beyond, the World. It would be, then, less a matter of art becoming realistic than of human history itself achieving itself in a creative and real way, that is, in a powerful realizing action. Historic characters would no longer play a role; they would no longer be representative figures; they would be the manifestation of the essential forces of social man as really present. The productive activity of men through history would cease being tragedy or comedy, as in a theater-world; it would become the one real stage for the full development of technique. The heroes of this stage would no longer have an official kind of appearance. They would wear no buskins on their feet nor crowns on their head; they would no more borrow masks, old or new, nor disguises. These non-heroic "heroes," heroes of the collectivity, would no longer speak words that overflow any content they may have; content would on the contrary overflow words and their form. All the superstitions and illusions, all the ideological poetry, and all the artistic forms that are a part of superseded tradition would have to be liquidated, so that the new social reality might take over and deploy itself without self-alienation in unproductive mediations. *The Eighteenth Brumaire of Louis Bonaparte* begins with this statement: "Hegel remarks somewhere that all facts and personages of great importance in world history occur, as it were, twice. He forgot to add: the first time as tragedy, the second as farce."[33] Within the perspective of universal reconciliation, the history of mankind becomes the real stage for the deployment of the practical energy of men. It no longer doubles itself into drama and comedy. The world of alienation begins with tragedy, and, having lost all *raison d'être*, just at the point of the full maturity of technique, it collapses into the ludicrous. Its gestures are vain and empty, its characters and events ridiculous, its actors only a mockery in their actions. Now that alienation has exhausted any meaning it had in coming into existence and no longer offers a source for poetry and art, mankind must decisively build its own future, and it must simply because now it can. This is the reconciliation that is to come about in a necessary, real, serious, and practical manner, not as a matter of the tragic, the comic, the poetic, the prosaic, the artistic, or the crude. The poetry of the past will be transcended. "History is thorough, and passes through many phases when it conveys an old form to the grave. The final phase of world-historical form is its comedy. The Greek gods, already once mortally wounded, tragically, in Aeschylus' *Prometheus Bound*, had to die once more, comically, in the dialogues of Lucian. Why does history proceed in this way? So that mankind will separate itself happily from its past."[34]

Life itself will absorb art, without becoming for all that *specifically* poetic, artistic, or aesthetic. The truth and beauty of the effort men make to conquer nature by technique will no longer need an ideal complement in the order of truth and poetic or artistic beauty. Marx holds fast to the idea that art is a sector of the division of labor, a technique that rests on the development of productive forces and the organization of society and trade; and he tells us what he thinks of the future, or rather the nonfuture, of art as art:

> The exclusive concentration of artistic talent in particular individuals, and its suppression in the broad mass which is bound up with this, is a consequence of division of labour. If, even in certain social conditions, everyone was an excellent painter, that would not at all exclude the possibility of each of them being also an original painter, so that here too the difference between "human" and "unique" labour amounts to sheer nonsense. In any case, with a communist organisation of society, there disappears the subordination of the artist to local and national narrowness, which arises entirely from division of labour, and also the subordination of the artist to some definite art, thanks to which he is exclusively a painter, sculptor, etc., the very name of his activity adequately expressing the narrowness of his professional development and his dependence on division of labour. In a communist society there are no painters but at most

190 Ideological Alienation

people who engage in painting among other activities.[35]

Art and poetry have asked their question of philosophic thought ever since its consolidation and systematization, that is, ever since Plato. The changeless and true idea of the beautiful, in separation (χωρισμός) from and yet with participation (μέθεξις) in the changing beauty of the sensible which but reflects it, begins to be a problem from the end of the Pre-Socratic dawn. We meet the problem again with the question of the links and relationships between Φύσις ("nature") and τέχνη ("art, technique"), ἀλήθεια ("truth") and ἔργον ("action, deed, work"). The language of philosophy, when it turns to these questions, has great difficulty taking hold of the language of poetry as ποίησις ("a making, a production") and ποιεῖν ("making, producing"). In the *Republic* and in the *Laws*, poetry and art draw the attention of the classic philosopher of the West in what is most fundamental to them, most problematic, and most fragile; poetry and art are in danger and at the same time are themselves dangerous powers. After a long progress, the last "systematic" philosopher of the West, Hegel, in his *Lectures in Aesthetics*, ponders once more, in definitive fashion, the truth of art. Art first, religion next, and finally philosophy constitute the three moments of absolute mind realizing itself in the history of the World. *Art* is the revelation of absolute truth in sensuous form, the sensuous representation of the idea, the casting of content into form; art, whose essence is poetry, is an expression of the divine. *Religion* expresses the absolute to a higher degree than art, relying on the representation of faith. Lastly, thought, in the form of *philosophy*, grasps the very reality of the truth of being-in-becoming through *Vernunft*, the concept and the idea. However, philosophy must also be superseded in and by true (absolute) knowing, the real knowing of *Wissenschaft*. Hegel thus saw that Spirit passes beyond art; he even saw that art could become something not absolutely necessary and that after a certain point it could belong to the past. Hegel did not mean that there would no longer be works of art, that all artistic movement and production would stop. His thought is that art is no longer an *absolute need* for us, that we no longer have this absolute need to incarnate and present a spiritual content in artistic, sensuous form. The sacred essence of poetry and art would no longer manifest itself; its constitutive and decisive truth would no longer be necessary in the world.

Certain passages in these lectures call for serious consideration, in that they pose but do not answer a really serious question, one

191 Art and Poetry

which we usually treat too lightly. In these passages we are brought face to face with the problem of art, not the theory of art, aesthetics, but *art*.

> Art has become for us an object, but just as our culture is not exactly characterized by a superfluity of life, and our mind and soul are unable any more to find the satisfaction that objects moved interiorly by a breath of life procure, it might be said that it is not by taking the point of view of culture, *our* culture, that we shall even have a mind for appreciating art and its right value, for taking account of its mission and dignity. Art no longer provides the satisfaction for our spiritual needs that other peoples sought and found in it. Our needs and interests have shifted into the sphere of conceptuality, and to satisfy them we have to call in reflection, thought, abstraction, abstract general concepts. Because of this, art no longer holds the place among what is truly living that it used to, and general concepts and reflection have taken over. This is why people today are given over to reflections and thoughts about art. And art itself, as we have it now, is only too well suited to becoming an object of thought.[36]

And, a few pages farther on, Hegel gives voice to these remaining questions:

> The work of art is, then, unable to satisfy our ultimate need for the Absolute. In our day people no longer venerate a work of art, and our attitude with regard to creations of art is much colder and reflective. In their presence we feel much freer than was once the case, when works of art were the highest expression of the Idea. The work of art solicits our judgment; we submit its content and the precision of its representation to reflective examination. We respect art, we admire it; only we no longer see in it something that cannot be surpassed, the intimate manifestation of the Absolute, we subject it to the analysis of our thought and we do this not with the intention of promoting the creation of art, but rather with the aim of recognizing the function of art and its place in the context of our lives. The fair days of Greek art and the golden time of the later Middle Ages are over. The general conditions of the present time are not favorable to art.[37]

192   Ideological Alienation

Finally, a few lines later, this man, whose ideas are echoed in Marx, whose thought perhaps embraced not only the "past" and "his own present" but could also look toward the future, has this to say:

> *In all these respects art is and remains for us, on the side of its highest possibilities, a thing of the past.* Herein it has further lost its genuine truth and life, and is transported to our world of *ideas* rather than is able to maintain its former necessity and its superior place in reality. What is now stimulated in us by works of art is, in addition to the fact of immediate enjoyment, our judgment. In other words, we subject the content, and the means of presentation of the work of art, and the suitability and unsuitability of both, to the contemplation of our thought.[38]

Marx, who paid special attention to Hegel's philosophy of art, offers his answer to the problem of art by saying what art is, what it has become in the modern world, and what it will become. From all that we have seen, it seems that in his eyes art and poetry constitute a *technē* that rests on the nondevelopment or underdevelopment of technique and that is part of the world of ideological alienation, the ideal world. As a result, the only possibility open to them, it seems, is abolition by the total, dealienating development of real praxis on the part of men working socially to appropriate the natural world for themselves through the process of transforming it. In the same way that θεωρία ("theory") is subordinated to πρᾶξις ("practice, praxis") without the basis of the distinction between the two being sufficiently explored, so ποίησις, as the essence of τέχνη, is subordinated to πρᾶξις, and τέχνη takes the sense of (productive) Technique. Τέχνη is thus severed from Φύσις, and ἀλήθεια is dissolved in the ἔργον, which is but the product of the productive practical energy of men who wish to satisfy their vital needs in the transformation of nature. There is no other actuality than reality as constantly actualized, as the realization and production of objects suitable for satisfying consumption. The brilliance of true beauty is absorbed by true reality, that is, by reality as the actual thing of power. Λόγος ("word, reason"), the common source of the languages both of poetry and thought, is silent; "Is it not inevitable that with the emergence of the press bar the singing and telling and the muse cease . . ."[39] Besides, language never really was a problem for Marx. Is it, too, to be superseded?[40]

Art becomes an object of aesthetics—whether aesthetics be taken as philosophic, scientific, technical, practical, or simply experienced —when it ceases to be art *in its ultimate purpose*. It accordingly addresses itself to the αἰσθησις, the senses, sensation, the emotions, the feelings; and men search at most for an intellectual significance— an intellectual sense—in all this enjoyment and consumption. Within the perspective of dealienation and dereification, every object can and should become an object of the senses, an objectification of the essential subjective-objective forces of social man. Having transcended alienation, thanks to the full deployment of the powers of technique, men will be able to enjoy with their senses and through the help of ideas in the imagination all the objects of the world— objects in the sense of "objects-for-subjects" rather than "objects objectively detached" (*objectal plus qu'objectif*). If nature is to be transformed from top to bottom through technique, even to the consummate loss of naturalness, utter desacralization, and the complete humanizing of everything that is in being, why should the production continue of such special works and objects as the creations of poetry and art, these special creative activities that are determined by the division of labor? Would not a sensuous consciousness that extended the senses suffice for the discovery of the sense of objects in that reality produced by man through his realization-labor? If the world becomes real, in what way will men still need an ideological complement, an ideal world?

Marx indicates the direction (*le sens*) in which his answers go, and he indicates also the sense, the direction (*le sens*) of current historical becoming. What he says is just what is being realized: art is losing its essence for the sake of gain on the part of technique. At the same time a whole host of literary and artistic *techniques* develops. Through all its manifoldness and its whole problematic condition, the new planetary reality shows its meaning: the *production* of the world of objects by human subjects. What is going on is the self-affirmation of man in the world, by man's objectifying himself in a *real* way.

> The manner in which [objects] become *his* depends on the *nature of the objects* and on the nature of the *essential power* corresponding *to it*; for it is precisely the *determinate nature* of this relationship which shapes the particular, *real* mode of affirmation. To the *eye* an object comes to be other than it is to the *ear*, and the object of the eye is another object than the object of the *ear*. The specific character of each essential power is precisely its *specific essence*,

> and therefore also the specific mode of its objectification, of its *objectively* actual living *being*. Thus man is affirmed in the objective world not only in the act of thinking, but with *all* his senses.[41]

The essential forces of objective subjectivity can fully objectify and assert themselves, in all their specificity, by means of the senses, by means of representation in the imagination, and by means of the thinking that corresponds to them. By conquering the natural world through technique, so that the world thereby becomes objective by the labor of men, human being no longer lives under the yoke of private property and the division of labor—which are the generating causes of alienation in its double nature—and so achieves itself totally in a social way. The complementary world of the ideal loses all reason for its existence. Precisely by the total praxis of triumphant labor man is freed from the narrow practical cares and the labor that made of him but a laborer, and so he will have complete "aesthetic" enjoyment of all objective, real wealth. He will no longer be alienated either in production or in the products of practical ποίησις. Art will no longer constitute one domain of the whole circle when everyone will be able to have "aesthetic" sensations, emotions, and ideas (*représentations*). Ideas (*représentations*) will emanate from concrete presences.

> Only through the objectively unfolded richness of man's essential being is the richness of subjective *human* sensibility (a musical ear, an eye for beauty of form—in short, *senses* capable of human gratification, senses affirming themselves as essential powers of *man*) either cultivated or brought into being. For not only the five senses but also the so-called mental senses—the practical senses (will, love, etc.)—in a word, *human* sense—the human nature of the senses—comes to be by virtue of its object, by virtue of *humanized* nature. The *forming* (*Bildung*) of the five senses is a labor of the entire history of the world down to the present.[42]

But can the senses and meaning wholly coincide?—Unless all question concerning sense disappears.

It would be quite legitimate to think that, beyond the death of poetry and art, beyond the death of the work of art and the poem, the dimension of the poetic and artistic will unfold as an "activity" that is not directly productive or technically organized. It would not be illegitimate to think that this poetic and artistic dimension might deploy itself, then, as Play.[43]

11.

PHILOSOPHY (METAPHYSICS)

AND THE SCIENCES

Philosophy is nothing other than religion (the "table of contents of the theoretical struggles of mankind") rendered into thought and developed by thought. Philosophy, that is, metaphysics, continues, crowns, and systematizes all ideological alienation: "The philosophic mind is nothing but the ⟨alienated⟩ mind of the world thinking within its ⟨self-alienation⟩—i.e., comprehending itself abstractly."[1] Abstract, metaphysical thought stands opposed to concrete, sensuous reality. Being itself generated by material oppositions, it generates in turn ideal oppositions, the materials of thought. In this way are born the illusory oppositions between the *in-itself* and the *for-itself*, *subject* and *object*, *mind* (*spirit*) and *matter*, *history* and *nature*, etc., etc. The act of the externalization of thought, nonetheless, continues to express the real externalization of alienated human activity, without recognizing it. Alienated thought, since it is in no way a truly operative instrument, only further alienates human practice through failure to consider it in all its true reality. The "spectacles" through which philosophy looks at the world prevent it from seeing the empirical facts, which are what all the deep metaphysical problems come down to. "Philosophy and the study of the actual world have the same relation to one another as onanism and sexual love,"[2] says Marx in quite categorical terms.

The species being of man, far from being confirmed and asserted in the world of thought, is invalidated and alienated in the mystical feeling that imbues speculative thought and its formal, abstract logic. The world of thought is alienated and alienating because it is only the ideological half of the real world, which is itself actually and materially alienated. Whether it work by way of intuition or reason, philosophic thought fights against the tree of reality on which it itself has grown only in order to leave it behind. "The *mystical* feeling which drives the philosopher forward from abstract thinking to intuiting is *boredom*—the longing for a content. The man ⟨alienated⟩ from himself is also the thinker ⟨alienated⟩ from his *essence*—that is, from the natural and human essence. His thoughts are therefore fixed mental shapes or ghosts dwelling outside nature and man."[3] Philosophic thought is thus condemned as having mysticism and religion for its source, as metaphysical, as nonhuman and inhuman.

This whole reign of abstract thought expresses the ruling power of those who hold the means of production. Philosophy may well be an abstraction, but it has not left off being bound to the concrete. The dominant class, which produces ideology, uses philosophy as a weapon. The character of generality and universality that seems to clothe philosophic thoughts serves admirably the particular objectives of the dominant class. As history progressively universalizes itself through the development of technique, through the increase in actual numbers of the dominant class, and through the convergence and extension of world relations, leading to the illusion that common interests exist for all members of society, the dominant ideas become more and more abstract. Ideas increasingly take the form of universality; for, inasmuch as the dominant class in each instance makes its domination a reality upon a base much broader than was the case for the preceding dominant class, it presents its interest as an interest that is common and universal. In this way it invests the ideas that inspire it and represent it with the form of universality, defending them as the only intelligent ones and the only ones universally admitted. Now, thought is not the theoretical form of something whose active, living essence is common, real being, social being; rather it "becomes" "universal" consciousness, that is, a generalized abstraction that is the enemy of real, concrete life. The individual no longer lives his human existence as such but represents it as a philosophic existence. On the other hand, correlatively, the individual can concretely manifest his life only in complete contradiction to abstract thought. More than that, thought systematized in the form of philosophy hides and masks this whole contradiction. "The *philosopher* sets up himself (that is, one who is himself an abstract form of ⟨alienated⟩ man) as the *measuring rod* of the ⟨alienated⟩ world. The whole *history of the alienation process* and the whole *process of the retraction* of the alienation is therefore nothing but the *history of the production* of abstract (i.e., absolute) thought—of logical, speculative thought."[4]

Man, this natural "and" human being, lives and produces in society. Nature taken in the abstract sense, nature in itself, is nothing for man; a nature anterior to human history has no meaning for man. Neither the philosophy of nature, that is, of nature as external to man and independent from his activity and industriousness, nor the philosophy of the history of man, which overlooks the natural, though human, needs of men, nor philosophical or abstract dialectical logic, in which thought separates from that of which it is thought, takes into account the basic situation: the sensuous activity of living individuals as powers that constitute and transform the sensuous world. Man from the outset is situated in a historical nature and a natural history, and he has all that constantly before him. This basic, indissolubly natural, historical, social, and human reality is precisely what is abandoned by philosophic thought as alienated metaphysics. The consummate ideologist, the philosopher, whether spiritualistic, as is Hegel, or materialistic, as are Feuerbach and some Neo-Hegelians, "does not see how the sensuous world around him is, not a thing given direct from all eternity, remaining ever the same, but the product of industry and of the state of society; and, indeed, in the sense that it is an historical product, the result of the activity of a whole succession of generations, each standing on the shoulders of the preceding one, developing its industry and its intercourse, modifying its social system according to changed needs. Even the objects of the simplest 'sensuous certainty' are only given him through social development, industry and commercial intercourse."[5] There is, of course, a nature that is antecedent to the natural history of man; yet man has never *had anything to do with* that nature, since he meets only the nature that is discovered in the course of his active history. Neither man's thought nor his practice meets this nature that is distinct from man. "Of course, in all this the priority of external nature remains unassailed," Marx admits.[6] But *that* nature does not concern man. "For that matter, nature, the nature that preceded human history, is not by any means the nature in which Feuerbach lives, ⟨it is⟩ nature which today no longer exists anywhere (except perhaps on a few Australian coral-islands of recent origin) . . ."[7] Does the thinking behind this statement betray an extraordinary lack of a sense of the cosmic? Is one to say that the Cosmos itself and Nature the all-embracing do not, properly speaking, "exist"? Do stars, minerals, plants, and animals exist only by virtue of having a place in human affairs? The passage just quoted that speaks of the world of the senses continues in these words: "The cherry-tree, like almost all fruit-trees, was, as is well known, only a few centuries ago transplanted by *commerce* into our zone, and therefore only *by* this action of a definite society in a definite age it has become 'sensuous certainty' for Feuerbach."[8] Marx turns his back to Nature as such and concerns himself only with the process of the transformation of nature into History. His humanistic naturalism is interested only in the total denaturalization of nature, which is accomplished by praxis in its aim to satisfy natural needs. Nature *is* only to the extent that man and his technique take hold of it. Φύσις ("nature" in the Pre-Socratic sense) is no longer present for Marx; the sensuous activity of men replaces the all-governing Heraclitean lightning.

As a result, all *philosophy of nature*, whether abstract or intuitive, idealistic or materialistic, mythological, theological or naturalistic, remains alienated from human practice by the fact that it gives a privileged status to the nature that "preceded human history."

On the other hand, all *philosophy of history and man* developed by the philosophic spirit is no less alienated. It too fails to recognize the true motor agency for the historical development of mankind; it too conceals both material needs and the real alienation of human beings. The philosophy of history does not know the starting point, that is, the action of men modifying all that is natural; it therefore does not proceed from real premises: "[The] premises are men, not in any fantastic isolation and rigidity, but in their actual, empirically perceptible process of development under definite conditions. As soon as this active life-process is described, history ceases to be a collection of dead facts as it is with the empiricists (themselves still abstract), or an imagined activity of imagined subjects, as with the idealists."[9] Since historical becoming has only been a progressive development of alienation and since what men do amounts only to their self-externalization in practical, social labor, it is clear that the theoretical grasp—be it empiricist or rationalist, materialistic or spiritualistic—of this process was and remained ideological, that is, alienated philosophically.

Equally set in alienation was the enterprise of thought that turned back to take hold of thought itself, the construction of *logic*. "Logic (mind's *coin of the realm*, the speculative or *thought-value* of man and nature—and hence their unreal essence) is ⟨*externalized*⟩ *thinking*, and therefore thinking which abstracts from nature and real man: *abstract* thinking."[10] Whether scholastically formal, transcendental, or abstractly dialectical, philosophic thought in the form of logic immobilizes or fixes the movement of thought inside a schema, severing it from real, concrete movement. Logic blocks negation and contradiction, and in so doing it is hardly able to provide an actual basis for negativity, because it situates negativity on the level of the idea, the concept, spirit, consciousness, thought. Marx's reproach against the philosophers who preceded Hegel, against Hegel himself, and against the ideological thinkers who followed him is that they remained imprisoned in the world of "logic" instead of setting off to battle it with its own weapons. He criticizes the ideological critic Bauer for staying on the plane of theoretical and ideological criticism rather than passing over to a material, effective, and actual plane: "How little this consciousness came into being even after the act of material criticism, is proved by Bauer, when, in his *The Good of Freedom*, he dismisses the brash question put by Herr Gruppe—'What about logic now?'—by referring him to future critics."[11] Marx's answer to this imprudent question amounts to this: Enough of philosophic logic—abolish it by plunging into praxis! The rational comprehension of praxis can and must supplant all abstractly rationalist logic. Every "deep" philosophical problem simply reduces to some empirical fact or to a totality of real facts, and so it is not consciousness and logical thought that can get to the resolution of the problem: "The solution of theoretical riddles is the task of practice and effected through practice, just as true practice is the condition of a real and positive theory . . ."[12]

Marx shows no more indulgence toward first or general philosophy, that is, metaphysics. Metaphysics with its ontological pretense in no way examines the sense (the *logos*) of the being of totality. Whether spiritualistic or materialistic, metaphysics from Plato to Hegel remains abstract, contenting itself with speculative mysteries. The shadow of dualism weighs upon all metaphysics. Metaphysical thought confuses the abstract and the real; it takes the abstract for the real and translates the real abstractly. There is no ontological truth in an answer that passes beyond the horizon of human sensuous, practical activity. The question of being has no meaning. What transcends concrete sensuous experience is alienated in becoming abstract: "Abstract spiritualism is abstract materialism; abstract materialism is the abstract spiritualism of matter."[13] The basis of everything that is, the primordially first *archē*, does not lend itself to being grasped either as spirit or as matter. Spiritualism and materialism are both abstract since they are both only theoretical. There is no fundamental questioning that can concern itself with something that would precede or found real, material human activity. Marx's materialism is *practical*, and in the name of this "materialism" he condemns all philosophy as speculative. Metaphysics has never pictured the World as the material of human labor, as nature's conversion into history (*devenir-histoire de la nature*), and so it has betrayed nature, man, and history.

Thinkers, philosophers, and theoreticians, being one and all metaphysicians and ideologists specializing in general ideas and the universal, are the "heroes" of this alienation. They put themselves forward as the standard for the world, imagining themselves to be the true makers of history, the council of protectors, the dominant force. But their alienation consists exactly in this: they only express and systematize the ideas of the dominating classes, their speculative constructions are conditioned by material realities, and they never get to a correct formulation of the way practice is a coming-into-being. These specialists in mental labor all only reflect the division of labor and the alienation of technique; the nondevelopment and then the underdevelopment of productive forces and the trammels put on this development by the relations of development determine the production of this whole sublime, sublimated wealth of ideas and ideology. There could not be, therefore, a *history of philosophy*, a history of an (autonomous) development in philosophic thought, in which philosophies, in detachment from their real bases, would be linked up to each other. There could not be a succession of philosophic systems bound together by a mystical connection, by an abstract dialectic that masked their social determinations and ties, their historical conditioning. Philosophic thoughts and all contemplative systematizations have no historical development belonging properly to them, for it is men, in the development of their material production, who, through the modification of objective reality, change as well their way of thinking and the products of their way of thinking. With his extraordinary sense of consistency, Marx radically destroys the whole edifice of metaphysics, philosophy, and speculative thought, not deeming it a basic, all-encompassing construction and therefore attacking it as superstructure. In his grasp of the "sense" of modern history and from his study of the meaning of planetary technique, Marx does not recognize philosophy as a historical power, and so he hauls it before the sovereign tribunal of productive, transformative material action; *philosophy dies under the blows of technique*. The tasks of philosophy become the tasks of historical praxis. Philosophy has one last thing to do: to unmask alienation, blazing the trail for the passage to the total development of productivity and abolishing itself in bringing itself to full realization. "It is the task of history, therefore, once the other-world of truth has vanished, to establish the truth of this world. It is above all the task of philosophy, which is in the service of history, to unmask human self-alienation in its secular form, once its sacred form has been unmasked. Thus, the critique of heaven is transformed into the critique of the earth, the critique of religion into the critique of law, the critique of theology into the critique of politics."[14]

Marx's philosophy, then, undertakes the task of pitilessly criticizing all alienations and finally abolishing philosophy. The abolishing of philosophy is to mean at the same time the realizing of philosophy. It is indeed by its realization that philosophy abolishes itself and transcends itself. Empiricists call for the negation of philosophy; pure theoreticians want philosophy to realize itself. The first do not see that one cannot abolish philosophy without realizing it, and the second do not understand that it is impossible to realize philosophy without abolishing it. The negation of philosophy means "the negation of philosophy as it has been up to now, i.e., of philosophy as philosophy."[15] The *Aufhebung* of the ideal complement of the alienated world is to coincide with its practical realization (*Verwirklichung*) within the universal reconciliation that will have transcended and abolished all alienation. This movement by which philosophy becomes world-order reality (*devenir-monde de la philosophie*), however, raises some questions that we shall take up later.

For the time being, our point is to clarify to a certain extent the Marxian criticism of philosophic alienation, the trial to which philosophy is to submit itself. This criticism is principally addressed to German philosophy in its consummation in Hegel's philosophy and the constructions of the Hegelians.[16] German philosophy has been able to develop only thanks to the nondevelopment of German history. Poor in actions, Germans have been rich in thoughts. "Just as ancient peoples lived their past history in their imagination, in mythology, so we Germans have lived our future history in thought, in philosophy. We are philosophical contemporaries of the present day without being its historical contemporaries. German philosophy is the ideal prolongation of German history."[17]

Hegel is the German philosopher who tried to grasp the totality of historical reality through philosophic thought. He pursued logically and "historically" the ideal genesis of all philosophic ideas, summing them all up and grouping them in his history of philosophy, which is the soul of universal history. Philosophic ideas thus stand all united to compose a whole, and this alienated philosophic world now calls for criticism. The negation critique of philosophy as philosophy must confront Hegel, with whom the "becoming" of philosophic thought as coming to be aware of itself, as becoming the spirit of the world thinking itself, reaches its culmination. "Therefore, that which constitutes the *essence* of philosophy—the ⟨*externalization*⟩ *of man in his knowing of himself*, or ⟨*externalized*⟩ *science thinking itself*—Hegel grasps as its essence; and he is therefore able to gather together the separate elements and phases of previous philosophy, and to present his philosophy as *the* philosophy."[18] With Hegel, all philosophy of history coincides with the history of philosophy; the World thus becomes Thought. Yet this process by which the world becomes philosophy (*ce devenir-philosophie du monde*) and by which at the same time philosophy becomes the world (*devenir-monde de la philosophie*) is not total and unitary. The world as a totality remains dismembered into a philosophy that thinks within alienation and abstraction, on the one hand, and on the other a reality that is torn apart, alienated and alienating, concrete and contradictory. Philosophy is itself doubly contradictory.

The critique of philosophic alienation is a dealienating enterprise. Its target is not so much some determinate philosophic abstraction as it is the "totality" of the historical, systematized "becoming" of philosophic thought. Hegel, the culmination of this becoming, is consequently to be radically criticized. Marx rises up against his "master" and, under the inspiration of his own philosophically antiphilosophic passion, begins the battle. It is a fight to the death not in the name of the love of Wisdom ($\varphi\iota\lambda o\sigma o\varphi\iota\alpha$) or in the name of the truth of $\Lambda\acute{o}\gamma o s$ ["word, reason" in a Pre-Socratic sense] but in the name of the will for the realization of *Praxis*, in the name of *practice as a conquering force*. Hegel is charged by Marx with a series of convergent "errors," all because of a false basis. He is accused of maintaining the illusions of metaphysical speculation and of deploying a mystical, abstract thinking. He is accused of making use only of an apparent criticism and a false positivism, being neither sufficiently critical nor radically negative. In a word, Hegel is charged with justifying that which is by pronouncing the real to be rational and the rational real. Marx's criticism of Hegel *wants* to take its attack to the center of Hegel's thought and to all its dimensions and consequences, as well as to its presuppositions. Everything, then—logic and philosophy of nature, theory of labor and political philosophy, phenomenology of spirit and philosophy of art and religion, history of philosophy and philosophy of history—everything is attacked, and we have already seen the way this Marxian critique works in regard to labor, civil society, the State, and man's self-consciousness. The basis for this effort to overturn Hegel, the meaning of this negation critique, amounts to this: a charge that Hegel did not discover the motor agency for and the riddle in the movement of the universal history of mankind, that is, the development of real productive forces, the movement of technique as an active power. "He has only found the *abstract, logical, speculative* expression for the movement of history; which is not yet the *real* history of man—of man as a given subject, but only man's *act of creation*—the *story* of man's *origin*."[19] In contrast,

> since for the socialist man ... the *entire so-called world history* is only the creation of man through human labor and the development of nature for man, he has evident and incontrovertible proof of his *self-creation*, his own *formation process*. Since the *essential dependence* of man in nature—man for man as the existence of nature and nature for man as the existence of man—has become practical, sensuous, and perceptible, the question about an *alien* being beyond man and nature (a question which implies the unreality of nature and man) has become impossible in practice.[20]

The God who lives hidden within the Hegelian thinking and Hegel's religious philosophy draws Marx's fulminations. Religious alienation, the center of all ideological alienation, is continued by Hegel, and the real existence of religion is not abolished. After a whole long dialectical journey (and detour), the divine being, transcending nature and human history, and religious man find again a final confirmation in the ultimate philosopher. For his part, this philosopher has taken into account only the spiritual and ideal essence of religion, not its alienated reality and its conditioning determinants. He confuses the empirical existence of religion with the speculative philosophy of religion and the religious existence of man with the philosophico-religious representations that he makes of this "existence"; the result of this is that he does not get to what is essential in man. Even when he tries to transcend religion, religion is "abolished" only for the sake of a further philosophic abstraction: absolute knowledge.

Hegel stands charged with maintaining the root and all the forms of double alienation, of alienation both practical and real and theoretical and ideological, since in his eyes all forms of real alienation are only forms of the alienation of self-consciousness. His dialectic is not a dialectic of reality but remains a dialectic of pure thought. This dialectic continues alienation, because it posits man as self-consciousness; man is thus an abstract, entirely theoretical being, and the whole *Phenomenology* does nothing but study the spiritual appearances (*phenomènes spirituels*) of the alienation of real, action-performing human being. The sensuous reality of sensuous activity is thus negated, since the object is constituted only by self-consciousness and activity. Total reality, the conversion of nature into history (*le devenir-histoire de la nature*) through human labor, exists only in function of knowledge, Spirit, concept, idea, thought, consciousness, and self-consciousness; and knowing replaces real, material life, which is objectness (objectivity). "The way in which consciousness is, and in which something is for it, is *knowing*. Knowing is its sole act. Something therefore comes to be for consciousness in so far as the latter *knows* this *something*. Knowing is its sole objective relation."[21]

Hegel posits and maintains alienated being and only introduces dialectical movement into thought being. *Negation* never negates alienation, for it remains a logical movement, confronting only essences in thought. The *negation of negation*, which pretends to be the absolute positive, ends up in fact with total abstraction, infinite spirit, and thus maintains alienation definitively. "In Hegel, therefore, the negation of the negation is not the confirmation of the true essence, effected precisely through negation of the pseudo-essence. With him negation of the negation is the confirmation of the pseudo-essence, or of the self-alienated essence in its denial; or it is the denial of this pseudo-essence as an objective being dwelling outside man and independent of him, and its transformation into the subject. A peculiar role, therefore, is played by the act of *superseding* in which denial and preservation—denial and affirmation—are bound together."[22] Despite the enormous revolutionary importance of the Hegelian dialectic, in which negativity is the principal, creative determinant, it is a dialectic that remains "logical" and ideological in its substitution of the becoming of spirit for the becoming of the reality of all that is in a material way. Not fully grasping the negativity of becoming as historical time, it conceives time only as self-relating negativity. Thus it can in no way shatter and transcend, that is, productively negate, the externalization, the alienation of the essential forces of man. In short, it is a dialectic that remains formal and empty, not having taken hold of revolutionary transformative practice, which is content-filled essential activity.

> The supersession of externalization is therefore nothing but an abstract, empty supersession of that empty abstraction—*the negation of the negation*. The rich, living, sensuous, concrete activity of self-objectification is therefore reduced to its mere abstraction, *absolute negativity*—an abstraction which is again fixed as such and considered as an independent activity—as sheer activity. Because this so-called negativity is nothing but the *abstract, empty* form of that real living act, its content can in consequence be merely a *formal* content begotten by abstraction from all content.[23]

Hegel's whole ontological, metaphysical philosophy culminates in the kingship of (divine) absolute spirit, the truth of the idea, and it does so because it began with just that. Spirit engenders the concept and the idea of Thought (Logic), spirit becomes alien to itself and alienates itself in Nature (Philosophy of Nature), and spirit returns to itself through law, morals, art, religion, and philosophy in the course of the universal history of mankind (phenomenology of spirit and philosophy of history). Gaining awareness of itself and its process, in the end it requires that philosophy itself be transcended in the interest of absolute knowledge. The being-becoming of totality is this becoming of absolute spirit culminating in the grasp of total truth by absolute knowledge. Hegel's statement, "... what is *thought is*, and ... what *is*, only *is* in so far as it is a thought,"[24] identifies, in a manner like that of Parmenides, thing and thought, the subjective and the objective, the real and the ideal. Marx sums up Hegel by quoting him: *"The absolute is mind*. This is the highest definition of the absolute."[25] A few pages farther on, we find these lines of commentary with which the manuscripts of 1844 come to a close: "Until and unless spirit inherently completes itself, completes itself as a world spirit, it cannot reach its completion as self-conscious spirit. The content of religion, therefore, expresses earlier in time than (philosophical) science what spirit is; but this science alone is the perfect form in which spirit truly knows itself. The process of carrying forward this form of knowledge of itself is the task which spirit accomplishes as actual History."[26]

Marx sets off to war against Hegel in the name of the total liberation of human praxis, in the name of a kind of thought that wishes to be only a guide for action. This is not a pure and simple denial of Hegel. In the heat of battle he spares no blows and frequently gets carried away. For example, at one point he writes: "It is and remains an old woman, faded, widowed *Hegelian* philosophy, which paints and adorns her wrinkled and repugnant abstraction of a body and ogles all over Germany in search of a wooer."[27] Marx's aim is the dialectical reversal of Hegel. Turning toward him and against him, one genius as it were confronting another, Marx tries to inaugurate a passage leading from the light of spirit to the reality of action. No philosopher has ever been *refuted* by another philosopher, and so it is with the "dialog" between Hegel and Marx. Nevertheless, Marx speaks in his own voice and aims to trace a new path.

In the afterword to the second German edition of *Capital*, we can—and should—read the following passage, which will perhaps help us to better understand the sense of the criticism that both the young and the older Marx make:

> My dialectic method is not only different ⟨in its base⟩ from the Hegelian, but is its direct opposite. [Recall that Marx never works with the logico-ontological distinction between *contraries* and *contradictories*.] To Hegel, this life-process of the human brain, i.e., the process of thinking, which, under the name of "the Idea," he even transforms into an independent subject, is the demiurgos of the real world, and the real world is only the external, phenomenal form of "the Idea." With me, on the contrary, the ideal is nothing else than the material world reflected by the human mind, and translated into forms of thought.
>
> The mystifying side of Hegelian dialectic I criticised nearly thirty years ago, at a time when it was still the fashion. But just as I was working at the first volume of "Das Kapital," it was the good pleasure of the peevish, arrogant, mediocre Ἐπίγονοι who now talk large in cultured Germany, to treat Hegel in same way as the brave Moses Mendelssohn in Lessing's time treated Spinoza, i.e., as a "dead dog." I therefore openly avowed myself the pupil of that mighty thinker, and even here and there, in the chapter on the theory of value, coquetted with the modes of expression peculiar to him. The mystification which dialectic suffers in Hegel's hands, by no means prevents him from being the first to present its general form of working in a comprehen-

> sive and conscious manner. With him it is standing on its head. It must be turned right side up again, if you would discover the rational kernel within the mystical shell.
>
> In its mystified form, dialectic became the fashion in Germany, because it seemed to transfigure and to glorify the existing state of things. In its rational form it is a scandal and abomination to bourgeoisdom and its doctrinaire professors, because it includes in its comprehension and affirmative recognition of the existing state of things, at the same time also, the recognition of the negation of that state, of its inevitable breaking up; because it regards every historically developed social form as in fluid movement, and therefore takes into account its transient nature not less than its momentary existence; because it lets nothing impose upon it and is in its essence critical and revolutionary.[28]

Marx's critical and revolutionary dialectic wants to liquidate philosophy. It fixes its eyes on material movement, the real becoming of human history, and looks at the movement of thought (and of philosophy) only as a reflection, translation, and betrayal of real movement. "Materialism" and "realism" here mean historical and humanistic materialism and realism, not something "ontological." "Corresponding" to the *material* movement of productive forces, to the *real* development of technique, are the *spiritual* movement of ideology, the *idealist* thought of philosophy, the alienations of consciousness. This dialectic is, therefore, neither ontological—since it has no wish to know some fundamental being such as matter itself or spirit itself—nor logical or gnosiological—since it does not tarry very long over reality itself or ideality itself. Not set within the alternatives of spirit and matter or of the real and the ideal, the Marxian dialectic aims to be a dialectic of human activity, of real, material, sensuous, practical, powerful, concrete activity. It wants to have done as much with philosophic, ontological, and metaphysical alienation, with its abstract preoccupation with being, spirit, and matter, as with logical and gnosiological alienation, that pretends to examine, no less abstractly, knowledge, reality, ideas, and thought. The language that the Marxian dialectic speaks is neither "cosmological" nor "conceptual"; its intent is to express the actual "*logos*" of human history and discover the meaning of human activity. Idle speculations about the primacy of *matter* or of *spirit*—ontological materialism and ontological spiritualism—about the opposition between *reality* and *idea*, being and thought, experience and reason—realism and idealism, logical empiricism and logical rationalism—do not concern Marx, supposing for the moment that they have ever been of interest to thinking worthy of the name.

"The same spirit that builds philosophical systems in the brain of the philosophers builds railroads by the hands of the workers," Marx writes in the *Rheinische Zeitung*, July 14, 1842.[29] Seeing technique—and not philosophy—to be the motor of becoming, he takes material technique above all to be that which his dialectic will dealienate and revolutionize. To achieve that, he has to shatter the contradictions of the world of alienation. The Marxian dialectic often separates what the Hegelian dialectic united, and it fights against any intervention by mediation; for mediation is that through which the being-in-becoming of totality is achieved in self-knowledge as universal and absolute Spirit. Marx, most assuredly, does not plumb the essence of the dialectic or work out the concepts of dialectical thought. He does not spend time on the distinction between realities and notions that were *contrary* and those that were *contradictory* or on the question of the *unity* or the *identity* of conflicting opposites; and he is not burdened by the problem of the ties binding "subjective" dialectic (that of thought) with "objective" dialectic (the real). Marx's summary and laconic dialectic is a resounding dedication to the task of resolving the historical contradiction of modern society, namely, the dynamic contradiction that "unites" the two classes in conflict. Those opposites that are the concern of his dialectic are the *proletarians*, who bear negativity, and the *capitalists*, who retain the productive forces and are the "positive" support of alienation.

"Actual extremes cannot be mediated with each other precisely because they are actual extremes. But neither are they in need of mediation, because they are opposed in essence. They have nothing in common with one another; they neither need nor complement one another. The one does not carry in its womb the yearning, the need, the anticipation of the other."[30] Thus Marx, in his *Critique of Hegel's "Philosophy of Right,"* blames Hegel for the *conciliatory* dimension of his dialectic. For him, opposing realities, real extremes, actually existing differences, remain irreconcilable; just as, for example, human sexuality and nonhuman sexuality can in no way be coupled, they remain fundamentally distinct. On the other hand, two opposite determinations lying in the same being, in the same entity, can meet. Differences that appear within one common entity can unite, and only these; for example, the male and female sexes show as extreme differences within human being, and so they attract each other and couple.³¹ As a result, every philosophic thinking that keeps affirming the unity of the world, of this world split in two by alienation, maintains and justifies alienation. "Hegel's chief mistake consists in the fact that he conceives of the contradiction in appearance as being a unity in essence, i.e., in the Idea; whereas it certainly has something more profound in its essence, namely, an essential contradiction."³²

Metaphysics can only miss the essential contradiction, because it itself depends upon alienation. Since it seeks to *reconcile* everything within mind, its "reconciliation" of reality and idea is abstract. Reality, nonetheless, remains torn apart and full of antagonisms; and philosophy, being total and unified, can only conceal this dislocation in the world. The lie told by philosophy resides in its very principle. Philosophy by its thinking has given expression to the world of alienation, and all that remains for it is to merge itself in the real becoming of the world.

It is not only philosophy and metaphysical thought that are alienated. *Science* is no less so. Scientific activity does not constitute an autonomous process that has its own internal logic. The real becoming of productive activity determines the "history" of scientific knowledge and activity; there is no *history* of science, as something independent of material productive activity. The very matter of scientific activity is provided for the scientific worker as a social product. What he does, despite his feeling that he does it on his own, he does socially—that is, through technique and society—but, since technique in every society has been up till now alienated, how could science not be?³³

In struggling with nature to extract from it the goods they need for life, men also set up the *natural sciences*. In the course of their history they work out the *historical* sciences. Their economic activity is expressed in the science of *political economy*. The practical relationships that join men to nature and among themselves in a unity of opposition are materially at the base of any scientific theory. Yet these relationships are alienated; and as a result their theoretical and scientific expression is alienated as well. A nonalienated science of an alienated reality is not possible.

The sciences are all the more alienated by their having been developed within the division of labor. For it is the alienating division of social labor that in general and according to particular modalities is the condition for the work of science, which is itself divided.

One can see that intellectual workers, members of the dominant class, are the only ones who do science, while scientific activities themselves are divided up in such a way that the splintering divisions among them prevent effective communication. In this way the whole sphere becomes autonomous, loses its relationship to other domains, and engenders yet more regions that themselves become autonomous through a specialization that leaves universality far behind. What is true for all areas of alienation is true for the "total" sphere of scientific alienation as well as for all its particular subspheres: through them the species being of man is alienated and Totality is atomized. "It stems from the very nature of ⟨alienation⟩ that each sphere applies to me a different and opposite yardstick ... for each is a specific ⟨alienation⟩ of many and focuses attention on a particular round of ⟨alienated⟩ essential activity, and each stands in an ⟨alienated⟩ relation to the other."[34] In leaving practice behind, though they issue from it, scientific knowledge and the many and varied scientific techniques suppose that they constitute a proper realm with its own laws. Yet the reign of scientific alienation is the expression of general alienation, the division of labor, and the interests of the dominant class. Moreover, even the different branches of the tree of science remain without real mutual communication, each one being cut off, both individually and as an ensemble, from true natural, human, and historical universality.

The encyclopedia of the sciences, the total system of scientific knowledge, absolute science—none of these has reached full constitution. Knowledge has remained divided and splintered, each scientific worker deeming his discipline and his method the single true one by which he judges all others, while at the same time not taking the trouble to relate his particular truth to even that restricted reality that corresponds to it. Isolated individuals live in "atomistic" society, unable in any way to complete the construction of the edifice of science. No doubt scientific progress has been achieved, resulting, in general, in a partial transcending of the watertight compartments that separate scientists. But these results remain meager.

> In astronomy, people like Arago, Herschel, Encke, and Bessel considered it necessary to organise joint observations and only after that obtained some fruitful results. In historical science, it is absolutely impossible for the "Unique"[35] to achieve anything at all, and in this field, too, the French long ago surpassed all other nations thanks to organisation

of labour. Incidentally, it is self-evident that all of these organisations based on modern division of labour still lead only to extremely limited results, representing a step forward only compared with the previous narrow isolation.[36]

The sciences remain divided into the *natural sciences* and the *sciences of human history*. It is an alienating division, for man is never involved with nature as something outside history or with a history independent from nature. There is not one basis for science and another for life; this scission is introduced by alienation, and with it science takes leave of the terrain upon which it rises and which ought to be its foundation, namely, the sensuous activity of men. "Only when it proceeds from sense-perception in the twofold form of *sensuous* consciousness and of *sensuous* need—that is, only when it proceeds from nature—is it *true* science."[37] Nature in this sense is in no way separable from human history, for "the nature which develops in human history—the genesis of human society—is man's *real* nature; hence nature as it develops through industry, even though in an ⟨alienated⟩ form, is true *anthropological* nature."[38] Nature, History, and Man hardly constitute separable entities, and so there could not be separate sciences. It is by his productive labor that man, in his anthropological, historical nature, enters into contact with nature to transform it through history. "*Industry* is the *actual*, historical relationship of nature, and therefore of natural science, to man."[39] Scientific abstraction, whether abstractly materialistic or abstractly idealistic, does not know how to grasp this process in which nature becomes history thanks to the labor and industry of man living in society. Alienated science, consequently, remains unnatural, antihistorical, and inhuman. Scientific alienation leads us to think that on one side are the natural sciences and on the other the sciences of human history, that actual reality and the constructions of knowledge are different, and that there are specific and particular truths. In betrayal of the sensuous needs of man (that is, of his natural human needs), scientific alienation, as the integrating part of ideological alienation, expresses the real alienation of man.

Cut off from each other and with no link to the totality of things, which is indissolubly natural, historical, and human, the sciences lose as well any actual linkage to philosophic thought. Philosophic alienation, the speculative and abstract systematization of scientific alienation, moves, so to speak, into scientific alienation, which in turn sets itself up as autonomous.

> The *natural sciences* have developed an enormous activity and have accumulated an ever-growing mass of material. Philosophy, however, has remained just as alien to them as they remain to philosophy. Their momentary unity was only a *chimerical illusion*. The will was there, but the means were lacking. Even historiography pays regard to natural science only occasionally, as a factor of enlightenment, utility, and of some special great discoveries. But natural science has invaded and transformed human life all the more *practically* through the medium of industry; and has prepared human emancipation, although its immediate effect had to be the furthering of the dehumanization of man.[40]

Marx sees in technique—in practical, sensuous activity—the motor of the historical development of mankind and of nature's transformation into the material of social labor. He sees technique's most highly developed form, *industry*, as that which prepares the dealienation of man, the freeing of his activity (though right now it consummates alienation), and the satisfaction of the totality of his natural, human, and social needs. And so Marx puts the association of *technique* with *science* beyond question; *productive, industrial technique* is even inseparable from *scientific technique*. Nevertheless, as we have already seen, "there is no history of politics, law, science, etc., of art, religion, etc."[41]

Even though he sees science as one particular mode of alienation (that is, a mode of basic alienation and a mode of the derivative alienation which ideology is), Marx does not seem to consign science to abolition, as he does politics, the State, religion, art, and philosophy. Science, linked to technique and constituted as *knowing* that rests on *doing*, can survive the transcending of all forms of alienation; and we shall try to see what becomes of science within the prospect of universal reconciliation. Still, science—as it has hitherto been conceived and practiced—remains alienated, and it is not this kind of science that will develop after the abolishing of alienation, for that abolishing will include scientific alienation.

We should not expect to find in Marx a theory of science or a clarification of the problem of the relationship between philosophy and science. We should not expect either to see Marx offering his own thinking as *science*. His metaphysical and historical thought, his scientific—and economico-political—analysis and theory are indissolubly linked, even though his scientific activity remains founded

213 Philosophy (Metaphysics) and the Sciences

by his philosophic thinking. The word *science*, under Marx's pen, does not have quite the sense it has for us, as we shall see, perhaps, in what will follow. Nevertheless, what seems certain is that science, as it has developed, constitutes only the systematized and technicized extreme point of alienation.

Like any real radical alienation and like all the other modes of ideological alienation, scientific alienation separates theory from practice. In addition, within scientific theory it separates the different domains into watertight compartments. Alienated scientific activity also separates nature from history and science from life. Scientific alienation cuts totality into sections according to different points of view, each of which deems itself the single true one; and thus it is an alienation that rests upon the division of labor and alienated technique.

All those things by which the Hegelian Spirit revealed and realized itself in the history of the world—namely, labor, the family, politics, law, morality, self-consciousness, art, religion, philosophy, and science, all that which brought the being-in-becoming of totality to the light of *logos*, the concept, and the idea, and, finally, to the light of *absolute knowledge*—all this is regarded by Marx as constituting so many forms and forces of the externalization and alienation of human being. In the *phenomena* that spirit grasps, Marx discovers the real *alienations* of man.

The totality of alienations must be transcended actually and not simply in thought. The *double* alienation, of practice and of ideology, must be abolished really and not simply in consciousness. The very core of the totality of alienations and of its *double aspect* must be rooted out. Private property must be annihilated, for economic alienation is the root of all alienation: ". . . economic alienation is that of *real life*; its transcendence therefore embraces both aspects [practical reality *and* consciousness]."[42] It is not philosophic thought or the thinker who can lead mankind to universal reconciliation, although reflective grasp plays a large role in Marx's thinking. In the last lines of Hegel's preface to his *Philosophy of Right* (which is the last work of his to appear while he was still alive), we find this passage:

> One word more about giving instruction as to what the world ought to be. Philosophy in any case always comes on the scene too late to give it. As the thought of the world, it appears only when actuality is already there cut

214 Ideological Alienation

and dried after its process of formation has been completed. The teaching of the concept, which is also history's inescapable lesson, is that it is only when actuality is mature that the ideal first appears over against the real and that the ideal apprehends this same real world in its substance and builds it up for itself into the shape of an intellectual realm. When philosophy paints its grey in grey, then has a shape of life grown old. By philosophy's grey in grey it cannot be rejuvenated but only understood. The owl of Minerva spreads its wings only with the falling of the dusk.[43]

Echoing these words, the following lines from Marx accuse Hegel of a twofold inadequacy:

The philosopher is only the organ through which the creator of History, the Absolute Spirit, arrives at self-consciousness *by retrospection* after the movement has ended. . . . so that the philosopher appears *post festum*.

Hegel is doubly inconsistent: first because, while declaring that philosophy constitutes the Absolute Spirit's existence he refuses to recognize the *real philosophical individual* as the *Absolute Spirit*; secondly, because according to him the Absolute Spirit makes history only *in appearance*. For as the Absolute Spirit becomes *conscious* of itself as the creative World Spirit only in the philosopher and *post festum*, its making of history exists only in the consciousness, in the opinion and conception of the philosopher, i.e., only in the speculative imagination.[44]

The whole point at issue here, namely, finally to have done with alienation and its ideological extension, and thus to have done with philosophy, so that productive forces can develop freely and totally, so that human praxis may know no more constraints, is summarized in Marx's statement in the eleventh thesis on Feuerbach: "The philosophers have only *interpreted* the world, in various ways; the point is to *change* it."[45] And it is this total, radical change, this transcending of alienation and alienations, that will lead man to regain his being, to be reconciled with, through conquest of, the becoming of totality—but totality interpreted in a certain way.

Part VI.

The Prospect of Reconciliation as Conquest

Viewing things within the perspective of many-sided, *radical alienation* in no way constitutes the final term of Marx's thinking. The alienation of man is given direct treatment in that thinking in order that mankind may be led to transcend the alienation of material forces, thought, and consciousness. The philosophically antiphilosophical thought of Marx culminates in the prospect of universal and *total reconciliation*; it is even inspired from one end to the other by this vision. Universal reconciliation will mean the reconciliation and recognition of nature and history, of man and society, of the individual and the community, of needs, of planetary technique, and of full satisfaction. In its embrace, philosophy will be abolished in being realized, thought will become reality, and philosophy world. Reconciliation will mean *conquest of the world*, the world being what reveals itself and makes itself through human activity.

12.

THE PREMISES FOR TRANSCENDING ALIENATION

In order to talk of *externalization* and *alienation*, does one not have to presuppose a being or a reality that "precedes" the externalization and the alienation? In order for there to be a *transcending of alienation*, a *total reconciliation*, does there not have to be a "reality" that can become alien through alienation and with which there will be a *reconciliation*? But can something that externalizes and alienates itself, something whose whole history is only a history of dispossession, ever have existed in all the truth of its reality?

All human history has been only the history of alienation; no reality has preceded alienation. Alienation means alienating man's activity, robbing and dispossessing his being. Man has always up to now been alienated man, sometimes more, sometimes less. What, therefore, is the meaning of this *return* of man to his existence, of which Marx speaks; for "the positive transcendence of *private property*, as the appropriation of human life, is therefore the positive transcendence of all ⟨alienation⟩—that is to say, the return of man from religion, family, state, etc., to his *human*, i.e., *social* existence."[1]

One must repeat: man, the being indissolubly natural and social, endowed with objective essential forces activating his struggle and his productive activity, the being that aspires to the total satisfaction of his needs—*man in this sense* is the center of Marx's thought. Human being is itself a product of nature that produces and reproduces within natural history; and history is only a continuous transformation of human nature. There is no "absolute" starting point. The origin of everything that is revealed in man, the origin of the manifestation of man's activity, lies in a (dialectical) movement. The first (dialogical) natural species relationship is the relationship that unites man to nature, a relationship that at the same time unites men as social beings; and there is nothing that "precedes" this relationship. This relationship is the source of the truth of reality. Human being is the nature of man, and nature is the human being of man. The being of the totality in its becoming, *Nature*, *World*, only *are* from the moment that nature is manifest to the eyes of man, man standing in opposition to nature by his labor and doing so simply because that is his human nature.

We cannot keep from repeating: Marx nowhere explicitly raises the problem of what "precedes" human activity, the process in which nature becomes history (*le devenir-histoire de la nature*). The whole of ontological movement and the entire evolution of nature, all that leads to man, remain beyond his vision. For him, everything seems to begin with the social activity of natural man; everything begins to exist for man at the moment when his natural being begins to work on and in nature in order to satisfy his natural, human needs. Nature is unveiled and, properly speaking, is born into human history, and it is the nature of man that makes man from the outset a social being. "History itself is a *real* part of *natural history*—of nature developing into man,"[2] writes Marx. Yet this "real part" is the Totality, for it is through this part, and in relation to it, that everything is, shows itself, and comes to be. Human history begins when nature culminates in man, and this culmination is *the* beginning. History is indeed "natural," but nature exists only as nature made-into-history (*nature historialisée*). "The *human* essence of nature first exists only for *social* man; for only here does nature exist for him as a *bond* with *man*...."[3]

Men were not *created* but appeared thanks to spontaneous generation; and the question of who, in nature, engendered the first man is a pure product of abstract speculation. Neither nature nor man was created or produced by a creator or a grand worker. They exist on their own account,[4] according to Marx, who holds that "*generatio aequivoca* is the only practical refutation of the theory of creation."[5]

There is, then, no *logos* or no dialectic that governs and penetrates the totality of that which is, if it is not the labor of man, his sensuous, meaningful activity; for by this the nature of man is manifested and "produces" itself. The World in this way becomes human history, the humanly natural history of all that is in its becoming. The earth itself, for example, that on which man moves and into which he returns, exists for man only by his naturally antinatural labor: "and land only exists for *man* through labor, through agriculture."⁶ The productive activity of man constitutes the act that originates all that, in *historical* becoming, is. Man begins to become visible in a "socialized" nature, and this nature begins to exist for man as soon as he appears as the animal that makes tools. There is just no kind of alien and superior being located above nature that then manifests itself in human history. There is no "being" that transcends the essence of man as that essence resides in its natural history, the motor agency of which is the social activity of men claiming the material world and seeking to appropriate it in order to satisfy their life needs. Human being has roots that are natural, and, exerting its species forces under the impulse of its natural drives, it builds social becoming. The human nature of man expresses the action by which nature becomes man (*le devenir-homme de la nature*), and this nature is the source of those essential, objective forces that push him toward his own externalization; man is from the beginning a natural and socially active being that seeks to satisfy the totality of its needs through labor. The origin of man is nature, his nature is human; the Nature with which he is involved is always social, and its becoming is historical. (Cosmic) Nature and (human) nature, (social) technique and (historical) becoming are therefore inseparably bound and manifest themselves together from the very beginning. The visible beginning of all that is, the originating act of the World, is human history, for "history is the true natural history of man,"⁷ and "only naturalism is capable of comprehending the act of world history."⁸ The originating act of all that is, the place from which it all can be grasped, is this point of intersection of the "humanism of nature" and the naturalism of man.

The social relationship that unites man with nature and men with their fellows is the real, fundamental relationship, and it is this relationship that is alienated from the very origin of historical development. The being of man and the nature of things are alienated from the beginning. For man, in the course of his natural history, performs actions only as self-externalization in self-alienation. By his social labor he creates a whole world of objects which is nevertheless foreign to him, having no part in his being. Natural drives and essential, objective forces urge human beings toward the objects of their needs; yet this reign of objects implies the reification of everything there is. The activity of man, which by its essence is to be natural and human, stands as neither *natural* nor *human* in that it continues to be reifying and alienated. Natural objects and produced, humanly made objects do not become objects suitable for the true satisfaction of human subjects. The essential forces of man indeed objectify themselves in this process, manifesting their creative power and making nature be for man. Yet this whole objective world full of "useful" realities remains foreign and stifling, alien and alienating. Man manifests and realizes himself in labor; yet he alienates himself in realizing himself and in accomplishing his works. Private property and the division of labor render man alien to himself and to the nature of things, to the world and to other men. All that man creates in externalizing himself remains external to him. From the beginning the history of social activity is a history of alienation; from primitive subproduction to capitalist superproduction, man alienates himself in realizing himself. Man has alienated himself from his true nature, from the true reality of his *essence*, and continues to do so, for nonalienation has never yet been a historical reality. What makes man a human being and nature historical—Technique—has from all time been alienated and alienating. Moreover, men have never gained reflective awareness of this state of things, and their self-consciousness, as well as their theoretical thought and science, has been illusory and ideological. Reflective grasp of alienation is one of the preconditions for transcending alienation, but it alone is almost nothing. The oppositions and antagonisms, the contradictions and conflicts that pit men against men and human beings against the world develop on a terrain that is alienated in a real way, and it is on this real terrain that the battle must be taken up.

Everything that appears to be cut in two actually is so, because of alienation. "The ⟨alienation⟩, which therefore forms the real interest of this ⟨externalization⟩ and of the transcendence of this ⟨externalization⟩, is the opposition of *in itself* and *for itself*, of *object* and *subject*—that is to say, it is the opposition, within thought itself, between abstract thinking and sensuous reality or real sensuousness."[9] Universal reconciliation means the abolition of these contradictions, the unification of thought and sensuous reality; it means the conquest that founds the unity of the totality, and not a reconquest of a lost state.

The premise for universal reconciliation is given by the true nature of man, his essence, by something that has acted up until now only as a self-alienation. In the course of history man has reached only an imperfect self-realization, since his realizations have been his reification. Nevertheless, man at the same time created the conditions for transcending alienation. Man can, then, (re)gain his essence, "regain" meaning: gain by discovering what constituted the

221 Premises for Transcending Alienation

hidden sense of his being and becoming, of his human nature and his natural social essence. It is not a matter of regaining a lost state of paradise, because that state has never yet existed. It is not a matter either of returning to a primitive, undeveloped simplicity, to a pretend original unity, or of regaining a nonnatural simplicity for man. What is involved is that men are to act in such a way that their being and their making dealienate themselves—for the first time in the history of mankind. It is a matter of "the *positive* transcendence of *private property*, as *human ⟨self-alienation⟩*, and therefore . . . the real *appropriation of the human* essence by and for man . . . therefore . . . the complete return of man to himself as a *social* (i.e., human) being—a return become conscious, and accomplished within the entire wealth of previous development."[10] In this way will be realized the "reintegration or return of man to himself, the transcendence of human ⟨self-alienation⟩ . . ."[11] The transcending of basic alienation, the positive abolition of private property, will lead to the actual appropriation of human being and of the world, by and for man. Abolishing the alienation of man himself will make it possible for the reintegration and *return* of man to himself and universal *reconciliation* to be realized. In this reintegration, return, and reconciliation, the antagonism between man and nature and between man and man will dissolve. Within reconciliation the secret origins, the nature of man and things, and the veiled meaning of all historical activity will be able to show forth. Man will thus reintegrate his being, his nature, and his essence; he will return to himself, he will *regain* a place that he has *never yet occupied*. The natural, human, and social essence of man, his species being, constitutes *the* premise for universal reconciliation. It is this essence that man, and mankind, must reintegrate, without implying a return to some position or situation that already really existed.

The history of mankind contains a riddle: What is the sense of its becoming? Because all human activity has so far been, indeed, sensuous and industrious, yet alienated, the riddle remains unsolved; and all history is only the history of the development of productive forces and a process of preparation for the solving of the riddle. The natural, historical becoming of mankind has created the real, material conditions for the reconciliation of man within himself, his labor, the products of his labor, and the world. The progressive, progressing evolution of technique makes both possible and necessary the revolution that will dealienate workers and labor. Private property, the division of labor, capital, and mechanization have allowed man to externalize and realize himself while reifying, unnaturing, and dehumanizing himself. What remains to be done, on the basis of all that has been gained, is to abolish that which alienates men. "Precisely in the fact that *division of labor* and *exchange* are embodiments of private property lies the twofold proof, on the one hand that *human* life required *private property* for its realization, and on the other hand that it now requires the suppression of private property."[12]

Technique served up till now to provide a partial, select, and fragmented satisfaction of human needs, but that all took place within the world of private property wherein subjects were separated from objects. The abolition of private property will permit man to regain his human, that is, social, existence in the satisfaction of the totality of his needs in a human way.

> The transcendence of private property is therefore the complete *emancipation* of all human senses and qualities, but it is this emancipation precisely because these senses and attributes have become, subjectively and objectively, *human*. The eye has become a *human* eye, just as its *object* has become a social, *human* object—an object made by man for man. The *senses* have therefore become directly in their practice *theoreticians*. They relate themselves to the *thing* for the sake of the thing, but the thing itself is an *objective human* relation to itself and to man, and vice versa. Need or enjoyment have consequently lost their *egotistical* nature, and nature has lost its mere *utility* by use becoming *human* use.[13]

The transcendence of alienation, the return of man to his human nature, to his social essence, universal reconciliation, this whole reintegration of man for the first time able to become reality—all this means transcending simple *egocentricity* (*egoité*) and *subjectivity*, transcending the reign of *utilitarian* and *egoistic* need and enjoyment. But, correlatively, it also means transcending simple *objectivity* and *otherness*, transcending *reification*. Men and things will behave in a human way. All this can actually take place because human nature, by its essence, allows and even requires it, although it has never yet been a reality. Man will therefore recover his total humanity, which is something that has never yet been fully manifested. But this humanity of man is the premise for transcending alienation, that is, for universal reconciliation. While wishing to go way beyond individualism, subjectivism, and egoism (and, correlatively, objectivism and utilitarianism), Marx's thought remains anchored in *man*, who is something more than those things; its intent is to humanize and socialize all that is while refusing to give recognition to any being whatsoever that would stand beyond or beneath man's being.

The solution to the riddle of the world—that is, to the riddle of the history of men, the dialectic of becoming in this history, the realization of its meaning, and the reconciliation of both the sensuous activity and the cognizable meaning found in labor—consists in the conquest-appropriation of the world and of man by and for man, an appropriation that will have left *private* property far behind. The truth of mankind's historical becoming has been manifested only negatively up till now, by alienation. The abolition of alienation will, consequently, allow the appropriation by man both of human being itself and of all that is. The abolition of both subjective and objective alienation will make possible the appropriation of the (at once subjective and objective) being of both man and things. Man will only thus appropriate what by essence belongs to him, though it has never yet actually belonged to him. So it is a "re"-appropriation, a taking possession by man of the totality of properties that, though naturally belonging to him, have never actually been his right up to now. It is the *claim* to real, material human life and the conjoined *claim* to the multiple wealth of the natural, social world that require the abolition of private property, which is the source of all alienation. The society of men can and must appropriate the totality of the world in a natural, social, and human fashion, since the essence of man is indissolubly natural, social, and human. The essential, objective powers of the human subject can and must be realized subjectively and objectively within the kingdom of reconciliation, in which neither egoistic needs nor reifying realities will any longer be found. This can all actually take place in historical time, because the subjective and objective nature of man, the human essence of all that is, both allows it and requires it.

According to Marx, the World is only the totality of being as that totality is manifested and produced through human activity; and so the human essence of all that is and the human essence of man are, at one and the same time, both what is alienated through history and what makes possible the transcending of alienation. Nevertheless, this human essence has never been fully manifested; it has not even empirically existed in its fullness and totality. It is the essence of the being of man and the being-becoming of totality, and it awaits its accomplishment. It is a potentiality awaiting actual being, for the practical energy of men has hitherto not realized it. All history is the history of the development of technique and alienation, and at the same time it is the history of the preparation of universal reconciliation. The whole needful, industrious life of men entailed alienation, produced it, and produced as well the conditions for its transcending. "We have before us the *objectified essential powers* of man in the form of *sensuous, alien, useful objects*, in the form of ⟨*alienation*⟩, displayed in *ordinary material industry* (which can be conceived as well as a part of that general movement, just as that movement can be conceived as a *particular* part of industry, since all human activity hitherto has been labor—that is, industry—activity ⟨alienated⟩ from itself)."[14] Once these essential forces are reified, an era of nonalienated labor must be inaugurated to rehumanize what has never yet been truly *human*, although it derives by its very nature from the being of man. "All history is the preparation for '*man*' to become the object of *sensuous* consciousness, and for the needs of 'man as man' to become [natural, sensuous] needs."[15]

Marx moves neither in the direction of providing an explanation of the ontological premises for human nature nor toward giving a metaphysical basis for the splitting of totality into a world of subjects and a world of objects; his interest is solely in the historical becoming that makes nature come to be for man thanks to the sensuous activity of technique. He does not put any effort into trying to grasp as truth some total history of the world: "*the entire so-called history of the world*,"[16] as he puts it and himself italicizes it. His passionate interest, the whole impetus of his theoretical and practical interest, lies in the movement of production (the production of man by human labor and the production of the world of wealth) and the possibilities of a full satisfaction of human needs. He wants to trace the true origin of man and to master the essential truth of human being in order that man may be able to accomplish the self-reintegrating movement of return to that place that constitutes his proper abode, even though this abode has never yet been a place he has lived in. The essence of man, his true natural, social, and human nature, constitutes that never-yet-lived-in abode; human beings, *without a country and without roots*, are now to put their being and their doing into action in order to achieve this (re)turn, this (re)integration, this (re)conquest of that which is the (metaphysical) precondition for transcending alienation. The movement of a descent to origins is achieved by going forward into the future: it is an ascent and a conquest, an integration and reconciliation of man with all that he is and all that is, a resolution without precedent and radically revolutionary, an entirely new answer to a challenging problem, an undertaking that may even itself provoke violent challenge.

The essential forces of man's species being have up to now externalized and alienated themselves in creating the works of human history. Yet these forces constitute the precondition for the possibility of reconciliation, and what is at issue is that everything be put to work to blaze a trail of approach to true human reality. The essential, objective forces must *become* what they are, namely, the motor agency for the development of social humanity, the link between the members of human society. Atomistic individualism can be transcended, since man's nature is in essence social and panhuman. "The highest point reached by contemplative materialism, that is, materialism which does not comprehend sensuousness as practical activity, is the contemplation of single individuals and of civil society.—The standpoint of the old materialism is civil society; the standpoint of the new is human society, or social humanity."[17] So run Marx's ninth and tenth theses on Feuerbach.

The fundamental premise that allows the transcendence of alienation, namely, the essence of man (something that has never yet been empirically found), is *metaphysical* in nature. And it is metaphysical in the traditional sense of that term, since it goes beyond the data of experience. Marx has never been able to establish the empirical existence of this natural, social, human, species essence of man—an essence whose whole history is but the history of alienation and which will show itself for the first time in the kingdom of universal reconciliation. Marx himself does not notice the metaphysical dimension of his thought. He writes: "The premises from which we begin are not arbitrary ones, not dogmas, but real premises from which abstraction can only be made in the imagination. They are the real individuals, their activity and the material conditions under which they live, both those which they find already existing and those produced by their activity. These premises can thus be verified in a purely empirical way."[18] But can the species essence of man, natural, social, human being, which is its own foundation, an essence never yet realized but nevertheless the thing that makes possible and necessary the transcending of alienation— can this basic metaphysical presupposition be verified in a purely empirical way? Can the prospect of radical dealienation, faith in the possibility of the total transcending of all alienation, and hope in future universal reconciliation be sustained on the basis of the data of experience? Are all those things implied in the wondrous development of technique as liberated from all impediments? Voicing his thoughts here, Marx writes that "every profound philosophical problem is resolved . . . quite simply into an empirical fact."[19] However, neither the ontico-anthropological premise for the transcending of alienation (that is, the glorious, richly endowed nature of man) nor the culmination of historical becoming, the kingdom of reconciliation, is an empirical fact. This premise in Marx's thought and the limited greatness of his vision are metaphysical, and they even constitute the culminating point of Western metaphysics; for this metaphysics is what gives birth to science and technique with their readiness to conquer the entire planet in the name of man laboring to satisfy his needs.

The metaphysical dimension of Marx's philosophical thinking, while wishing to get beyond subjectivism and objectivism, idealism and spiritualism, realism and materialism, retains as a central axis both human subjectivity, though a "subjectivity" that is socialized, and a certain "materialist" conception of reality. His thinking takes its point of departure from the natural, sensuous, real, material drives and needs of man, man considered as sensuous, real, material activity and the kingdom of reconciliation conceived as a state of harmony between the manifestation of human *senses* and the uncovering of the *sense* of the being-in-becoming of totality. In accord with this view of things, then, anything that would underlie as foundation for, lie beyond, or transcend the transformation-working activity of man would be *nonsense*. "Spirituality" has no true part in man's essence, and so it will not find a place in reintegration after alienation is transcended. In the metaphysical vision of *human nature*, the history of which is the history of alienation and to which man will return after the radical abolition of all that alienates him, there will be no place for any kind of "metaphysical" power. What Marx finds striking is the "materialistic connection of men with one another, which is determined by their needs and their mode of production, and which is taking on new forms, and thus presents a 'history' independently of any political or religious nonsense which would especially hold men together."[20]

The *negative* metaphysics of Marx wants the human essence to be positively achieved in *social physics*, with "history" itself being in some way transcended. The beginning, the act that originates human being, is certainly history; "man too has his act of origin—*history*—which, however, is for him a known history, and hence as an act of origin it is a conscious self-transcending act of origin."[21] Since man has been alienated from the moment of his historical origin, even though he possesses a complete essence, since the whole of history is the history of alienation, and therefore since "history," "so-called universal history" is only the history of the preparation for the total transcending of alienation, the point now is for man to realize his being, something that never yet happened in the course of the process by which Nature becomes man (*dans le devenir-homme de la Nature*). The history of the world has been a history of the radical alienation of man and of world, and "universal" history has never been universal. Marx turns his back decisively on any kind of anecdotal historiography and even on historiography that recounts historical events. For him, historical becoming is in no way an abstraction. His aim is to take hold of the World as the totality of what is shown and done through human labor and so have man achieve himself integrally in his transformative labor. In this, nature, history, mankind, and technique obey a common rhythm, since they are all united in their common essence.

The issue, then, is that men (re)gain the original sense of unity for all that is, a sense never yet positively manifested. Negating negation, negating whatever denies his being, man can achieve himself using all his senses by setting himself up, through that negation of negation, in the position that will create and found the era of total reconciliation. In discovering the (hidden) sense of what he is and of all that is, and disencumbered of all metaphysical nonsense, man can bring to reality that which constitutes his essence, which was once covered up by alienation. And so, after the abolition of labor as alienation, on the one hand, and of private property, on the other, he can recover that which constitutes his inalienable property, namely, his panhuman humanity.

Marx, forging this prospect of total dealienation out of a metaphysical grasp of the true nature of man, Marx, foolishly optimistic about the possibility of the reconciliation of man with himself and with the world, Marx, the thinker sometimes transcended by his own thought, this same man also writes these lines: "The transcendence of ⟨self-alienation⟩ follows the same course as ⟨self-alienation⟩."[22] Perhaps we are not yet in a position to understand this thought from the founder of Marxism, this acknowledgment that *reintegration*, the *return* of man to himself, the *recovery* by man of his essence, constitutes in some way a *"resumption,"* a *"repetition,"* a *"return"* of the same alienation. The sentence quoted means, of course, that the movement of dealienation follows the same path as that taken by alienation, except in the reverse direction. But it also says that *the abolition of (self-)alienation follows the same path as (self-)alienation.*[23] In order for the hidden sense of a movement to reveal itself, in order for its meaning and direction to be manifest, that movement, surely, has to be completed. Before a trail is followed, it is not easy to tell where it is going.

13.

COMMUNISM: NATURALISM, HUMANISM,

AND SOCIALISM

Marx does not want to stop at a theoretical knowledge of historical movement. Nor is he any more interested in criticism for the sake of criticism. For criticism, far from being a "passion of the head," is on the contrary "the head of passion." "Criticism is no passion of the brain, but is rather the brain of passion. It is not a scalpel but a weapon. Its object is its enemy, which it wishes not to refute but to destroy."[1] The all-inclusive, social condition of alienation, which has long been partially understood, criticized, and refuted, draws the lightning of Marx's negation critique only because he wishes to prepare its radical and total abolition, for this alone is what can lead humanity to an open future. In this sense, "the weapon of criticism certainly cannot replace the criticism of weapons; material force must be overthrown by material force...."[2] Every theoretical theory was alienated because it did not grasp the truth of reality, the alienation of human activity. Although Marx's theory begins in the head of a thinker, it aims to be a practical and revolutionary instrument, the lever of the material emancipation of the proletariat and of mankind. The passage just quoted continues: "But theory, too, becomes a material force once it seizes the masses. Theory is capable of seizing the masses once it demonstrates *ad hominem*, and it demonstrates *ad hominem* once it becomes radical."[3] "To be radical is to grasp matters at the root. But for man the root is man himself."[4] A few lines farther on, Marx makes clear that, since "man is the supreme being for man," the *categorical imperative* to emancipate completely man's natural, social being can be stated only thus: "to overthrow all conditions in which man is a debased, enslaved, neglected, contemptible being."[5]

Man, the sad hero of alienation, must become the joyful hero of reconciliation. After the overthrow by material violence of all the relationships that alienate man, the reign of real *humanism*, of fully achieved humanism, can and must be instituted. This humanism will mean the accomplishment of the true *nature* of man, of his *social* essence, for man is in essence a *community* being. When philosophic thought recognizes the true social nature of man and exposes the alienation in human history, when it becomes a thinking that denounces the ideological character of all past philosophy, then it takes an active part in the coming of human emancipation; but it does so not as (pure) philosophic thinking. It is by transforming itself into an instrument of the revolution, by changing into material violence, that this thinking can actively contribute to the establishment of humanistic naturalism, which in turn will abolish it since it will have become reality. "Just as philosophy finds its material weapons in the proletariat, so the proletariat finds its spiritual weapons in philosophy."[6] For the proletarians, the most dehumanized of human beings, are the heroes of the emancipation of man and not simply laborers in its service. "The head of this emancipation is philosophy, its heart is the proletariat. Philosophy cannot be actualized without the abolition (*Aufhebung*) of the proletariat; the proletariat cannot be abolished without the actualization of philosophy."[7]

The abolition[8] of philosophy by its realization, the abolition of the proletariat by the emancipation of man, the abolition of private property by the communist movement, in short, the abolition of all forms of alienation, comprises the solution to the riddle of history and the end of tragedy. This radical and total solution can dissolve all ties, *economic*, *political*, *religious*, and *ideological*, only if it is truly revolutionary; for all these ties continue the alienation of man. It is not enough to eliminate some one of them: they must all be abolished together. "Naturally, the transcendence of ⟨alienation⟩ always proceeds from that form of ⟨alienation⟩ which is the dominant power."[9] Alienation, of course, takes different forms. Nevertheless, economic alienation is the root of all alienation and of all the forms that it may take. So it is against private property that the first engagement of the battle is fought.

The solution to the riddle of history does not, however, lie in a *crude, mechanical communism*, a communism that "abolishes" private property *by generalizing it*. Dealienation cannot be only in the sphere of the economic. Crude communism hardly abolishes the condition of the worker itself but, on the contrary, extends it to all men. "For it the sole purpose of life and existence is direct, physical *possession*. The task of the *laborer* is not done away with, but extended to all men. It wants to do away *by force* with talent, etc."[10] *Crude*, thoughtless communism[11] does not reject private property, but rather it is the positive, generalized expression of it; for as the general, full-blown form of private property it gives positive expression precisely to (private) *property*, which has only its *private* character eliminated. Private property thus becomes only community property, and we remain in the same world of alienation, except that the community continues to maintain a relationship of ownership with the world of things. "In negating the *personality* of man in every sphere, this type of communism is really nothing but the logical expression of property, which is its negation."[12] Envy, covetousness, and the need to possess foster a crude communism that is incapable of recognizing the true nature of man. Neither natural nor human, this communism does nothing but make a collective phenomenon of egoistic individualism and reified realities.[13] Perceiving the extent to which mechanical communism might be taken for the form and content of the movement of de-alienation, Marx lays particular stress on the crude and limited character of this communism that wishes to annihilate what cannot be possessed by all as private property. He discerns too, perhaps, the extent to which the risk of this kind of communism weighs upon the communist revolution whose coming he is preparing; and he vigorously fights it as a counterfeit of radical communism, as in truth only a generalized capitalism. Marx's wish is that man satisfy the totality of his natural, human needs, that he manifest all the wealth of his essential properties by appropriating the world in a human and nonpossessive manner. Marx, the visionary of a naturalistic humanism, denounces the perspective that confuses being with having: "The crude communism is only the culmination of this envy and of this leveling-down proceeding from the *preconceived* minimum. It has a *definite*, *limited* standard. How little this annulment of private property is really an appropriation is in fact proved by the abstract negation of the entire world of culture and civilization, the regression to the *unnatural* simplicity of the *poor* and *undemanding* man who has not only failed to go beyond private property, but has not yet even reached it."[14]

Appearing against the background of the wretched world of private property as a tendency to establish private ownership in the form of a positive communal having, this thoughtless communism is the universalization of capitalism; the human community only becomes the communal capitalist. "The community is only a community of *labor*, and of equality of *wages* paid out by communal capital—the community as the universal capitalist."[15] The shadow of the old world weighs upon the world claimed to be new, and Marx for a moment was able to take hold of this shadow covering the Promethean enterprise of man's emancipation. But his scientific faith in the possibility, indeed the necessity, of the complete and total abolition of all alienations prevented him from giving attentive thought to this particular opening.

The simple suppression of political alienation is again not a satisfactory solution to the riddle of history. Founding true communism cannot be and must not be a matter of either economics or politics, exclusively or even principally. Political communism remains extremely limited. In only suppressing the *State* and demanding civil equality for all members of a community, political communism, whether democratic or despotic in nature, does not succeed in abolishing private property and the alienation of man (notice we do not say "the alienation of the citizen"). Just as crude economic communism only generalized private property and the slavery of women, political communism only generalized civic equality. Marx repeatedly presses the point that the simple suppression of private property and true communism are not the same thing; neither are political and human emancipation. In "On the Jewish Question," he dedicates several perceptive pages to the insufficiency of political emancipation. "*Political* emancipation certainly represents a great progress. It is not, indeed, the final form of human emancipation, but it is the final form of human emancipation *within* the framework of the prevailing social order. It goes without saying that we are speaking here of real, practical emancipation."[16] Political emancipation is not the absolute and total mode of human emancipation, because in being politically enfranchised man is liberated through the intermediary of a particular reality. The State is not a universal reality. As an organ of alienation, it is only the intermediary between man and the freedom of man, and alienated man relies upon the State to realize his humanity. "It follows that man frees himself from a constraint in a *political* way, through the state, when he transcends his limitations, in contradiction with himself, and in an *abstract, narrow* and *partial* way. Furthermore, by emancipating himself *politically*, man emancipates himself in a *devious way*, through an intermediary, however *necessary* this intermediary may be."[17]

No more than private property can be actually abolished by political measures aiming for its annulment can the distinction of human being into *man* and *citizen* be transcended by political means. A communism that would be exclusively, or principally, political does not abolish man's *double* existence: his citizen existence as the public man within the political community and his existence as the private man living his particular life—or, in other terms, his general existence, in alienation, and his particular existence, in contradiction to the first, as an individual living his life. Political life may try as it might to set itself up as true, noncontradictory species life, but it will not do it, because it implies alienation. Political life cannot become man's real life whatever it does: "It can only achieve this end by setting itself in *violent* contradiction with its own conditions of existence, by declaring a *permanent* revolution. Thus the political drama ends necessarily with the restoration of religion, of private property, of all elements of civil society, just as war ends with the conclusion of peace."[18] So Marx was able to see equally well all the dangers involved in a *political* rather than a total communism, a communism that only restored all the powers of alienation under another political form.

Indeed, the generalization of political equality means the generalization of the rights of the citizen and not the full recognition of the rights of man. Economic communism makes all men *workers*, the communism that advocates the community of women makes all women *prostitutes*, political communism (democratic or despotic) makes all men *citizens*; and none of these "communisms" takes note of the fact that the worker, women as enslaved to man, and the citizen are but figures of human alienation. Marx urges us to notice that "the so-called *rights of man*, as distinct from the *rights of the citizen*, are simply the rights of a *member of civil society*, that is, of egoistic man, of man separated from other men and from the community."[19] The true communist revolution could not be political, as was the bourgeois revolution, but is human and social: "Human emancipation will only be complete . . . when [*man*] has recognized and organized his own powers (*forces propres*)[20] as *social* powers so that he no longer separates this social power from himself as *political* power."[21] It is in ceasing to be the alienated worker and the abstract citizen that man can become what he is: species man. It is in ceasing to be the egoistic individual that he can regain his community essence. The recovery by man of all his properties, the abolition of the worker and the citizen in favor of the real man, the regaining by species being of all activities: this is the meaning of human emancipation. Such was Marx's thinking when he tried to exorcise the ghost of political communism.

The solution to the riddle of history, which is something that is neither principally economic nor principally political, cannot be a matter of the simple elimination of religious alienation. Communism and the elimination of religious alienation are not the same thing, yet communism is constitutionally tied to atheism. "Communism begins from the outset ... with atheism,"[22] Marx says, and thus humanism is rooted in atheism. Nevertheless, he continues: "But atheism is at first far from being *communism*; indeed, it is still mostly an abstraction. The philanthropy of atheism is therefore at first only *philosophical*, abstract, philanthropy, and that of communism is at once *real* and directly bent on *action*."[23] Since atheism in its rejection of religious alienation does not go beyond the level of merely negating a being that would transcend nature and man, it is not sufficiently radical. In eliminating God and religion, in passing from the theological sphere to the anthropological problematic, theoretical atheism does indeed establish humanism, but it remains abstract; for the simple denial of God or the simple denial of religion as the negation of real man does not yet mean that the reign of man's true positiveness has begun. The "death of God," the murder of God by men, is not enough to found communism as the "absolute" humanistic position, in possession of its own foundations. The position—that is, the negation of negation from which communism is built up—is the grasp of natural man as being. "*Atheism*, as the denial of this unreality [the unreality of the nature of man, from the religious point of view], has no longer any meaning, for atheism is a *negation of God*, and postulates the *existence of man* (*das Dasein des Menschen*) through this negation; but socialism as socialism no longer stands in any need of such mediation. It proceeds from the *practically and theoretically sensuous consciousness* of man and of nature as the *essence* (*des Wesens*)."[24]

The communism that is essentially economic aims to abolish not property but private property and to emancipate man by the *mediation* of (generalized) property. The communism that crudely opposes marriage and monogamous love and advocates the rule of the community of women can realize its dream only by way of the *mediation* of universal prostitution. The communism that is essentially political tries to suppress a certain kind of State and wishes to free man by the *mediation* of the State, transforming all men into equal citizens. The communism that suppresses God and moves by way of the mediation of this denial to posit man fails to grasp the basic essence of man's nature. All these particular suppressions, all these negations of negations that deny the species being of man, still need mediation and still want the abolition of alienation, having failed initially to *posit* man's natural being as the basis of all that, by way of man, is. Yet, the radical elimination of private property remains *the* premise for all total emancipation and constitutes the precondition for the realization of practical and positive humanism, of that humanism that proceeds positively from itself to transcend what stood as preparation for it.

The communism that wants before all else to eliminate religious alienation seems to overlook the fact that the positive abolition of private property means the positive abolition of the root of all alienation and, therefore, of religious alienation too. The positive abolition of private property itself, however, does not yet constitute the full being of achieved communism. "Atheism, being the supersession of God, is the advent of theoretic humanism, and communism, as the supersession of private property, is the vindication of real human life as man's possession and thus the advent of practical humanism (or just as atheism is humanism mediated with itself through the supersession of religion, whilst communism is humanism mediated with itself through the supersession of private property). Only through the supersession of this mediation—which is itself, however, a necessary premise—does positively self-deriving humanism, *positive humanism*, come into being."[25] The abolition[26] of private property is the most necessary negation and the most necessary without being absolutely sufficient to found communistic humanism positively. The abolition of the State, the abolition of God, and the abolition of religious alienation flow from this first negation, and, while having to be themselves accomplished in their own proper way, these abolitions cannot alone constitute the meaning of humanistic communism. It goes without saying that, just as the point is not simply to generalize economic life (and private property) or the political life of citizens (and the power of the State), so it cannot be a matter either of simply generalizing religious life (and the divinity). *Man* does not have to become *god* after the abolition of God. Free of all that betrays his true nature, he has only to become what he is, real man relying upon himself, being his own basis for himself. Man does not have to become a *superman* either, once he will no longer pursue his development under the heavy sky of the gods; it was the gods who were conceived as supermen. Man, who is neither god, nor superman, nor subman, is to achieve himself through his humanity. "Man, who has found only his own reflection in the fantastic reality of heaven, where he sought a supernatural being (*Übermenschen*), will no

longer be disposed to find only the semblance of himself, only a
non-human being (*Unmenschen*), here where he seeks and must
seek his true reality (*wahre Wirklichkeit*)."[27]

Neither essentially economic, nor essentially political, nor essentially antireligious, the solution to the riddle of history is moreover not essentially ideological. The positive and radical abolition of alienation cannot be the result of a modification in consciousness or thought. To interpret what exists in a different way without actually changing it means to recognize and maintain it by means of one more theoretical interpretation; to forge new theories in view of a new and better future remains a labor that is ideological, alienated, philosophical, and abstract. The theoretical critique of the existing world or the philosophical construction of an idea that would prepare a new world can never replace real, material practice. The motor force of history is not theory, be it corroborative, critical, or abstractly revolutionary: it is actual revolution. Yet the practice of fully achieved humanism presupposes also a theory that offers a pitiless denouncement of the double aspect of alienation, namely, real alienation and alienation in thought and consciousness. The circle that binds thought (and consciousness and theory) to action (and practice and reality) is perhaps not only dialectical but vicious as well. Though Marx takes theory to be derivative, he continues to recognize that it also has creative power. His classification puts thought from the very beginning on the side of theory, so that, given a massive separation between theory and practice, he is uncertain what to do with it. Thought remains always stained with ideology, and this is true for revolutionary thought too. Nevertheless, reflective grasp, revolutionary theory, and thought guiding action are necessary preconditions for the communist movement, even though they be in no way sufficient. Indeed they found this movement, as a movement that gradually, as it reaches consummation, divests itself of those very conditions.

Perhaps a passage from the one book Marx wrote directly in French can help us to understand what revolutionary theory is and what becomes of communistic thought in the course of its realization. In *The Poverty of Philosophy*, we find these words:

> Just as the *economists* are the scientific representatives of the bourgeois class, so the *Socialists* and the *Communists* are the theoreticians of the proletarian class. So long as the proletariat is not yet sufficiently developed to constitute itself as a class, and consequently so long as the struggle it-

> self of the proletariat with the bourgeoisie has not yet assumed a political character, and the productive forces are not yet sufficiently developed in the bosom of the bourgeoisie itself to enable us to catch a glimpse of the material conditions necessary for the emancipation of the proletariat and for the formation of a new society, these theoreticians are merely utopians who, to meet the wants of the oppressed classes, improvise systems and go in search of a regenerating science. But in the measure that history moves forward, and with it the struggle of the proletariat assumes clearer outlines, they no longer need to seek science in their minds; they have only to take note of what is happening before their eyes and to become its mouthpiece. [But does not what happens before their eyes also happen under the prodding of their thought?] So long as they look for science and merely make systems, so long as they are at the beginning of the struggle, they see in poverty nothing but poverty, without seeing in it the revolutionary, subversive side, which will overthrow the old society. From this moment, science, which is a product of the historical movement [which itself is nevertheless also a product of science], has associated itself consciously with it, has ceased to be doctrinaire and has become revolutionary.[28]

So a moment will come when theory will be so completely associated with action that it will cease to be doctrinaire theory and will be fused with the revolutionary movement of social practice. The negation of ideologies from the past, the abolition of the alienation of thought and consciousness, and above all the construction of revolutionary thought on the base of revolutionary reality are necessary, while all the time they are but *premises* to practical, consummated humanism. For they are still *mediations* (still negations and abolitions) that will themselves be transcended by the establishing of communism as the fundamentally humanistic, practical, positive, and real *position*.

True communism will not generalize theory and philosophy or make all men *"thinkers."* Just as the issue here is not a matter of generalizing the economic life, the political life, or the religious life, so it is not a matter either of generalizing the contemplative or theoretical life. Communism will negate all the negations that negate man's species being, will transcend the negation of negation, and will be that new position that will have no more need of private property and the State, of the divinity, or of ideas. Far from wanting to generalize philosophy, even revolutionary philosophy, communism wants to abolish it in realizing it. We shall see a little later the serious problems that this abolition of philosophic thought presents.

The solution to the riddle of history, the transcendence of all alienations, the reconciliation of man with nature, with himself, with his fellow men, and with the totality of the world, is *neither* essentially *economic nor* essentially *political* or *philosophic*; it is rather a total thing that comes by way of the revolutionary negation of the totality of the existing world. "In reality and for the *practical* materialist [not the materialist philosopher], i.e., the *communist*, it is a question of revolutionising the existing world, of practically attacking and changing existing things."[29]

Revolutionizing the existing world means abolishing what is, since all that is, is alienated and alienating. To begin with, the most fully real basis of alienation must be abolished, namely, *economic life*: that is to say, private property, alienating labor, the very condition of being a worker, the division of labor. For reconciliation to be achieved, economic life *as such* must be abolished. This abolition is *the* premise for dealienation, though it is not yet at all sufficient to found communistic humanism in a positive way. Communism is not a movement that exhausts itself in the economic order. Nevertheless—nevertheless, Marx, whose very thinking addresses the phenomenon of economics in order to free man from its grip, this same Marx rivets his gaze in fascination upon this very object and says: "Its organisation [the organization of communism] is, therefore, essentially economic . . ."[30] And Marx says this not of a crudely economic communism but of true communism, of *his* communism. This man, whose intent is to transcend the primacy of the economic, boldly writes that the *organization of communism is essentially economic*. The revolutionary whose aim is to abolish all that is, here runs up against a reality it is difficult to get around. The ghost of economics is not so easily conjured away, and the economic problem, bound up as it is with that of modern technique, is extremely difficult to *solve*.

Nevertheless, Marx wants man, by abolishing the root of all alienations, by abolishing private property as human alienation, to rediscover, regain, reintegrate, and conquer his true nature, his indissolubly natural, human, and social being. Abandoning all his alienated abodes—the economic world, the political world, the family, religious life, ideological life—man is to accomplish a movement of return (and of reintegration: *Reintegration oder Rückkehr*) to his naturally human and humanly social existence, even though this existence and this essence are an abode he has never yet lived in; for history is altogether the history of alienation, of the loss of man. The abolition of private property will lead man to the non-possessive appropriation of nature as it is manifest through his needs and his activity and to the conquest of his origin, his being, and the totality of the world. He will thus be able to enter into the era of universal reconciliation, and all antagonisms between man and nature, between man and men, between subjectivity and objectivity, between freedom and necessity, and between reality and thought will have been abolished. Within this reconciliation, the relationships that will bind men to nature and to each other will be natural and socially human and not "economic." "This communism, as fully developed naturalism, equals humanism, and as fully developed humanism equals naturalism;[31] it is the genuine resolution of the conflict between man and nature and between man and man—the true resolution of the strife between existence (*Existenz*) and essence (*Wesen*), between objectification and self-confirmation, between freedom and necessity, between the individual and the species. [It is, therefore, *the* total and universal reconciliation.] Communism is the riddle of history solved, and it knows itself to be this solution."[32]

Since alienation is a radical thing, affecting the very roots of man, that is, his natural and social humanity, the resolution of the antagonisms by which Nature, Man, and Society appear as distinct powers in conflict can only be total; it would allow man to regain his essence, to show it forth in his existence, and to stop separating the totality of what is from being, that is, from the being of the world as indissolubly natural, human, and social. Man would thus be reconciled with the World, the world being, according to Marx, the totality of what is revealed (what becomes visible, as he puts it) and made thanks to the total activity of man, who is the essentially practical being, the one who brings about realization.

Communism, (re)discovering the common essence of what is, will be *humanism consummated*. Man will have become that which he is but has never yet been: the being by virtue of which all that is, *is*, all "things" becoming for him objects replete with humanity through his objective, social subjectivity. As consummated humanism, this communism will be *naturalistic*. In recognizing the totality of man's natural needs and aiming for their full satisfaction, this naturalism knows that nature is "nothing" without man. Nature is manifested as nature, and becomes man's nature and nature for man, through the social life and the historical becoming of men. Marx is indifferent to greater Nature, to *Physis*, to, we might say, cosmic nature; what he always has in view is the (social) nature of man.

> The *human* essence of nature first exists only for *social* man; for only here does nature exist for him as a *bond* with *man*—as his existence for the other and the other's existence for him—as the life-element of human reality. Only here does nature exist (*erst hier ist sie da*) as the *foundation* of his own *human* existence. Only here has what is to him his *natural* existence (*natürliches Dasein*) become his *human* existence (*menschliches Dasein*), and nature become man for him. Thus *society* is the unity of being of man with nature—the true resurrection (*Resurrektion*) of nature—the naturalism of man and the humanism of nature both brought to fulfillment.[33]

As a result, communism will realize the common essence of that which, by its essence, is nature-humanity-society, and in this sense it will be at once consummated naturalism, consummated humanism, and consummated socialism. It will be all that precisely as *active*, practical, and realistic humanism.

The dealienation of man is at the same time a dealienation of nature; socialism consummated will be a *resurrection of nature*. Nature begins to die as soon as it begins to exist, because human activity, that by which nature is, has been forever both alienating and alienated; and so Marx calls this coming of nature back to life resurrection. Yet the *resurrection* of nature, like man's *reintegration* or *return* to himself, is only the return to a life that has not yet been lived. In coming back, nature will reintegrate—or return to—its nature, a nature that has not hitherto existed except in alienation. The very terms that Marx uses—*Rückkehr, Reintegration, Resurrektion* (later we shall see him speak of man's *reconquest*: *menschlichen Wiedergewinnung*)—could lead one to suppose that at the end of an "eschatological" process a *lost Garden of Eden* or *Golden Age* would be regained, reintegrated, rewon, and that the movement in question would lead to a terrestrial paradise. And, in fact, Marx's communistic "prophecy," aspiring to the realization on earth of man's happiness, which in socialist society means the reintegration, the return, the resurrection, and the reconquest of a lost, alienated essence, does imply a kind of eschatological Jewish vision. The requirement of justice and universal peace, the idea of suffering resulting from man's uprooting, and the denial of a heavenly paradise all spring from it. Notwithstanding, Marx neither supposes nor believes in a lost paradise or a vanished golden age, any more than in an end to human history. Nor does he speak of a prehistoric and primitive communism that would be the *thesis* against which an *antithesis*, namely, the universal history of alienation, rises and that would be rediscovered and re-created at a higher level within fully developed communism (*negation of negation* and new position). According to Marx, all human history has been the history of alienation, with nothing preceding alienation historically or ontically. What was alienated was the social nature of man. The issue is, therefore, to recover that nature lost from its first appearance, but the movement of reconquest is something absolutely new and without precedent. This premise for reconciliation is "metaphysical," for the nature in question precedes all experience, but its accomplishment is to be physical, human, and historical.

The realization of man's naturalism along the axis of the humanism of nature can take place only by way of society and through the life and social activity of men. Nature must not be given a privileged position at the expense of man and society, nor man at the expense of nature and society, nor society at the expense of nature and man. Properly speaking, one must not even distinguish here these three aspects of the same being, of common being, of the *being*-in-becoming of totality. Indivisibly naturalistic, humanistic, and socialistic, communism will reconcile what was in conflict. Marx warns: "Above all we must avoid postulating 'Society' again as an abstraction *vis-à-vis* the individual. The individual *is the social being*."[34] For even death does not keep the individual from being species being, in the historical becoming of mankind; the determinate individual is mortal as individualized species being. Until after his death, does not the individual leave traces of his passage *on earth*? Is not the historical memory of men the earthly depository for the actions and dealings of transitory individuals?

Marx abolishes the three basic theses of theology and metaphysical anthropology: God, freedom, and the immortality of the soul. God is abolished by the position of human being as foundation. Freedom coincides with necessity in the kingdom of integral communism. The immortality of the soul is rejected because only the (species and individual) existence of the real man is actual, for after death a man's existence in the historical process leaves only traces, which, of course, express the species dimension of the individual rather than his determinate particularity. Marx works these abolishings in a fairly brutal and nonmeditative manner, and he does so in the name of *practical* humanism.

True communism, naturalistic, humanistic, and socialistic, abolishes all alienations and all distorted mediations in order to build the world of human emancipation, in order to allow man to deploy all his power and totally realize his will. It is a communism whose face is turned toward the future. Consequently, one must not impose on communism the needs, the limitations, and the intentions that have been created by the regime of alienation in the superseded past. It will not be envy, greed, the need for immediate, possessive, and utilitarian gratification, the sense of having, ideological morality, and the demands of conscience that will be the motives for human action when man, the living totality, will have found his unity within the unity of the being-becoming of the world. When man will have overcome his fragmentation and the duality that arises between the sensuous and material aspect of beings and things and their meaningful, spirit-order (*spirituel*) aspect, thereby finding the unified expression of the unity proper to his essence, then will he be able to appropriate his being in a universal way by appropriating at the same time the universal essence of all that is. Marx pays scant attention to the difference that separates any *original unity* from an *enterprise of unification*, and he does not let himself be bothered by the extreme difficulty involved in any attempt to *reunite* what was *separated*, when he conceives active humanism as generating total man and total society, and the total activity of men as inserting organically into Totality. "Man appropriates his total essence in a total manner, that is to say, as a whole man. Each of his *human* relations to the world—seeing, hearing, smelling, tasting, feeling, thinking, observing, experiencing, wanting, acting, loving—in short, all the organs of his individual being, like those organs which are directly social in their form, are in their *objective* orientation or in their *orientation to the object*, ⟨the appropriation of that object, the appropriation of *human* reality;⟩ its orientation to the object is the *manifestation of the human reality* . . ."[35] Total man, splendid in his totality-being, no longer lost in the object, no longer separating object from subject, and now actively manifesting his human reality in enterprises of realization, will express the very beginning of the totality of the world now revealed to man, because man is the root of all that is, he who makes exist that which, without his labor, would not be.

The positive and radical abolition of private property, the root of alienation, and not just its simple removal or generalization, will lead to the complete emancipation of all human properties and will even bring things to act in human fashion with regard to man. The new man will in this way show forth all the wealth of his natural, human, and social needs, his passion will become action, and his freedom will coincide with necessity. Human sufferings will have taken on a meaning altogether new in becoming themselves powers for the enriching of human life. Suffering (*Leiden*), passion (*Leidenschaft*), and inevitable negativity will be able to change into a motive for enjoyment and positive action. "It will be seen how in place of the *wealth* and *poverty* of political economy come the *rich human being* and the rich *human* need. The *rich* human being is simultaneously the human being *in need of* a totality of human manifestations of life—the man in whom his own realization exists as an inner necessity, as *need*. Not only *wealth*, but likewise the *poverty* of man—under the assumption of socialism—receives in equal measure a *human* and therefore social significance."[36]

Total man will live, of course, in a total society, man and society being only two different expressions of one being. This life—social, human, and natural (natural because the new society will be the resurrection of nature)—will give a new life and a new meaning to all man's works and all the world's phenomena. For, within universal reconciliation, things themselves will supersede their alienation and reification and will begin to exist fully and positively as man's products. The industrious nature of man, the consubstantiality of man (in the singular) and of society (man in the plural, men), and the wholesale transformation of nature into history will be realities in the new social situation in which sensuousness and meaningfulness, "materiality" and "spirituality," beings and things, reality and thought, and practice and theory will come to *coincide* (but, then, what will become of their difference?). Nonetheless, the *senses*, *material*, *real practice*, and *human being* will prevail over the spiritual (thought, contemplation, theory, and the meaning of the being of totality) even after the transcending of dualistic alienation. "Here we see how consistent naturalism or humanism distinguishes itself both from idealism and materialism, constituting at the same time the unifying truth of both,"[37] says Marx. He continues, however, to define the man of his real humanism as a *practical materialist* and to return all activity to *sensuous activity*. Although total unity ought to be realized within universal reconciliation, one of the powers of totality continually takes precedence over the others.

To the same degree that Marx's pessimism is radical and sombre regarding the past and the present his optimism about the new world is unbridled and rosy. His idea being that whatever is calls forth its opposite—in the world of alienation only—he sees "⟨alienation⟩ . . . hastening to its annulment."[38] His confidence in the *practical energy* of men, a productive and creative, transformative and reality-making energy, is without bounds. The resolution of the tragicomedy of universal history will thus be a complete happy ending. Since the whole "history" of philosophic thought has succeeded in resolving nothing, it will dissolve in practice and technique. "We see how objectivism, spiritualism and materialism,[39] activity and suffering, only lost their antithetical character, and thus their existence as such antitheses ⟨in the social situation⟩ [in fully achieved socialism, one should read]; we see how the resolution of *theoretical* antitheses is *only* possible *in a practical way*,[40] by virtue of the practical energy of man. Their resolution is therefore by no means merely a problem of understanding, but a *real* problem of life, which *philosophy* could not solve precisely because it conceived this problem as *merely* a theoretical one."[41] It is, then, the practical energy of men that constitutes the one power capable of leading them to the conquest of their being and of the world, thereby making the essential and objective forces of human subjectivity and the objective, real products of social activity coincide. But does not this practice too presuppose a true theory? Having set thought over on the side of theory once and for all in a sole concern for practical truth and activity, Marx, try as he might, does not succeed in reconciling *logos* and *praxis* in a single unity.

Although the reign of total reconciliation, according to Marx's assertions, transcends all contrary and contradictory positions and "isms," it still continues to give a privileged position to one set of terms at the expense of another, and it does so even on the level of the total, global, and integral unity said to be attained. Marx takes the hostility between the senses and the spirit to be necessary as long as the human feeling for nature "and" what he calls the "human sense of nature" are not yet produced by man's own labor.[42] The full emancipation of human labor must, accordingly, reveal everything to be the product of the social activity of men. *The totality of what is* is thus reduced to *the total activity of man*. And this ("total") praxis is material, sensuous, real, and objective; it is a production destined to satisfy the "totality" of human needs. Yet, even within humanistic and socialistic naturalism, the appropriation of the world by men is achieved by virtue of a labor and an enjoyment that are sensuous, real, objective, and material. The other side is hardly transcended in favor of a unity; it is simply done away with. But the other side, destined to disappear—though Marx never knows exactly what to *do* with it—concerns more than just what founds and transcends the sensuous, that is, the ideal, the subjective, and the spiritual, all metaphysical powers implying an alienating dualism; it also concerns nature itself, that which in the name of fully achieved naturalism is destined to cease to be nature and become but the product of human technique. The "resurrection of nature" means, then, the total abolition of nature, *the complete technification of what is "natural."*

Marx does not want to admit a possible flaw lying within reconciliation or imagine the possibility, much less the necessity, of a nonidentity between the senses and the meaning of the whole enterprise. He continues to proclaim that, after the transcendence of alienation, nature will rise again through the productivity of human subjects, who are at the same time objective subjects; for man, now total, will no longer be lost in the world of objects but will fully objectify himself in objects as produced things. What will determine the being of this total man, now reconciled with his being and with all that is, will no longer be a strange and alien power but his own essence, propelling him toward realization. For "the appropriation of what is alienated and objective, or the annulling of objectivity in the form of *alienation* . . . has to advance from indifferent foreignness to real, antagonistic alienation."[43]

Socialist man, wholly freed from all that would limit and escape him, will be a being living in a "solely" sensuous world, a being who will be the producer of all that is. Though producer, he is at the same time a product, a product of the process by which nature becomes history (*du devenir-histoire de la nature*), a product of communist society."⟨*Fully achieved*⟩ society (*die gewordene Gesellschaft*) produces man in this entire richness of his being—produces the *rich* man *profoundly endowed with all the senses (all—und tiefsinnigen Menschen)*[44]—as its enduring reality."[45] Marx rejects crude and thoughtless communism, conceiving the man of his naturalistic and humanistic communism as a being totally and deeply anchored at the same time in the sensuous and the meaningful (*dans le sensible et le signifiant*).[46] The meaning of man's becoming, however, remains reduced to sensuous activity, to *practical* energy. The new man and the new society *shall be* in a state of continuous actualization, of perpetual movement, at every moment realizing and actualizing some sensuous, global meaning, in a continual transcendence toward new realizations and actualizations, all in the same direction. Production as an act will have taken precedence over any particular product.

A great epoch of Western metaphysics, that is, of Greek, Judeo-Christian, and Modern metaphysics, reaches a culmination with Marx. The being-becoming of the totality of the world, the sense of all that is, no longer lies for him in the *logos* and ἀλήθεια ("truth") of a *divine, indestructible physis*, in an *idea* that *participates* in the sensuous but is *separated* from it, or in the *entelechy* and ἐνέργεια ("act") of a *divine Nous* and *first mover*, as we find it respectively for the Pre-Socratics, Plato, and Aristotle. Nor does it lie in the *actualitas* of an *absolute God*, who *created all that is from nothing*, who is *transcendent* and *all-powerful*, as for the Jews and Christians. Lastly, it does not lie in the encounter of the *ego cogito* (i.e., the *res cogitans*) with the *res extensa* in the *transcendental subject*, that which founds and grasps *objectivity*, or in the history of the world of absolute *Spirit* becoming *self-conscious* and *absolute knowledge*, as it is respectively in Descartes, Kant, and Hegel. Marx continues the whole of Western metaphysics—principally its third period, European, modern thought, the philosophy of subjectivity as thinker and agent (a philosophy which is as well, by implication, a philosophy of objectivity)—and he does so by radically transcending philosophy precisely as search for the truth, as thought, theory, knowledge, and science. He generalizes, socializes, and universalizes *human subjectivity*, which is found to be objectified, and takes hold of all that is from the angle of sensuousness (*Sinnlichkeit*), objectivity (*Gegenständlichkeit*), and reality (*Wirklichkeit*); and that (sensuous) world, objectivity, and reality are turned into a product actualized by practical realization-activity (*Wirkung, Aktion, Praxis, Verwirklichung*) on man's part. For it is the productivity of men that makes nature and the Totality be and become for man. The being-becoming of the totality lies in the indissoluble original and essential unity that binds nature, man, and society together; and that is what is manifested and made by the power-filled labor of human society. Abolishing all material and ideological alienations, revolutionizing the social structure from bottom to top, and doing away with superstructural powers, man is to come to express his being (once again, natural-human-social) by impressing upon all that is the signs of his technical praxis, by bringing what, without his labor, does *not* exist to be and become, through his power and will. Marx wants to found the epoch of world history, properly speaking, (*Weltgeschichte*) with the whole Universe becoming human history by virtue of the activity, the productivity, the technique, and the practical energy of human society. The human collectivity (i.e., human subjects objectifying themselves) becomes thus the basis—should we still call it metaphysical?—for *planetary technique*, the agency that produces all that is. Communism, naturalistic, humanistic, and socialistic, is the movement whose burden it is to accomplish the task of universal reconciliation and to allow the full satisfaction of the totality of human needs, needs that renew themselves endlessly as they are satisfied.

For this "program" to be actually realized, real premises are necessary, premises that are essentially practical (Marx never knows whether theoretical presuppositions are also essential and necessary). These premises, far from being abstractly deduced, are produced by the very movement of the productive forces of technique, by the global development of society, by the degree of intensity or the scale of the class struggle, by the depth of historical action, and by the dimensions attained by the rise of the masses.[47]

The first real premise, the first practical condition for the radical abolition of private property and every alienation, lies in the great increase in productive forces, the high development of technique. Without this increase, the true socialization and universalization of wealth would be impossible, and all that would be generalized is misery. Without this precondition, "*want* is merely made general, and with *destitution* the struggle for necessities and all the old filthy business would necessarily be reproduced."[48] Communism is, therefore, envisioned by Marx as constituting the future for countries that are technically developed, the countries of Western Europe, for communism consists of the socialization of wealth and not of wretchedness; communism ought to be what follows capitalist industrialization. Marx based his hopes on the communist revolution in Germany. "In Germany *no* form of bondage can be broken unless *every* form of bondage is broken. Germany, enamored of *fundamentals*, can have nothing less than a *fundamental* revolution. *The emancipation of Germany* is *the emancipation of man*."[49]

The high degree of development in productive forces, however, must not be only local and national but universal, or almost so. Only men who belong to universal history, who are "empirically universal" and not merely local individuals, only proletarians who link directly to universal history can give life to the communist enterprise. A local, national communism could not be communism and could not be sustained. "Empirically, communism is only possible as the act of the dominant peoples 'all at once' and simultaneously, which presupposes the universal development of productive forces and the world intercourse bound up with communism."[50] Consequently, communism must be a universal movement born of Western European society, bourgeois capitalist society, which is destined to embrace history everywhere by transforming history into universal history.[51]

The second practical condition for bringing about the revolution that would lead to communism flows from the first, and it is the existence of an enormous mass of mankind (namely, the immense majority) that is deprived of all property, that is radically alienated and in irreconcilable contradiction to the world of existing wealth and culture. Once alienation becomes a power that can no longer be tolerated, in that it expropriates the overwhelming majority of mankind, it will give birth to the revolt of this proletarian mass and lead them to the revolution, in which, according to the final words of the *Communist Manifesto*, "the proletarians have nothing to lose but their chains. They have a world to win."[52] Utterly alienated, practically the entire mass of mankind can only march toward the "expropriation of the expropriators," toward total revolution.

Deprived of all property and all satisfaction, the working masses, even though they activate productive forces and produce the wealth, take account of the fact that they have only to overturn those who hold the means of production and direct the relations of production in order to gain access to a human existence. Being reduced to a state of subhumanity and crushed by inhuman powers, the majority of mankind is the motor of the movement that leads to the appropriation of man by man and for man. The individuals that compose this class are no longer empirical, particular individuals. They are nothing and they have nothing in the present world. Products of the development of "universal" history, these individuals belong to universal history, and it is their task to realize that history in production. "The proletariat can thus only exist *world-historically*, just as communism, its activity, can only have a 'world-historical' existence. World-historical existence of individuals, i.e., existence of individuals which is directly linked up with world history."[53]

The old society, which generated all the productive forces it could contain and developed modern technique right to the limit of *its* possibilities, produced at the same time this overwhelming mass of proletarian workers. It has thus created the twofold practical condition for objective and subjective human emancipation. The conditions for the complete liberation of the exploited class and for the emancipation of mankind altogether are given by the development of the motor agency of history, namely, the productive forces and those who activate them. The revolutionary class itself possesses the greatest productive power of all productive powers.

The mode of production and the division of labor condition the two antagonistic classes. The class of the proletarians is the true activator of productive forces. But, as the relations of production are blocked by private property, the common labor of the proletarians, bound as it is to the development of productive forces, comes into violent collision with those who hold the means of production, the bourgeois and the capitalists, who put forward their particular interest as the general interest. Held in the tight grip of private ownership, the productive forces, instead of serving men positively, are changed into destructive forces and generate negativity. Property holders represent the conservative side, which in fact is the side destructive of society, while the proletarians are the force that would destroy that state of affairs. The proletariat carries all of society's burdens and has the enjoyment of nothing, so it stands opposed to all other classes. It comprises the majority of society, sets productive forces into motion, and produces wealth through the intermediary of the sale of its labor force; and through this sale it is transformed into a simple reified commodity. In wishing to abolish the contradiction that exists between the productive forces that it itself activates and the relations of production that are bound to private ownership, the proletariat does not want to become a dominant, consolidated class: it wants to abolish both private property (and proprietors) and the mode of alienating labor; it will abolish together itself, as proletariat, and its antagonist.

The mass of workers need class consciousness, the awareness of the radical necessity of a total revolution. This revolutionary consciousness comes to it "from outside," even though it is generated by its own situation; that is, it is theory of a kind that presupposes, but also founds, practice. This intellectual realization provides light for the struggle for emancipation. Without the active intervention of the proletariat, the old state of affairs could not be abolished. The *necessity* for a dissolving process which is inherent in bourgeois society converges with the *willed action* of the proletariat and gives birth to this action. "Indeed private property, too, drives itself in its economic movement toward its own dissolution, only, however, through a development which does not depend on it, of which it is unconscious and which takes place against its will, through the very nature of things; only inasmuch as it produces the proletariat *as* proletariat, that misery conscious of its spiritual and physical misery, that dehumanization conscious of its dehumanization and therefore self-abolishing. The proletariat executes the sentence that private property pronounced on itself by begetting the proletariat . . ."[54] A few lines farther on, Marx highlights the not godlike but human role of the proletariat, for the proletarians are quite the opposite of gods; and he makes clear what the essence of the proletariat is.

> It cannot abolish the conditions of its own life without abolishing *all* the inhuman conditions of life of society today which are summed up in its own situation. Not in vain does it go through the stern steeling school of *labour*. The question is not what this or that proletarian, or even the whole of the proletariat at the moment *considers* its aim. The question is *what the proletariat is*, and what, consequent on that *being*, it will be compelled to do. Its aim and historical action is irrevocably and obviously demonstrated in its own life situation as well as in the whole organization of bourgeois society today.[55]

The accomplishment of the proletariat's task—to discover its essence—comes by way of the economic, political, and ideological overturning of the class that is economically, politically, and ideologically dominant. This will not be for the proletariat just a repetition of one of those revolutionary experiments that have seen the light of day in the course of history, which has been the history of alienation. Many revolutionary upheavals have already taken place, but "all earlier revolutionary appropriations were restricted."[56] Only the high development of productive forces, of technique, and of big industry make possible and necessary the abolition of private property and class antagonism. Previous revolutionary appropriations were equally limited, because neither the totality of needs nor the totality of capabilities was yet fully developed. Only the modern proletarians, totally dispossessed and frustrated, can and must become the moving agency for the limitless development of dealienated social activity.

"The periodically recurring revolutionary convulsion" has never yet been strong enough to overturn the very basis of all that is, thereby making a total revolution, a revolutionizing of "the very production of life" as it has been up till now, an overturning of the total, alienated activity on which this production of life was based and remains based.[57] The *idea* of this upheaval has been expressed many times, Marx notes, as the history of communism proves. Communistic *thought* has dared to attack private property; but, *really* and *practically*, the material basis of the production of life and human activity remained almost intact. It is precisely this knot woven of private property, labor as alienating, and the division of labor that practical materialists, real communists, and revolutionary proletarians have to undo—by cutting it.

Yet the periodically recurring revolutionary convulsion known in history follows an evolutionary movement, undergoes a *development*. It sets going an ever-increasing mass of individuals; it each time enlarges the circle of social activity. "Every new class, therefore, achieves its hegemony only on a broader basis than that of the class ruling previously, whereas the opposition of the non-ruling class against the new ruling class later develops all the more sharply and profoundly."[58] History *tends* to become universal, but only the proletariat can really universalize all that is.

Marx is not only the scientific analyst of economic contradictions; he is also the "prophet" of total reconciliation and the technician of mankind's emancipation through the emancipation of the proletariat. He is at the same time an inspired *warning voice*, constantly putting others on their guard against all the false meanings that the communist movement could take on. His words almost *conjure up* that movement, and later on communism will adopt his thought as its basis. He writes:

> In all revolutions up till now the mode of activity always remained unscathed and it was only a question of a different distribution of this activity, a new distribution of labour to other persons, whilst the communist revolution is directed against the preceding *mode* of activity, does away with *labour*, and abolishes the rule of all classes with the classes themselves, because it is carried through by the class which no longer counts as a class in society, is not recognised as a class, and is in itself the expression of the dissolution of all classes, nationalities, etc., within present society.[59]

The proletarian revolution, therefore, is seen as overturning the entire infrastructure and superstructure of capitalist society and of every society that has alienation within itself, in order to begin with a new foundation, of a kind never before known, for the social life of men. Labor, private property, and the division of labor must be abolished; but Marx does not tell us how, after the abolition of private property and after the socialization of wealth, *labor* and the *division of labor* can be actually superseded in a society based on a spectacular development of technique and productivity.

Full of confidence about the possibility—and necessity—of this radical and total transformation, Marx asserts that the practical movement that is to culminate in this modification can take the form and have the content only of a *revolution*. Revolution is necessary both "because the *ruling* class cannot be overthrown in any other way" and "because the class *overthrowing* it can only in a revolution succeed in ridding itself of all the muck of ages and become fitted to found society anew."[60] The ruling class possesses the means of production and the State. In order to dispossess and dealienate society, the bourgeois State must as well be abolished, and in this revolutionary work the proletariat itself comes to maturity and experiences the conditions for the existence of the new communal society.

In this movement of progression toward total reconciliation the *revolutionary, communist* consciousness of the proletariat finds its expression; and this is a consciousness "which may, of course, arise among the other classes too through the contemplation of the situation of this class."[61] It is a consciousness which is accessible to individuals who, though belonging to the ruling class, detach themselves from that class to rally to the revolutionary class; they are able to do this by virtue of having "raised themselves to the level of comprehending theoretically the historical movement as a whole."[62] Only the dialectical bond between revolutionary *ideas* and communist *thought*, on the one hand, and the revolutionary *class* and the *real movement* of communism, on the other, will guarantee vigilant attention to the task of the radical abolition of private property, (alienating) labor, and the division of labor. Nevertheless, as we have already had occasion to note as a problem, the unifying link in this relationship remains ambiguous: sometimes it is the *material, real* situation of the revolutionary *class*, its *practical* experience, that gives birth to revolutionary ideas, awareness, and thought; sometimes it is *theory*, the *understanding* of the situation and of real movement, *reflective grasp*, that generates revolutionary action. The "vicious" circle that ties real practice and action (the material agency that creates theoretical notions) to consciousness and ideas (which are derived forms, but oh so powerful and basic!) may well turn "dialectically," but it remains all the same "vicious," for it is not clear what is *the* power that gives it its motion.

However that may be, the communist action that gives rise to communist ideas, while itself a product of revolutionary consciousness, is to contribute to a vast production of communist awareness, revolutionary thought penetrating the masses and guiding their action. For that, a massive change is necessary in the men who undertake the struggle against "the very production of life" as it has been up till now,[63] for these are the men who will accomplish this total transformation of man's basic activity. Men are destined to change into a new kind of creature in this undertaking which is to modify the roots of social life. "Circumstances make men just as much as men make circumstances,"[64] as this idea is put dialectically, if a little vaguely, by the resolute opponent of Hegel's dialectic. If the revolutionary masses do not pursue the task right to the end, bringing communist thought and the truth of universal history into reality in a practical way, only simple, nonessential changes will occur in society, and only the impotent *idea* of a regenerative upheaval will persist.

As Marx tirelessly repeats, the revolutionizing class is and is not a *class*. From the outset, though a determinate class, it stands as representing all society; its "particular" interest coincides with the true universal interest. Comprising the total mass of society in the grip of the one ruling, possession-holding class, it has not yet been able, owing to the pressure of social conditions, to develop into a particular class with particular interests. It aspires thus to a domination that will not be domination. Its victory will assure the emancipation of all members of society, and it will lead mankind toward that new form of society in which "it is no longer necessary to represent a particular interest as general or the 'general interest' [i.e., the abstractly universal] as ruling."[65] The final stage in class struggle will come to an end with the simultaneous abolition of the two ultimate antagonistic classes and will lead to a society without classes, a society of full technical development and *total planning*.

On the economic level, the proletarian revolution is to abolish simultaneously *separate economic life*, *private property* in all its forms, *labor* itself as man's externalization, the *division of labor*, and all the old conditions for *production* and *trade*. All that constitutes, quite obviously, a still negative task, to be followed by the positive construction of socialistic economy, the communal and communistic organization of economic life. While wanting to abolish economics, *the organization of communism remains essentially economic*.[66] Socialized economic life will take on another meaning while still being *economic life*.

Everything will be subordinated to all; individuals will subordinate to themselves the powers that once subjugated them. Human forces will make things subject to themselves. The human totality will be able to appropriate the totality of what is. The proletariat—and humanity on the whole—will appropriate the totality of productive forces by prodigiously developing them, by freeing all the powers of technique. This universal appropriation will take on a total character, altogether free of alienation, and neither the *conditions of labor* nor the *division of labor* nor the *instruments of labor* will be able to set themselves up as alien powers hostile to the members of socialist society. "In all expropriations up to now, a mass of individuals remained subservient to a single instrument of production; in the appropriation by the proletarians, a mass of instruments of production must be made subject to each individual, and property to all. Modern universal intercourse can be controlled by individuals, therefore, only when controlled by all."[67] Without taking the trouble to explain how all that will be possible in a world of high technical development, Marx continues to announce that the abolition of *private property* (*Privateigentum*) and the communist regulation of production will allow the *transcending* (*Aufhebung*) of all that is alien (*Fremdheit*) to the social nature of men, while the elimination of *alienation* (*Entfremdung*) opens the way to the free and total *appropriation* (*Aneignung*) of the world by man. "With the abolition of the basis, of private property, with the communistic regulation of production (and, implicitly in this, the destruction of the alien relation between men and what they themselves produce), the power of the relation of supply and demand is dissolved into nothing, and men get exchange, production, the mode of their mutual relation, under their control again."[68]

The communist regulation of economic life will free men from the tyranny of economic powers, for they will dominate these powers, controlling them and exercising *their power* over them, according to their consciousness and will—these being deemed in conformity with the rhythm of technical development. "*All-*

round dependence, this natural form of the *world-historical* cooperation of individuals, will be transformed by this communist revolution into the control and conscious mastery of these powers, which, born of the action of men on one another, have till now overawed and governed men as powers completely alien to them."[69] Social necessity and individual freedom, the production of objects according to the conscious will of subjects, and cooperation in interhuman activity will take on the appearance of a reality familiar to man. By appropriating productive forces and assuring a propitious industrial and technical development, by developing the totality of individual capabilities (for total appropriation advances hand in hand with the blossoming of the totality of capabilities), and by readying the conditions for the satisfaction of the totality of their needs, the members of the new society will be transformed into total individuals and will live in a historically universal way (*weltgeschichtlich*).

Homo oeconomicus will be transcended after the economic basis of present society is overturned; the man of the new society will no longer have an economic horizon that is egoistic, narrow, limited, and basely utilitarian. Through the struggle for the universal appropriation of what man is and of what he does, the proletariat will strip itself of all that still remains of its former social position and take up apprenticeship in dealienated life.

We can understand what Marx means by *radical abolition of private property*, although he does not seem to think this abolition is a basis sufficient to completely supersede all alienation. The socialization of the means of production (the appropriation by the members of the human collectivity of the totality of wealth, in that all that is made becomes the common good of men, in short, the common possession of all the "sources of life") is clear, though there are *serious difficulties* involved in it. We perhaps understand, not without effort, what Marx means by the *abolition of labor as it has been up till now*. Once the condition of the worker, as worker, is abolished, all men will work productively and a new era of work will begin: labor will become a manifestation of human life, being no longer a means to earn—that is to say, lose—one's living. It is when Marx speaks of the *abolition of the division of labor* that he becomes truly difficult to follow. "The transformation, through the division of labour, of personal powers (relationships) into material powers, cannot be dispelled by dismissing the general idea of it from one's mind, but can only be abolished by the individuals again subjecting these material powers to themselves and abolishing the division of labour. This is not possible without the community."⁷⁰ Can the division of labor and all the alienating specialization that it implies be actually transcended by a community that would give free reign to the development of industry, machines, and a revolutionizing technique? Is this man whose thinking focuses on the nature of planetary technique, this realistic analyst of the division of labor, turning idyllic and becoming a dreamer in advocating total labor? Does he imagine this division-free activity performed by the members of communist society in the form of some "natural," "primitive," or craftlike occupation? Reflecting with a kind of tenderness on the medieval craftsman, whose situation was always determined by the feudal economic structure of society, he writes: "Thus there is found in the medieval craftsmen an interest in their special work and in proficiency in it, which was capable of rising to a narrow artistic sense. For this very reason, however, every medieval craftsman was completely absorbed in his work, to which he had a contented, slavish relationship, and to which he was subjected to a far greater extent than the modern worker, whose work is a matter of indifference to him."⁷¹ There could be no question of a return to medieval craft labor on the technical plane of the communist era. But how does Marx conceive this total activity, totally new and without precedent?

It must not be forgotten that the communist regime will have transformed all that is natural into a product of the social activity of men, for a complete utilization of natural forces for humanly social ends will have taken place. *Fully achieved naturalism* means *fully achieved technicism*, in that socialistic, communistic, humanism will bring about the "casting-off of all naturality."⁷² With God abolished and nature subjugated and transformed into a human and social reality, the practical activity of men will no longer be involved with a nature that is independent of them. When the communist regime abolishes private property and the division of labor, it will achieve at the same time "the abolition of the antagonism between town and country," this abolition being "one of the first conditions of communal life."⁷³ What makes one man a restricted "town-animal" and another an even more restricted "country-animal," what feeds the antagonism between them, will have passed out of existence owing to the total development of productive technique, the complete industrialization and socialization of nature. And so, at this level, the division of labor into material and manual work, on the one hand, and mental and intellectual work, on the other, will as well disappear.

The question again arises: How is the deployment of the power of division-free labor set out in concrete terms? We ask this of Marx himself, who, after having underscored the "fact" that man's activities will truly depend upon his will, so that no stabilization and rigid fixing of social activity would intervene to disturb the harmony of total praxis, has recourse to a kind of illustration for his thesis. It is Marx himself, the implacable *negator* of reality as it exists, who makes the *positive* statement about the new reality that we are going to quote. He tells us, first of all, that in *natural society* (*naturwüchsigen Gesellschaft*), in technically underdeveloped society, in which activity is not voluntarily but naturally divided, man's own act becomes an alien, inimical power that subjugates him. Activities at this stage, then, being too close to nature, are not chosen by man's free will but are imposed on him by the nature of things: man is a hunter, a fisherman, or a shepherd, and he is forced to remain that. There could be no question, consequently, of a reassumption of these natural—but not at all voluntary—activities on the level of a naturalism transformed into humanistic socialism. Here, then, is what will happen in *communist society*: "In communist society, where nobody has one exclusive sphere of activity but each can become accomplished in any branch he wishes, society regulates the general production and thus makes it possible for me to do one thing today and another tomorrow, to hunt in the morning, fish in the afternoon, rear cattle in the evening, criticise after dinner, just as I have a mind, without ever becoming hunter, fisherman, shepherd or critic."[74] In the world of planetary technique and the most powerful development of the totality of productive forces, in the world that will have completely transformed nature into history and stripped all that is of its natural character, in the world that will no longer know the distinction of labor into mechanical work and spiritual work nor the opposition of city and country, the communist man will be able, following his will and good pleasure, to give himself to these primitive, ancient, and medieval activities; he will be able to *hunt*, *fish*, *raise livestock*, and *criticize*. Is not this anticipatory vision of total man's activities in total society, such as it is expressed by the fanatic opponent of all ideology and every utopia, itself ideological, touching as it may also be in its naïveté and its rural, idyllic tones? At any rate, it *makes us think*; it makes us reflect how the abolition of economic life and every economic alienation by an arrangement that is essentially economic and by a communistic regulation of production raises problems that Marx did not anticipate. The transcendence of labor and the division of labor, such as they have existed, by the prodigious, universal development of productive technique will culminate in a reality far more *problematic* than that within which man might hunt, fish, farm, or criticize, at his good pleasure and at any hour of the day. Even as an imaginary projection, does not this prevision remain terribly flat? Does it not, under its rather idyllic colors, show us the transcendence of "differences" into a world of generalized *in*difference? Or, rather, will not man's activity, in its polytechnical character, be of the order of Play? Does *play* constitute the "sense" of human action once it has surmounted the search for meaning, without for all that foundering in the absurd and the meaningless? Marx allows us to play with this "hypothesis" without explicitly formulating it. Will there be a global human activity in which productive and creative work, poetic activity and recreational activity will be fused in their singleness, and will this one activity be Play? Speaking of labor as producing reified values, Marx denounces the inherently uninteresting and unattractive character of that labor, and he says that the worker does not enjoy his activity as the "play (*Spiel*) of his bodily and mental powers."[75] But he says nothing further of the possibility of a future opening into the dimension of play.

The total freeing of man by the liberation of the revolutionary proletariat is to take a political path, destined to end with the abolition of political alienation and the political life itself. First of all, the proletariat is to overturn the existing State, which is the organized and organizing power of the existing state of things. "Further, it follows that every class which is struggling for mastery, even when its domination, as is the case with the proletariat, postulates the abolition of the old form of society in its entirety and of domination itself, must first conquer for itself political power in order to represent its interest in turn as the general interest, which in the first moment it is forced to do."[76] Because individuals seek only their particular interest, the general interest, which expresses their real interests, appears "alien" to them at first; but they will be educated by—and through—their own revolutionary action.

The overturning and annihilation of the existing State, by the revolution, is the proletariat's very first task. For the proletariat cannot purely and simply take over the machinery of State as it already is, in order to make it work in another way. The old state apparatus must be annihilated, and the worker class, having taken power, must not continue to manage things with the former bureaucratic machine. In a famous letter to Kugelmann, Marx writes: "If you look at the last chapter of my *Eighteenth Brumaire* you will find that I say that the next attempt of the French revolution will be no longer, as before, to transfer the bureaucratic-military machine from one hand to another, but to smash it, and that is essential for every real people's revolution . . ."[77] The passage from capitalism to socialism cannot be made peacefully. Revolution is necessary, and it will set up *the dictatorship of the proletariat* which will allow the proletariat to organize, begin building socialism, and annihilate its enemies. "Between capitalist and communist society lies the period of the revolutionary transformation of the one into the other. There corresponds to this also a political transition period in which the state can be nothing but *the revolutionary dictatorship of the proletariat*."[78] The dictatorship of the proletariat and the State corresponding to it are strictly provisional, temporary, and bound to disappear. Emerging from the primordial economic struggle of the proletariat and closely bound to it, its fight on a political level aims to win political power. Society is the basis for the State. This is true for bourgeois, capitalist society and for that society which leads to socialism and communism; the State found in this transitional period will be entirely subordinated to society and will not acquire an independent reality. Moreover, it is destined to disappear. The same is true of class struggles, the mainspring of becoming for "every society up till now" and especially for history in recent centuries. This class struggle leads to the proletarian revolution and the revolutionary dictatorship of the proletariat, powers taking mankind to the classless society. The dictatorship of the proletariat sets the final note in class struggle and abolishes class antagonisms.[79]

Marx is the accuser of the *State, law,* and *political life,* the furious theoretician of the radical abolition of the State, law, and political life, and the seer of a society without classes and without a State, a society that no longer separates the sphere of social, socialistic activity into economics and politics; and yet he knows that the State, (coercive) law, and politics cannot be abolished right away. Conceiving the notorious transitional period as an inevitable reality, he writes: "What we have to deal with here is a communist society, not as it has *developed* on its own foundations, but, on the contrary, just as it *emerges* from capitalist society; which is thus in every respect, economically, morally, and intellectually, still stamped with the birth marks of the old society from whose womb it emerges."[80] The new society continues certain realities from the old, although it intends its negation. Negation itself remains affected and infected by what it negates, though Marx has never taken too seriously this affecting, infecting feature. His confidence in the absolutely positive future of fully achieved communist society is unshakable, and he does not let his vision be obscured by the dark shadows that the old world casts upon the new.

Marx willingly admits that the law that will be used by transitional society will remain "a right of inequality, in its content, like every right."[81] Furthermore, he has never let himself be blinded by the idea of equality; he has no wish to see inscribed on the banners of socialist society the formal terms *Freedom, Equality, Brotherhood* (*Liberté, Egalité, Fraternité*). His goal is the abolition of *class differences*, not the establishment of a reign of *abstract equality*. It is a matter of abolishing the inequality of material conditions and the conditions of individual development and of annihilating class differences: simply wishing for the establishment of a rule of indistinction, a kingdom of "equality," which is something unrealizable, is not enough. Even fully achieved communism will not, and is not supposed to, be under the imperial rule of "equality."

This provisional and transitional society must be taken for what it is: a society that violently negates capitalism, without however being fully socialistic and communistic. The proletariat will *dominate*, but *not* as a *class*. Although in possession of state political power, it will not, properly speaking, institute a new governmental power. During its campaign of conquest, the working class will set up a communal association in place of the old civil society, which is a framework within which political power summed up the antagonism between the exploiters and the exploited. Marx raises the question about the new political power and answers it by disposing of it.

> Does this mean that after the fall of the old society there will be a new class domination culminating in a new political power? No.
>
> The condition for the emancipation of the working class is the abolition of every class, just as the condition for the liberation of the third estate, of the bourgeois order, was the abolition of all estates and all orders.
>
> The working class, in the course of its development, will substitute for the old civil society an association which will exclude classes and their antagonism, and there will be no more political power properly so-called, since political

> power is precisely the official expression of antagonism in civil society.[82]

However, the problem of political power in the first phase of communism is not so easily disposed of. The *class struggle* that necessarily leads to the *dictatorship of the proletariat*, which, as *revolutionary*, constitutes a *transition* leading through the abolition of all classes to the classless, state-free society, is without question the keystone of Marx's *political* thought. This crucial component of political thought remains, however, highly problematic. The proletariat constitutes a class; it organizes as a political party; it overturns capitalist domination economically and politically; it sets itself up as the ruling class and takes over political power. "The first step in the revolution by the working class is to raise the proletariat to the position of ruling class, to win the battle of democracy. The proletariat will use its political supremacy to wrest, by degrees, all capital from the bourgeoisie, to centralize all instruments of production in the hands of the State, i.e., of the proletariat organized as the ruling class; and to increase the total of productive forces as rapidly as possible."[83] This we read in the *Communist Manifesto*, which continues a few paragraphs later as follows:

> When in the course of development class distinctions have disappeared and all production has been concentrated in the hands of a vast association of the whole nation, the public power will lose its political character. Political power, properly so called, is merely the organized power of one class for oppressing another. If the proletariat during its contest with the bourgeoisie is compelled, by the force of circumstances, to organize itself as a class, if by means of a revolution it makes itself the ruling class and, as such, sweeps away by force the old conditions of production, then it will, along with these conditions, have swept away the conditions for the existence of class antagonisms and of classes generally, and will thereby have abolished its own supremacy as a class.
> In place of the old bourgeois society, with its classes and class antagonisms, we shall have an association in which the free development of each is the condition for the free development of all.[84]

Does this mean that the questions concerning power, politics, and the State are thereby resolved? Marx has refused to predict the concrete forms that "the future State" will take in a communist society; but what he has said about it remains problematic. This association in which and by which freedom will have become possible and actual, so that each man's personality develops freely within the union of consensus among individuals, remains far too idyllic. For Marx, who wanted to transcend the artificial contractual ideas of the *Social Contract*, the *conscious*, *voluntary* union of individuals which exercises its control over the totality of the development of productive forces and over all the social development of individuals will become reality only in the later phase of communism; but it will be on the way to realization in the transition period in which the nonstate State will itself be preparing its own withering away. No bureaucratism is to insinuate itself into the new social organization, according to Marx, for he is the determined adversary of all bureaucracy. Social functions are in no way to become concentrated, consolidated, and reified in the hands of some rigid civil service or administration.

The fate of the State during this march of conquest, however, remains unclear, and neither Marx nor Engels has seen how to exorcise this ghost. In a letter dealing with the Gotha program, written March 18-25, 1875, to August Bebel, Engels tries to make things more specific—but does he? He writes:

> A free state is one where the state is free in relation to its citizens and is therefore a state with a despotic government. The whole talk about the state should be dropped, especially since the Commune, which was no longer a state in the proper sense of the word. The *"people's state"* has been thrown in our faces by the anarchists too long, although Marx's book against Proudhon and later the *Communist Manifesto* directly declare that with the introduction of the socialist order of society the state will dissolve of itself and disappear. As, therefore, the "state" is only a transitional institution which is used in the struggle, in the revolution, in order to hold down one's adversaries by force, it is pure nonsense to talk of a "free people's state"; so long as the proletariat still *uses* the state, it does not use it in the interests of freedom but in order to hold down its adversaries, and as soon as it becomes possible to speak of freedom the state as such ceases to exist. We would therefore propose to replace the word "state" everywhere by

the word *Gemeinwesen* [Community], a good old German word which can very well represent the French *commune*.[85]

The problem of the process that leads to the withering away of the State and of the forms taken by the administration of beings and things is, however, far from being solved.[86]

Must one say that this problem is solved in theory but not in practice? Does Marx, who hardly wished to accept the distinction between the truth of (abstract) theory and that of (concrete) practice, fall victim to the alienating power of this dualism? For theory continues to proclaim the truth of communism without saying too much about the practical means that will turn it into reality. We are told that "*social* organs develop in the *form* of society; thus, for instance, activity in direct association with others, etc., has become an organ for expressing my own *life*, and a mode of appropriating *human* life."[87] What *organizational forms* will these *organs* for the expression of life take? Through what administrative *organisms* will the intention to appropriate human life in a global fashion be shown?

Far from answering these questions, Marx stresses the point that this appropriation is conditioned by three things: by the degree of development in productive forces, by the appropriating individuals, and by the manner in which it is to be done. "This appropriation is first determined by the object to be appropriated, the productive forces, which has been developed to a totality and which exist within a universal intercourse. . . . This appropriation is further determined by the persons appropriating. . . . This appropriation is further determined by the manner in which it must be effected."[88] The *individuals appropriating* and the manner in which naturalistic, humanistic, socialistic, and communistic appropriation is effected have, therefore, an important role to play, and the fate of this appropriation, to a very great extent, depends upon these things. The role of *leaders* called to direct the movement of liberation is not to be neglected either. Leaders can be positive in accelerating or negative in slowing down the evolution of socialism-communism. Marx, writing again to Kugelmann, says this: "World history would indeed be very easy to make, if the struggle were taken up only on condition of infallibly favourable chances. It would on the other hand be of a very mystical nature, if 'accidents' played no part. These accidents themselves fall naturally into the general course of development and are compensated for, again, by other accidents. But acceleration and delay are very dependent upon such 'accidents' which include the 'accident' of the character of those who at first stand at the head of the movement."[89]

Marx does not, therefore, lack at least a measure of perspicacity regarding accidental factors, good or bad, in political evolution. He recognizes, occasionally, that the splendid, total victory of the revolutionary workers is an extremely difficult thing, that one stage of victory for the workers is offset by a new struggle that ends with their defeat, and that one class may make a revolution so that it may benefit another class. Yet, knowing all that, Marx remains confident in the coming of total human emancipation.

In order to bring about this emancipation, the communist movement must accomplish a number of things preliminary to it: the violent overthrow of the very basis of all the old conditions of production and trade, management and consumption; the annihilation of the political and administrative forms of the old State bureaucratic apparatus; the removal of any would-be natural character from all those items that pass for the natural and that as such are a hindrance to the creative power of the advancing proletariat. Everything must be subjected to the *consciousness*, the *will*, and the *power* of individuals united; what seemed to be able to exist independently of individuals is bound for annihilation, since all that exists is a product of human activity. Communism inaugurates an era that is entirely new: ". . . for the first time [a movement] consciously treats all natural premises as the creatures of hitherto existing men, strips them of their natural character and subjugates them to the power of the united individuals."[90] New men, practical materialists, disburdened of old structures and superstructures, in their practical endeavors treat "the conditions created up to now by production and intercourse as inorganic conditions, without, however, imagining that it was the plan or the destiny of previous generations to give them material, and without believing that these conditions were inorganic for the individuals creating them."[91] In the evolutionary course of the movement of dealienation, organization will no longer come to superimpose itself on the organic, and all that is "organic" and "natural" will exist by virtue of technique freed from all restriction in its sphere of action. Personal activity will no longer be limited "labor" but will coincide with the fullness of material life. History will be completely transformed into universal history, and all that is will be universalized. Particular individuals, freed of all particularity and all economic, political, national, and ideological barriers, will be "brought into practical connection with the material (*as well as intellectual*) production of the whole world and be put in a position to acquire the capacity to enjoy this all-sided production of the whole earth (the creations of man)."[92]

This whole militant process of dealienation and appropriation will lead mankind to the second phase of socialism, namely, communist society properly speaking and the reign of triumphant reconciliation. All oppositions will have been transcended and members of the collectivity will participate in benefits no longer *according to their labor* but *according to their needs*. Need, the motivating force for human activity, will be entirely satisfied in fully flowered communist society. For, in the first phase of socialism-communism, in the society that is "still stamped with the birth marks of the old society from whose womb it emerges,"[93] the individual can consume only in proportion to the social labor he provides, that is, in proportion to the individual quantum of social labor. Marx invites us to imagine for once a community of free men working with socialized means of production and dispensing their many individual forces as one same, single, collective force in accord with a common plan. Within this communal social order,

> Labour-time would, in that case, play a double part. Its apportionment in accordance with a definite social plan maintains the proper proportion between the different kinds of work to be done and the various wants of the community. On the other hand, it also serves as a measure of the portion of the common labour borne by each individual, and of his share in the part of the total product destined for individual consumption. The social relations of the individual producers, with regard both to their labour and to its products, are in this case perfectly simple and intelligible, and that with regard not only to production but also to distribution.[94]

Although this order will have abolished private property, it is not yet the communist order, either economically or politically. It merely inaugurates it. Here again we have the famous first phase of communism. Let us review for a moment the stages of the process that leads to total reconciliation, to communism fully formed as naturalism-humanism. From the very bowels of *capitalist society* is born the *trend toward emancipation* for the proletariat (and for man). To abolish private property and shatter the old bureaucratic and military apparatus of the State, the proletariat is obliged to resort to *revolution*. At its victory, the revolution installs the *revo-*

lutionary dictatorship of the proletariat and creates a *new State* that is a non-State. The worker class thus wins democracy, *socializes the means of production*, and creates the first phase of *socialism*-communism, within which each may work according to his capabilities and share in the consumption of goods according to his productive labor.

> In a higher phase of communist society, after the enslaving subordination of the individual to the division of labour, and therewith also the antithesis between mental and physical labour, has vanished; after labour has become not only a means of life but life's prime want; after the productive forces have also increased with the all-round development of the individual, and all the springs of co-operative wealth flow more abundantly—only then can the narrow horizon of bourgeois right be crossed in its entirety and society inscribe on its banners: From each according to his ability, to each according to his needs![95]

The full, and doubtless planetary, development of productive forces, the harmonious development of the entirety of each individual's capabilities, the abolition of labor and of the difference between town and country, the total transcendence of private property and bourgeois law, the withering away of the State (and of democracy) shall all, therefore, lead mankind to communism, to the solution of the riddle of history, to the reign of satisfaction for the totality of needs. "Economics" and "politics" will have thereby been transcended, and social evolution will no longer be a matter of political revolution.

The circuit is thus completed. The point of departure is rediscovered in the point of arrival. Man, who works to live, struggles against nature to produce his life, and aspires to the satisfaction of his natural, human, and social needs, remains robbed, frustrated, unsatisfied, and alienated, all from the first act of human "history" right up to the first act of the beginning of true history. In communism, as at once naturalism, humanism, and socialism, man is reconciled with all that is; the means of satisfying his needs, namely, labor, is no longer labor, properly speaking, no longer divided labor, and becomes his life need par excellence. Labor is thus abolished, so to speak, by its generalization. To total production correspond a satisfaction of needs and a consumption of collective wealth that are no less total. And so the circle is closed: needs, labor, production—labor as need, satisfaction of needs, consumption according to needs. The individual becomes an organic member of the social, socialist body, man rules over things, and all naturality is abolished by virtue of humanistic technique. From the Renaissance on, since the grasp of the *ego* of the *cogito* as operative upon the *res extensa*, man as human individual (and not simply as a psychological subjectivity), individualism, begins to be posited as a quasi absolute. But individualism is consummated in its self-extension, taking on a planetary dimension in the "absolute" collectivism, in the community that makes it possible to realize the individual's needs, the needs of individuals.

The transcendence of alienation can, according to Marx, be actually realized; such is the hypothesis (*Voraussetzung*) of socialism-communism. But this hypothesis has real premises. "Communism is not for us a *state of affairs* (*Zustand*) which is to be established, an *ideal* to which reality [will] have to adjust itself. We call communism the *real* movement which abolishes the present state of things. The conditions of this movement result from the premises now in existence."[96] Marx *wants* to keep an equal distance from "fatalism" and "voluntarism"; historical fate and human activity do not exist and do not work without each other. Necessity itself calls up action; real movement itself generates the forms and forces that move forward. What *is truly* necessary can *become reality* only through a movement which, being practical, acts to impart becoming and being precisely to that for which the premise conditions are actually found to exist. For communism is an essentially practical movement, and one cannot overemphasize its practical character. Marx categorically asserts "that communism is a highly practical movement, pursuing practical aims (*praktische Zwecke*) by practical means (*mit praktischen Mitteln*), and that only perhaps in Germany, in opposing German philosophers, can it spare a moment for the problem of the 'essence' . . ."[97] Communism is therefore a product neither of fate nor of some free will. It is not a rigid state that is to be established or the theoretical construction of an ideal. It exists through the practical movement of negativity, and it pursues practical ends (the full satisfaction of the real needs of men, for these are the motives for action) by practical means (total, planetary technique, the mobilization of all productive forces). Communism exists, that is, it becomes, through this mobilization. It is in movement and its "essence" is mobile. It posits negation as negation in order to negate the world as it exists, the world that nevertheless provides it with its preconditions; and it comes to posit the negation of negation as a new and unprecedented position, as the point of departure for the movement that moves toward the total, practical appropriation of all that *really* exists. Communism, pursuing the real movement that sets it itself moving, is "the actual phase necessary for the next stage of historical development in the process of human emancipation and rehabilitation (*der menschlichen Emanzipation und Wiedergewinnung*). *Communism* is the necessary pattern and the dynamic principle of the immediate future, but communism as such is not the goal of human development—which goal is the structure of human society."[98]

Not only is communism not the final goal, the ultimate term and definitive end in the becoming of human history, but also henceforward it implies its own limits. Though it is the movement of victory over Totality and of the (re)conquest of the essence of human totality, it hardly escapes limitation. Communism will be in turn transcended, but not before it has been completely realized. No one has yet paid attention to Marx's words here. Are we even ready to hear and understand them, and is the historical situation now still just as distant from them? For in regard to "*actual* communist action," the "movement" of communism: "History will come to it; and this movement, which in *theory* we already know to be a self-transcending movement, will constitute *in actual fact* a severe and protracted process. But we must regard it as a real advance to have gained beforehand a consciousness of the limited character as well as of the goal of this historical movement—and a consciousness which reaches out beyond it."[99] We once more find thought and consciousness transcending the real movement whose process of actual development is at the same time a process of self-abolition. Difficult as it is to grasp and acquire, this consciousness, never altogether real and almost always illusory and alienated, being the reflection of practice *and* the condition for all true practice, makes its appearance once again; and here it does no less than transcend, from this point forward, the real movement of communism itself, which is able to escape neither limitation upon itself nor its own future supersession. Communism's theoretical master, therefore, knows—or does he merely suspect?—that *practical*, *real* movement (for communism, among other things) is limited, and its goal can hardly be the ultimate end of historical becoming. What is more, this movement finds itself from the beginning transcended by *thought* and *consciousness*. All Marx's efforts to confer more reality upon the *movement of reality* than upon the *dialectic of conscious-*

ness do not prevent the latter from being at work from the beginning and from having in the end a position of overview with regard to historical becoming.

Socialism-communism, as naturalistic humanism, begins from consciousness and ends with a consciousness that transcends it. Have we not seen Marx write that socialism "proceeds from the *practically and theoretically* sensuous consciousness of man and of nature as the *essence*"?[100] Communism begins, therefore, from man's theoretical *and* practical self-consciousness, though this may be in principle something epiphenomenal. This consciousness, of course, develops in the course of the practical history by which man extends nature and brings it "to be." Nevertheless, the Marxian conception of thought, knowledge, and consciousness, though wishing to be "materialistic" and "empirical," does not avoid "intellectualism." Consciousness is and remains in a state of advance beyond the actual situation. For "consciousness can sometimes appear further advanced than the contemporary empirical relationships, so that in the struggles of a later epoch one can refer to earlier theoreticians as authorities."[101] What seems to be indicated here is that thought does not allow of being entirely absorbed into real communism. It should, however, integrally coincide with the movement by which things are brought to realization. Does this mean that thought—when it is truly thought—moves in the dimension of the future and that at any actual moment it is never exhausted? Marx thought it was not enough that "thought strive to become reality"; "reality must itself strive toward thought."[102] This striving *toward*, however, cannot annul the tension proper to thought itself.

14.

THE ABOLISHING OF PHILOSOPHY

BY ITS REALIZATION

Abolishing the foundational phenomenon of alienation and its epiphenomenal forms is to lead necessarily to the transcending—and abolishing—of philosophy in and through its realization. What was theoretical and alienated will be practical and real. The whole heaven of ideas will have become useless on the earth of men. Practical energy will resolve all problems. In the kingdom of total reconciliation, philosophic thought can be abolished as such only if it is to be realized, because it cannot be actually transcended before its full realization has been achieved.

In his doctoral thesis the young Marx already wrote: "For the world to become philosophic amounts to philosophy's becoming world-order reality; and it means that philosophy, at the same time that it is realized, disappears. What philosophy struggles against on the outside is its own inner deficiency. Precisely in its struggle it falls itself into the wrongs that it combats as wrongs in its opponent, and it abolishes (*aufhebt*) these wrongs only by falling into them."[1] The theoretical truth of philosophy, which is the abstract truth of practical becoming, can be dealienated only in being made reality; philosophy is abolished, therefore, in order to be realized. In combating the evils of the world, philosophy itself falls victim to these evils and to some of its own, for the enemy it fights takes up habitation within it and begins breaking it up. To abolish the evils of the world and its own, the only thing it can do is to realize itself in a self-abolition. The real movement of the world was raised to the rank of philosophy; the world became more and more philosophical. Correspondingly, the movement of abstract thought impressed itself more and more into the flow of concrete reality; philosophy became world. The world of theoretical, philosophic truth was alienated, just as the world of practical, worldly reality. Consequently, the point is to abolish the real alienation of the world and to realize the truth of philosophic thought by abolishing the latter as such, since it remained untrue, that is to say, unreal and detached from transformative will and practical energy. Once the (real) world has become an ideological world in a general alienation, once the ideological world governs reality by a treason against it, once the concrete has become abstract and the abstract determines the concrete, there remains but to abolish the *double* aspect of alienation. Thus "the theoretical mind, having been freed itself, becomes a practical energy, passes out of the shadow kingdom of Amenthes, and turns toward worldly reality already there without it."[2] However, in transforming itself into *practical energy* and *will, theoretical mind* and *thought* find their transcendence in their loss. "What was an inner light becomes a devouring flame that turns outwards."[3] And this flame will light up praxis reconciled with its destiny, which is the conquest of the real world.

We must be careful here, however. The task consists *neither* in "actualizing (realizing) philosophy without abolishing-transcending it" *nor* in "abolishing-transcending philosophy without actualizing (realizing) it," as Marx warns us in his "Contribution to the Critique of Hegel's 'Philosophy of Right': Introduction."[4] Each of these two demands operating alone remains one-sided, one falling victim to the worship of sterile theory, the other succumbing to the cult of unenlightened practice. The task consists in transcending the alienating contradiction between the reality of the world and the half-truth of philosophy. Philosophy can hardly be made reality and transcended without the abolition of the proletariat, and the proletariat cannot be abolished without the realization of philosophy.[5] In the world of alienation the proletariat is nothing, while it should be everything. In the world of the transcending of alienation, theoretical thought, which puts itself forward as everything, will be reduced to nothing in its scission from practical energy. Reconciliation will mean, among other things, conciliating truth with passion and passion with truth.

Nurtured by earth but seeking escape in the clouds of heaven, philosophy both expressed and betrayed its origin, losing itself in the pursuit of a distant goal. By returning to earth it resolves itself into the movement of the real world; no longer standing opposite to the world, it becomes world. What was of the order of abstract thought, speculation, pure theory, consciousness, and knowledge—

that is, all that which Hegel encompassed by total, absolute knowledge—is now, in dealienation, found to be absorbed by practical action, life-energy, social experience, productive labor, power-filled will, transformative action, total praxis. Yet the "philosophy of philosophy" that Marx proposes is extremely dialectical, ambivalent, and ambiguous. Philosophy is an ideological form of alienation *as well as* an abstractly true theory; it derives from practice, but it leaves practice in thinking about it. Philosophic thought presupposes empirical reality, natural, historical, human reality; *but* it as well provides real action with its lights and weapons. It abolishes itself in realizing itself, and *yet* ("philosophical") consciousness transcends historical movement and is in advance of the concrete situation. Does this mean that "philosophic" *thought* remains *present* in its *absence*, that it turns off and takes another path?

Marx did not fail to note that "modern philosophy has only continued the work begun by Heraclitus and Aristotle."[6] He was, therefore, aware of belonging to a tradition, a tradition of philosophy which he wished to bring to a final close—this itself being a most philosophic temptation. Philosophy is mind alienated within the alienated world, is intellectual onanism, is a strange activity foreign to sensuous reality; and yet philosophy can, by the real grip consciousness can take on things, by the understanding of historical movement, be an effective guide to the practice that will realize it in abolishing it. It is for the purpose of illuminating practice that revolutionary philosophy wants to be implacably *critical* of all that exists, without fear of the radical results of this criticism and without respect for any existing power. Criticism is indeed not the goal but the means, the means for transcending alienation and reaching integral communism, which does without any philosophy.

The philosophic thinking that prepares its own transcending does not have to construct the future; its task is to put itself at the service of historical becoming, while having a comprehensive overview of it. Mankind has long dreamt of its emancipation. Analytic criticism gives it the awareness of the possibility of actual liberation, and so criticism, far from setting itself up in a static, abstract present, dynamically binds up the past and the future as a participant in the coming of the new event. The past (which, having passed, is passed beyond, transcended [*le passé (dépassé)*]) is indeed to be irretrievably abolished; *yet* revolutionary thought can but put into practice the ideas from the past—by dealienating them, by realizing them, and by abolishing them. In its movement forward mankind always continues what it has begun, and by its critical reassumption of the old world it opens itself up to the new. The action of true philosophic thought is critical and therapeutic. Marx almost sketches out a "psychoanalysis" of the ideas and actions of human history in a letter to Ruge from September 1843: "The world has long dreamed of something of which it only has to become conscious in order to possess it in actuality. It will be evident that there is not a big blank between the past and the future, but rather that it is a matter of *realizing* the thoughts of the past. It will be evident, finally, that mankind does not begin any *new* work but performs its old work consciously."[7] And a few lines farther on he continues: "It is a *confession*, nothing else. To have its sins forgiven, mankind has only to declare them for what they are."[8] By awakening the world from the dream in which it was immersed, by explaining to it its thoughts and acts, one lifts the weight of its sins from it, on the condition, to be sure, that this explanation and this awakening lead mankind to the abolition of alienation.

Revolutionary thought, then, revolutionizes former thought, that is, destroys it, and makes metaphysics become historical (i.e., naturalistic, humanistic, and socialistic) physics; for the "humanity of nature" and "nature generated by history" will shine forth in the light of man's works in communist society. Revolutionary thought wants to be radical, empirical *analysis* of the existing world; it wants to be naturalistic, humanistic, and socialistic *criticism*; it wants to be positive thought; it wants to *contribute to action as the agency of realization*. Its premises are material and real, sensuous and practical, empirical and universally concrete; its premises flow from the observation and confirmation of that which is. This is the way, at least, that Marx's revolutionary thought interprets itself; but its true, deeper presuppositions do not reveal themselves to his eyes. Marx's thought does not have enough acuity to see that it is an extension of Greek metaphysics, Judeo-Christian theology, and modern philosophy. To begin with, it comes after the "dissociation" of *logos* and *physis*, of *idea* and *phenomenon*, of *theory* and *praxis*, of *social law* and *natural law*, all "dissociations" that hardly suffice to clarify the problem of *technē*. In the second place, Marx's thought comes after the positing of an absolutely first creative Act, the divine Act of human creation, and after the revealed history of *God made man* and *dying* in order to lead man *toward the final Redemption of his sins*—Redemption being a repetition of creation and signifying the *total victory of Spirit over Nature*. Marx's thought, finally, comes after the discovery of the *ego cogito*, that is, of the *res cogitans*, which, operating upon the *res extensa*, prepares the limitless unfolding of man's *will-to-power*, *ratio*, and *consciousness*, all of which transmutes into *science*, *technique*, and *productive action*. As we have already indicated, Marx's thought is above all an extension of this third era of Western thought, the institution of Subject.

The absolutely productive subject, in the form of the subjects who devote themselves to total social praxis, makes up the (metaphysical) foundation of Marxism and of the planetary technique in the course of which all that is comes to appear more and more as a product, as the result of a *production*, the world itself being simply that which is disclosed through *human activity*. Marx appeals at the same time to the movement of reality, that is, the basis of thought, and to effective thought, that is, the agency that orders reality. Without asking himself what it is that one calls reality, his thinking is the organic continuation of bourgeois thought, even though he negates it, that is, even though he *wishes* to negate it; and he is caught in the grip of Hegelianism and nineteenth-century materialism. Without taking too great cognizance of what is implied by consciousness, without knowing very well what to do with it, Marx in his thinking brings a movement to its completion and remains affected by and infected with something that he radically rejects. Do not Marx's whole thinking and his communistic perspective remain afflicted by the principles and realities of bourgeois society? Does not the shadow of private property and the possession of material goods lie upon communism, which, in the end, does not know whether it is or is not identical with the abolition of private property? The movement whose foundation Marx is laying passes from the individual to the collectivity, but it remains based on the same fundamental elements upon which the modern world rests, that is, upon which it totters.

This new practical seizure of totality that is to be accomplished, for the sake of abolishing the untrue truth of philosophy through realizing it in material, productive action, is it truly a total grasp of all that is? Or is it the grasp of one of the stages of the total pyramid of being, one *aspect*, to be sure, of all that is, but not the one, integral physiognomy? Like a cord running through all reality without embracing the being-in-becoming of the totality in all its fullness, the truth of Marx's thought, though professing itself free of any presuppositions, nonetheless sets itself out in the world of dualism. The one, single ground, not something made over into a unity or voluntaristically united, is still cruelly lacking. Marx empties the world of all meaning that does not allow of reduction to sensuous activity and then gives brilliant expression to this world empty of meaning, a world where meaning, all meaning, is absent, totality broken, and unity dismembered.

There is deemed to be an interaction between structure and superstructure in its various forms: between reality and thought, between sensuous, material movement and conscious, spiritual movement, between transformative practice and the understanding of this practice. Marx professes to transcend the opposition that sets spiritualism and idealism against materialism and realism. "The spiritualism vanishes with its opposite materialism," as Marx puts it in one place.[9] The material and the real, the sensuous and the active, the practical and the empirical, however, do not lose a position of privilege, at the expense of dialectical interaction and the claim of unity. From this it results that, once sacrificed, the spiritual and the ideal, mentally grasped meaning (*la signification*) and thought, theory and consciousness as transcended, all *reappear* and hang suspended in mid-air.

Physical life-needs and the practical movement of economy leading dealienated technique to the complete satisfaction of needs comprise the circle within which Marx's thought, the final terms of his analysis, and the aims of his perspective all move. The true connection between men—as well as between men and all that is—is *materialistic*.[10] The hero of the Marxist adventure is, therefore, not "the total man" but "the *practical* materialist, i.e., the *communist*."[11] Does not materialism, therefore, vanish together with its opposite, spiritualism?

All transcendent meaning is to be abolished, since it constitutes absurdity and nonsense. Its transcendence is a transgression against and a betrayal of sensible reality. The abolition of philosophy is consequently necessary, and its *Aufhebung* goes together with the true, real realization of true, real reality. What is truth and what is reality? Does not Marx, in his fight against alienating fetishism, reification, and "thingification" (*chosification*), succumb to the very thing against which he is struggling? Are not the (objective) subjects who produce objects (whose destiny is a relationship to subjects) for him *the* true, real heroes of historical becoming? But if this world of subjects-objects is a world transcended by a thinking and a consciousness—a communistic consciousness—is it surprising that consciousness, whether denigrated or lauded, should be found to have no foundation? Does not the dualism between real being (*Sein*) and conscious being (*Bewusstsein*) persist, together with the opposition of subject and object?

Human history, which is the extension of natural history, which is nature becoming history (*le devenir-histoire de la nature*), *brings* nature *to be* for man; for nature as preexistent to human history is empty of any interest, and even natural beings such as "sheep or dogs in their present form certainly, but *malgré eux*, are products of an historical process."[12] The foundation, therefore, of Marx's thinking and of communism does not lie in a grasp of the very being of Nature. Its *naturalism* consists in the naturalistic vision of materially natural needs, as the motor agency in the conversion of nature into history (*du devenir-histoire de la nature*) through the work of men. Where, therefore, must we look for the foundation of the being-in-becoming of the totality of all that is through becoming? In Man? Man is doubtlessly the center of *humanistic* naturalism; but who is this man, this natural, social being who both displays a sensuous, real activity and is endowed with consciousness? Is Marxian humanism sufficiently grounded? Marx tells us that man is "the subjectivity of *objective* essential power, whose action, therefore, must also be something *objective*."[13] Does the subjective-objective being of man, grasped in an anthropological, humanistic way and taken as the "reality" toward which everything converges and to which it reduces, simply do without any grounding? Marx indeed has also written that "man's *feelings*, passions, etc., are not merely anthropological phenomena . . . but truly *ontological* affirmations of being (of nature) and . . . are only really affirmed because their *object* exists for them as a sensual object . . ."[14] Yet ontological movement, the logos of the becoming of being, is consistently ignored in favor of historical human process, in favor of being here taken as that (nature) which *is* only owing to the action of men. And man's economic, historical activity—that is, the action of the subjectivity of essential objective forces—has only to do with objects, the objects of some kind of production whose destiny is to be consumed by subjects. Though he wants to transcend at once the isolated subject and the alienated object, Marx hardly gets beyond the conception of man that makes him Subject and opposes to him an object-world in such a way that man becomes himself an object (though a subjective object).

Whether they be created, fashioned, produced, or made, whether they be consumed, destroyed, or abolished, objects continue to hold supremacy. For, even "wherever the sensuous affirmation is the direct annulment of the object in its independent form (as in eating, drinking, working up of the object, etc.), this is the affirmation of the object."[15] The ontological being[16] of passion, that is, of human action, is itself moved by man's natural needs and does not go much farther than the fashioning of objects and sensuous enjoyment. All segments of the totality become materials for labor.

Philosophy, abolishing itself to realize itself, will thus come to realization in the world of humanistic naturalism, in communist society, whose apotheosis begins the reign of the objective subject who produces and consumes objects; and here all that is will have become practical and material. Capitalist industry has created the technical, material conditions that, once private property is abolished, permit the appropriation of the totality of being by men. "Only through developed industry—i.e., through the medium of private property—does the ontological essence of human passion come into being, in its totality as in its humanity."[17] The transcending of this mediation will make possible the total appropriation of "ontological being" (*l'"être ontologique"*).[18] Yet the question arises: Does the socialist, communistic community continue to lie under the sign of (bourgeois) technique and property? The problem of the positive, radical abolition of private property, the source of alienations and, through its abolition, the fundamental means for dealienation, remains terribly embarrassing. Without calling into question just the statement that all alienations come down, reductively, to the empirical existence of private property (so that all alienations are eliminated through its being abolished), we ask rather whether the reifying power of property does not affect the world of fully realized philosophy, that is, the world of fully achieved communism, even across the abolition of its bourgeois, capitalist form? Did not Marx write: "The meaning of private property—apart from its alienation—is the *existence of essential objects* for man, both as objects of gratification and as objects of activity"?[19] Despite his fight against any conception of communism that makes it the reign of generalized private property, does he not, when all is said and done, remain in the dimension of property? Aiming for expropriation of the expropriators as it does, does the Marxian project get beyond the horizon of appropriation?

Though it takes itself to be a philosophic thinking that already transcends philosophy, Marx's thought is not very attentive to the problems it contains and to its wavering in regard to certain crucial points. It does not wonder about itself but questions and puts into question what it calls the real in order then to give sharply assertive answers. Little concerned with its own path of movement, it gives no more time to the matter either of its own deeper presuppositions or of its own final consequences. It is a thinking that wishes to get on quickly, with little appreciation for questions that can only remain open ones.

With a kind of naïveté, Marx wishes to begin from *sensuous activity* alone, as if there were a sensuous activity and a materiality that, *as such*, would precede all thought. Natural, social, human sensuousness and materiality are, in the eyes of Marxian thought, at the basis of every true scientific mode of apprehension: the physical, the biological, and the historical. Will philosophy, in seeking to transcend itself, become scientific knowledge? Will the action by which philosophy becomes science (*ce devenir-science de la philosophie*) coincide with its abolition in and through the humanistic naturalism of communism? We shall leave this problem without an answer for the moment and try to reach the essence of what Marx means by true dealienated science.

The practical, technical, and historical relationship of man with nature constitutes the basis of science. Since nature exists only by and for man, there is not "autonomous" movement, dialectical or other, for nature or matter. Man is a natural being and nature becomes "human" through historical becoming, which alone is "dialectical." The *relationship*, therefore, that unites and opposes man and nature is dialectical. The separation of the sciences into *sciences of nature* and *sciences of man and his history* loses its meaning. The science of nature is the terrain for the development of human science, while the latter encompasses the former. Technique is and remains the sovereign basis for science. "Industry is the *actual*, historical relationship of nature, and therefore of natural science, to man. If, therefore, industry is conceived as the *exoteric* revelation of man's *essential powers*, we also gain an understanding of the *human* essence of nature or the *natural* essence of man. In consequence, natural science will lose its abstractly material—or rather, its idealistic—tendency, and will become the basis of *human* science, as it has already become the basis of actual human life, albeit in an ⟨alienated⟩ form."[20] Since there is not one basis for the sciences and another for life, in that science is itself a real part of the history of nature, and since the nature of Nature lies in sensuous activity, while sensuous activity at the same time constitutes the true ontological nature of man—that is, his anthropological, historical nature—then is there not necessarily but *one* Science? Science in fact takes its starting-point in nature, but to do that means to begin with human history, because "man is the immediate object of natural science; for immediate, *sensuous nature* for man is, immediately, human sensuousness . . ."[21] Correlatively, "*nature* is the immediate object of the *science of man*: the first object of man—man—is nature, sensuousness; and the particular sensuous human essential powers can only find their self-understanding in the science of the natural world in general, since they can find their objective realization in *natural* objects only."[22] The scientific knowledge that man can acquire encompasses the reflective awareness of his intimate relationship with nature, the one science, that is, the science of the action by man through which nature becomes history for man, and his own self-consciousness. Thus, after alienation is transcended (through the complete development of *technique and consciousness*), there will be only one Knowing, with its basis in practice. "Natural science will in time incorporate into itself the science of man, just as the science of man will incorporate into itself natural science: there will be one science."[23] It will be possible for this to take place in the kingdom of reconciliation, since it is only alienation that separates what was and is identical. "The *social* reality of nature, and *human* natural science, or the *natural science about man*, are identical terms."[24] The movement of natural-and-historical becoming, becomes visible and transparent beginning with social man and for him. As for socialist man, when he will have transcended all alienation and been reconciled with his inseparably natural, human, and historical nature, that is, simply put, with nature, he will have gained reflective awareness of global movement and will possess a science as solid and fluid as becoming itself.

Communistic reconciliation actually and scientifically transcends the opposition of nature and history, on the one hand, and of the individual and society, on the other. Natural being and the perspective of the sciences of nature, historical being and the perspective of the sciences of history, and human being and the perspective of the sciences of man merge into a universal, single, and common becoming. The resulting being-in-becoming (which is also being as having-become), this natural-historical-human being-becoming (which is also being as coming-to-be), is grasped only by a unitary, global, and holistic science. The natural dimension, the anthropological dimension, and the historico-social dimension no longer stand opposed to each other, not in the interest of some kind of accommodation but in order to form a really indissoluble unity, something that can be done, since nature, man, and community *are* inseparable from one another. There will be but one universal unity, with no drama of subjectivity and no problem of a psychological order interjecting itself askew into the historical process. Marx has no desire to give consideration and recognition to existential dramas and anthropological questions in all their glorious and wretched particulars. He wants to bring together the human totality and the totality of all that is as it is within becoming. The human totality should not be separable from Totality, neither in reality nor in thought. And so he declares: "Man, much as he may therefore be a *particular* individual (and it is precisely his particularity which makes him an individual, and a real *individual* social being), is just as much the totality—the ideal totality—the subjective existence of thought and experienced society for itself."[25] Human social totality will coincide with socially natural totality.

The unitary, dealienated science that Marx advocates is not considered specifically naturalistic, specifically anthropologistic, or specifically sociologistic, although biologism (the biologism of natural needs), historicism, and sociologism do not seem to be truly transcended. We find consistently that the founder of Marxism never lets up proclaiming that the full reality of nature is social and the true reality of historical man is natural, that the science of nature is obliged to become human and the science of man natural, and that all this knowledge is to unfold within one single science based on practice and technique.

But how will reunification come about between nature and history, history and man, being and thought, real being and conscious being, once they are severed, and what is the one basis upon which total science could be developed? Will the unitary ground be found to transcend all duality, all splitting up and substitution? "*Sense-perception* (see Feuerbach) must be the basis of all science. Only when it proceeds from sense-perception in the two-fold form both of *sensuous* consciousness and of *sensuous* need . . . is it *true* science."[26] Does not this *twofold form* shatter the foundation put forward as one, single, unitary, and universal? Alongside the natural movement of sensuous need proceeds consciousness, something *different* from material, sensuous reality, even though itself characterized as sensuous. Consciousness is highly problematic and hardly permits total absorption by sensuous, material, actual, practical reality. It is and remains different and even comes to soar in transcending survey over empirical movement. Marx once more, as always, falls back on the consciousness of man the (objective) subject; that is the site of the drama, almost more truly so than the material space within which human activity is loosed. He again and again comes upon it because it—consciousness—exists from the very outset, and from the very outset it is *separate* from what is not it. And the triumphant science of communism seems unable to absorb consciousness.

The consciousness that grasps the alienating character of alienation interprets that which is with a view to its radical and revolutionary transformation, functions as the light for the fight for liberation, accompanies the communist movement, and finally moves beyond it. It does not resolve altogether into real history, material practice, or the science of history-nature. The history of mankind's becoming is surely not essentially a matter of the process of the development of consciousness. Yet it would not be historical becoming if it were not illumined by consciousness. In the eyes of communistic thought, "the entire movement of history is, therefore, both its *actual* act of genesis (the birth act of its empirical existence) and also for its thinking consciousness the *comprehended and known* process of its *becoming*."[27] Thinking consciousness and theoretical thought do not in fact come to lose their power for the sake of what is believed to underlie them as their basis. Every deep theoretical problem, Marx remarks ironically, may well come down to being some kind of practical activity, but it remains no less true that theory illumines and guides practice.

In the course of historical becoming, thought (philosophic thought), knowledge (theoretical knowledge), and consciousness (thinking consciousness) *must* become science; they are obliged to become science because they will have become historical. History will triumph both as the universal historical becoming of mankind and as the scientific grasp of this becoming. The one science, at once science of nature and science of man, will be finally and essentially *historical*. Real history embraces at the same time the history of how nature becomes man (*du devenir-homme de la nature*) and the history of human society (for these "two" histories condition each other); and so the one, global, total science can only be a science of the historical becoming of all that exists, a science that can never stop, that can hardly acquire a static rigidity, that instead possesses a rhythm in conformity with and adequate to the rhythm of actual becoming. "We know only a single science, the science of history. One can look at history from two sides and divide it into the history of nature and the history of men. The two sides are, however, inseparable; the history of nature and the history of men ⟨mutually condition⟩ each other so long as men exist."[28] Never taking leave of the earth and real terrain, the science of history will truly explain the real genesis of all natural and human phenomena by the historical action of men, through which all that is, is manifested. The space of this new science, as well as its time, will be historical, since space and time constitute the place and rhythm of historical becoming, they themselves having existence by this becoming. This dealienated science will therefore explain the development of all man's practical and theoretical activities by material, reality-making praxis; it will be one with praxis and serve, by turns, as its master and servant.

The path taken by philosophy to bring about its realization in its abolition is consequently also the path of movement toward science. In ceasing to be abstract truth, that is, to be *errance*, philosophy becomes world, which means that it becomes the science of a world based on technique. In becoming the planetary science for a planetary world, does it thereby cease to wander in error? Marx seems to think it does. Real, positive, unitary, fundamental, essentially historical science thus takes the place of philosophy and constitutes the truthful grasp of the historical process of practical development for men struggling to take in nature by annexation. Philosophic thought and knowledge, theory, and speculation are abolished and realized by changing into real, scientific knowledge. All that is will be finally grasped by science, to be transformed by technique, since science itself is only a theoretical technique. All that is will be finally brought back to "human practice and the comprehension of this practice," as the eighth thesis on Feuerbach requires.[29] For what is at issue here is not transcending (abstract) thought in order to make it a little more contemplative but abolishing it by making it real in human practical, sensuous activity. Marx praises Feuerbach for wanting to seat science on and in the sensuous, but he reproaches him for not conceiving sensuous materiality as a practical sensuous activity. The fifth thesis on Feuerbach makes this cardinal point clear: "Feuerbach, not satisfied with *abstract thinking*, wants *contemplation*; but he does not conceive sensuousness as *practical*, human-sensuous activity."[30]

Will the science that succeeds philosophy be but the faithful reflection of practice, with no grounding task of its own? Will human praxis *and* the understanding of this praxis totally coincide? Marx's philosophic theses are so laconic, and too often so summary, that they do not develop either the question or the horizon for the answer. Even so, despite all his effort to derive theoretical science from material practice and thinking consciousness from sensuous reality, thus making all that is come down to man's industry and activity, the problem of the role and fate, the importance and repercussions of these second-rank powers continues to be obscure. And this obscurity is due primarily to the fact that the derived powers seem sometimes to be quite a bit more determinative and fundamental than would first appear. Still, it seems that inside the world of communism the practical, the real, the material, the sensuous, the active, the powerful, the actual, and the realizational continue to hold the first place. Because of its program of bringing down, not to say reducing, the "higher" to the "lower" and making the latter the "higher" thing (which does not prevent the dethroned power from returning again or from being present in its absence), the Marxian perspective remains embarrassing; no one knows quite how the causal link between cause and effects, between the sensuous base and light-filled, or at least light-shedding, conceptual meanings (*significations*) is to be made in a precise way. Is the only causality a one-way causality, and, if not, whence does a reciprocal causality get its moving force? Where Marx no longer sees clearly, he appeals to consciousness, but consciousness is not itself preeminently clear. Marx, whether he likes it or not, continues a whole era of modern European thought as a faithful, though rebellious, child of this phase of becoming; in this way he is a prisoner, and *doubly* so: he is prisoner to practice *and* to consciousness, that is, to their *duality*.

In being launched upon this course that would lead to a simultaneous, joint abolition and realization, will thought (philosophic thought) have no more bearing upon the understanding of human practice, which will all the same demand reflective grasp—reflective grasp that perhaps transcends science? Will thinking philosophy disappear totally? Philosophy as formerly done—autonomous philosophy—will come to an end; it will be replaced by technical science and scientific technique in pursuit of their own ends. But will transcended philosophy vanish without a trace? Will it be integrally absorbed by science, will it actually become Science; or will it still subsist by changing itself into a science a little different from the science of history, one and global as it will still be? Will (philosophic) thinking become in some way the science of sciences, the consciousness of science, or will it constitute a kind of general method for scientific procedure? In the world of dealienated subject and objects, can philosophic thought still be an affair of subjects at grips with objects? Marx does not have clear vision here, yet he tries to cut the Gordian knot. "Where speculation ends—in real life—there real, positive science begins: the representation of the practical activity, of the practical process of development of men. Empty talk about consciousness ceases, and real knowledge has to take its place. When reality is depicted, philosophy as an independent branch of knowledge loses its medium of existence."[31]

What will follow upon the death of philosophy? Marx's answer is at once brutal and unsure, and remains problematic, for the text we just quoted continues in the following words: "At the best its place can only be taken by a summing-up of the most general results, abstractions which arise from the observation of the historical development of men. Viewed apart from real history, these abstractions have in themselves no value whatsoever. They can only serve to facilitate the arrangement of historical material, to indicate the sequence of its separate strata. But they by no means afford a recipe or schema, as does philosophy, for neatly trimming the epochs of history."[32] Former speculative philosophy, philosophy as philosophy, autonomous philosophy, the philosophic movement of abstract thought, will no longer be the mediation; it will lose its medium of existence by abolishing itself in its realization in practice and technique, which in turn will themselves need only real knowledge, positive science, the concrete exposition of the development of the sensuous activity of men. Philosophy can in no way locate itself above science. *At best*, Marx adds, there will still be a place for "a summing-up of the most general results, abstractions which arise from the observation of the historical development of men." This will replace the philosophic syntheses that are now at an end. The theoretical summing-up of the most general results, in the form of abstractions from the observation of the historical development of men, will still, of course, be tolerably abstract, but it will have no value if separated from real history, from practical activity.

Let us state it openly: in being realized, philosophy is *abolished*; it reaches its end, that is, its goal and final achievement. It ceases to be. It will be replaced by the real knowledge of positive science, which will still include, at best, a summing-up of the most general results. It is consequently nonsense to talk of a communist philosophy. *Nevertheless*, thinking consciousness will always be at work, without being utterly used up in the practical works of men in action; for conscious thought seems unable to keep from soaring in survey over real movement, and it is resistant as well to being totally absorbed by science.

The men of the era of transcended alienation will be rid of the presence of any transcendence. Does this mean that the power of absence will no longer be manifest there? Such an idea does not enter Marx's mind. He sees men dealienated. These men are going to produce and reproduce, fashion objects and nonspeculative theories, and fully enjoy what they produce; they will get a comprehensive hold on their practice and development in total, historical science. With all alienation annihilated, they will find their being, that being which grounds their history "independently of the existence of any political or religious nonsense which would especially hold men together."[33] No meaning and no problematic of a metaphysical kind will come to disturb their life and put it into question, since, by the very essence of things, "mankind thus inevitably sets itself only such tasks as it is able to solve, since closer examination will always show that the problem itself arises only when the material conditions for its solution are already present or at least in the course of formation."[34] The great questions, those which man asks at moments of great importance, those which confront man even when he has no answers, will have become dumb, better still, will cease to demand of him some reckoning. To every *why?* will correspond a *because*, all on a practical level; question and answer will no longer have any metaphysical character. The question will no longer be to interpret the world (philosophically), but continuously to transform it (i.e., on a practical level).

Modern philosophy has, to be sure, pursued in its own way the task already begun by Heraclitus and Aristotle, as Marx was careful to point out. Having run its course, it finally reaches fulfillment in abolition by becoming reality and transmuting into Technique. Heraclitus tried to search the *logos* of the being-becoming of the totality of the world, to inquire into the rhythm and sense of the One-All. As a thinker-poet, he grasped the harmony and war that were conjoined at the very heart of the unity of totality; here all that is was seen as in time, which was "a child playing a game of draughts," and "the kingship [was] in the hands of a child."[35] Heraclitus gave expression to supreme Play through his enigmatic language, which, echoing the words of the Sibyl and of the oracle of the Delphic god, "neither speaks nor conceals, but indicates."[36] Aristotle, the Alexander of Macedon of Greek philosophy, as Marx puts it, once more poses the supreme problem. If we take the trouble to do so, we can read these words in his *Metaphysics*: "And indeed the question which was raised of old and is raised now and always, and is always the subject of doubt, viz., what being ($\tau\iota\ \tau\grave{o}\ \check{o}\nu$) is, is just the question, what is ⟨beingness⟩ ($\dot{\eta}\ o\dot{v}\sigma\iota\alpha$)?"[37] After the

287 The Abolishing of Philosophy

realization-annihilation of philosophic thought, this metaphysical thinking, this metaphysical question, and this metaphysical language will no longer have any meaning. These problems will no longer arise. Man will have become the producer of his being and the producer of all that is. The being-in-becoming of the totality of the world will no longer be a problem. Productive activity, positive action, and labor will become what they are, namely, the ground of the being of beings. For "this activity, this unceasing sensuous labor and creation, this production, [are] the basis of the whole sensuous world as it now exists . . ."[38] The world will no longer be divided into a sensuous (physical) world and a suprasensuous (metaphysical) world, and meaning or sense will no longer be separated from the senses. Sensuous activity will constitute the *raison d'être*, but not the metaphysical meaning, of the sensuous world, which will have liquidated the question of *the* meaning as something to be spiritually conceived (*la signification*). The world will not have a (transcendent) sense any more than it will be senseless (absurd). The point will be to "conceive the sensuous world as the total living sensuous *activity* of the individuals composing it."[39] Can, therefore, philosophy have anything to do in this one-moded world? Will man, as he who undertakes, by means of technique, the conquest of the *natural, historical, human* world, have need of philosophic thought?[40]

Religion, art, poetry, and philosophy will, consequently, no longer have a place in the world of total reconciliation. Material production—material but without reification—will be the creative power and will provide the key to the "secret" of all that is. All this appears highly consistent. And yet we have already heard Marx tell us that the men of the communistic world, as national and local history gradually is completely transformed into world history, will be "brought into practical connection with the material (as well as intellectual) production of the whole world and be put in a position to acquire the capacity to enjoy this all-sided production of the whole earth (the creations of man)."[41] What kind of intellectual production will, then, come to be added over and above material production or combined with it, after the annihilation (as advocated) of the very existence of the opposition between the material and the spiritual? Is it a matter of the spiritual production of times past, that is yet deemed to be radically transcended? Is it a question of a new kind of *spiritual production*? After so radical a condemnation of the spiritual—by means of the weapons of consciousness, it is true—after the proclamation of the unification of the world in the dimension of the sensuous, indeed, after the full realization of philosophy through its annihilation, after the absorption of thought by real science, will there still be something *more*? Will the power of transgressing all that is real, the power of transcendence, persist in showing itself? Will the force of metaphysics impose intellectual realities right in this sensuous world? Will what was to have been annihilated launch an appeal, the appeal of absence? Marx does not quite close the circle, his mind free of uneasiness. It all goes on as if he suspected, at times, that the production and reproduction of life, the fashioning and using of objects, the limitless production and enjoyment of sensuous riches, the universalization of practical life, and the technical seizure of all that is do not in any way suffice as grounding principle for the world. In a fully achieved communistic world the spiritual ought not exist; *all* is to have become *one* and differences are to have been resolved by and in unity. But the continual transcending of all that is, is always at work, and the questions to which mankind can give no answer lie waiting for mortals, whether they wish it or not.

CONCLUSION.

OPEN QUESTIONS

One would have to gain an understanding of the path taken by Marx's thought in its unity, grasp the center of the Marxian schematic, and enter into dialog with the sense and rhythm of this movement that leads to the need for a reflective grasp (by thought) of real process (the process of actual dialectic), that is, of the development of Technique, in order thereby that the motor of historical development might develop fully.

Marx does not begin with some form of the being-in-becoming of the totality of the world, some form of first and ultimate being, nor does he wonder whether the original *archē* is of material or spiritual, natural or divine order. He does not begin with some original matter in movement whose spatiotemporal becoming would lead to man as one who thinks about this process and acts upon it. Nor does he begin with Nature, whose evolution, that is, whose dialectical movement, would lead from matter to life and from life to human consciousness. There is no ontology in Marx, no first philosophy, either spiritualistic or materialistic. Marx rejects precisely all ontology and all metaphysics, though he is unable to divest himself of a kind of implicit "ontology" and succeeds only in rejecting metaphysics in realizing it—"metaphysically?"—in technique. Must one say as well that there is in Marx no philosophy of nature and no cosmology, that he does not talk about the evolution of nature or of the order or becoming of the Universe?

Marx begins with men, with natural, human, social beings who labor to satisfy their life needs and for that purpose set technique in motion. Men are moved by their drives. To satisfy them, they struggle against nature, extracting the goods destined to satisfy their needs, which in turn are renewed and increased in the degree to which they are satisfied. The drives (which are natural, human things), the needs, are the motive agency for the (social) activity of men. The movement of the drives toward their satisfaction meets up with the instinct for production that seems to dwell in man—indeed it does much more than just meet up with it. Man, therefore, produces his living and is himself a product of production. Producing his life thus by productive activity, by labor, by technique, he is always linked with other men. The economic development of productivity, the technological development of productive forces, of the instruments of technique, constitutes the real basis for the development of man and for the structure of his history, and activates social life. In the course of the transformation of nature into history by social labor, other forces as well develop as the expression of the weakness of man and as signs of the insufficient development of technique; these forces are legal and political relationships, the State, religion, art, and philosophy. These forms of idealistic superstructure are the illusory reflection of real development, being powers that organize it by giving it restraining channels and sublimate it by disguising it.

The very production of life, and this basically, is alienated. Men have a strange relationship with their labor, and the product of their labor remains alien to them. Men and things are reified and lose their essence. Men put productive forces into motion, but these forces do not belong to the totality of the members of the community. The regime of private property alienates man from labor; his being and his making are cut off from each other and both crushed by the inhuman power of having. Men have become laborers and workers, but they do not collectively possess the means of production, which, held by a class of individuals, allow the exploitation of socially provided labor. In consequence, man loses his being in gaining his livelihood. Political and ideological alienation, the disguised expression of the conditions of real existence and of real alienation, completes the derealization and reification of men. The second-order existence they are condemned to lead in politics, religion, art, and speculative thought prevents them from properly directing their efforts toward actually realizing, on a real, unitary, and total terrain, the ensemble of their needs and their open totality. On the level of bourgeois, capitalist society, the antagonism between the two classes in conflict—the producers and the proprietors of the means of production—reaches its high point. Only the proletarian revolution, utterly overturning the former state of things and, after its victory, installing the temporary revolutionary dictatorship of the proletariat, can lead mankind to socialism-communism. After the radical abolition of private property and the alienating superstructural powers, socialism-communism, at the peak of its development, will allow each man to have enjoyment according to his needs in a society without classes or State, until communism also is transcended. Man, then, can and should be (re)conciled with both nature itself and with his own nature, with both mankind itself and with his own humanity, now that all has become fully open to man's activity. Communism has the responsibility of realizing this conquest; yet Marx is careful to note that communism is not as such the final goal of human history. It is the figure of the near future, the energetic principle for the transformations that are necessary.

It would be tempting to make Marxian thought more complete in its totality-mindedness (*plus totaliste*). Its point of departure would then be matter in movement, the being-in-becoming of the totality of the world, which, through a unitary and total dialectical process, would become inanimate nature, then life, then finally give birth to man endowed with thought. This social, historical man would develop productive forces and superstructural forms until, in the kingdom of communism, he would (re)gain awareness of the total movement and take it in hand. The circle of things would be thus finally closed by virtue of the dialectic of movement in original, material being, which goes from nature to history in order to become expressed in the dialectical thought of men. This schema, however, is not Marxian, though it is vaguely Marxist. It is largely attributable to Engels; and no Marxist to this day has taken the trouble to reflect on the materialist "ontology" that, under the name of dialectical materialism or materialist dialectic, asserts matter in movement to be primary, with no knowledge of what this dialectic of matter (or of energy) is and how it implies the logos of its becoming.[1]

Marx, however, has nothing to do with this kind of matter. *He* begins and ends with the matter of human labor, the materials with which social life is built, material activity which is producer of this world. By an immense reductive effort, he brings back (and narrows down) to production all that is and comes about—and what *is* here is what is *made*.[2] He thus indeed does get to what is and comes about; for everything is found to be actually reduced to ceaseless and omnipotent production: the production of material goods and cultural goods, of use-value and exchange-value, of ideas that have value and those that are worthless. Marx states the truth of the modern and contemporary world and announces what will take place in the planetary world of tomorrow which will be realized by Marxism through being abolished and transcended.

Are we henceforward to understand Marx better than he understood himself? This question concerns every attempt to understand a still living thought, and it must remain open. To *understand* Marx is still a highly difficult task, for no one knows how to approach this thinking that wants to see thought absorbed by practice and that yet knows thought to transcend real movement.

There is in Marx an extraordinary passion for nothingness, an unrestrained will for transcendence, that no one dared see. Marx's whole thrust concentrates on this point: abolish, annihilate, transcend. The goal is the abolition of all that is, whether in the order of reality or of thought; the annihilation of all that exists, since it exists as alienating and alienated; the transcendence of every obstacle so as to pave the way to the praxis and the technique to come. In the world of reification, the past rules the present, the present is sordid, and the future is blocked. Through the communist movement it is the future that must speak, and communism itself will be abolished, annihilated, and transcended in the process of its realization. When the totality of beings will have been reduced to nothing, since it is henceforward a total nothingness, then alone can there arise a new, open totality, that will itself doubtlessly sink into nothingness. Marx rejects all that is; everything is alienated and alienating and bars the path to the future that can be established only by its annihilation. The Marxian vision is an advanced form of magnificent nihilism, a *planetary nihilism*, and yet, in consequence of this, it grasps *planetary technique* as the one lever that can put the world, this *errant star*, into motion, by abolishing the putrefied world and its "errance."

Nothing resists Marx's negating critique: not production as it has taken place, nor the relations of production, nor political and public life, nor family and private life, nor religion, nor poetry and art, nor philosophy and the sciences, nor the communist movement itself. Man's economic existence and his civil existence, his familial, religious, artistic, and theoretical existence are all alienated, impeding him precisely from really existing as man. Existence is alienated both because the burden of the material weighs upon it and because, to flee the material, it must take refuge in the world of idea and ideology. The world of ideas, however, cannot be a remedy for the absence of world.

Does Marx's thought grasp the truth of what it combats? His

thinking offers itself to being gripped by that which is. The truth of what is, is only a form of *errance*, and Marx wants to expose this errance, without, however, daring to go all the way in grasping errance as such. In his wish to put a final conclusion to alienation, Marx wants to put a stop to an errant becoming.

Placing Marx's thought on several levels makes us grasp its polyvalence, but it creates the risk, despite everything, of disguising its unity. An economic problematic, a political problematic, a political, anthropological, ideological, or "futurist" problematic cannot be dissociated, and yet each does raise specific difficulties.

The Economic Problematic

Marx's economic work is far from being wholly understood. The economic horizon of the Marxian enterprise is not easily seen. The crude economics of the Marxists too often prevents us from understanding the economic thinking of the man who founded Marxism.

The infernal cycle that ties production to consumption and the endless turning of this cycle have to be understood as regulated by the development of technique and by its rotational movement. There is a lot of talk, following Marx, about productive forces and about their development as determining all the rest. One would nevertheless have to understand these forces in depth and see technique as the motor of their development. Even more than the development of productive forces, it is, let us call it, technological development that counts first. The organization of labor and the division of labor are determined by the technological process. "Labour is organized, is divided differently according to the instruments it disposes over. The hand-mill presupposes a different division of labour from the steam-mill."[3] The hand mill presupposes, that is, determines, feudal society, the steam mill capitalist society. "The hand-mill gives you society with the feudal lord; the steam-mill, society with the industrial capitalist."[4] Do not automatic machines go hand in hand with socialist society? Machines, industry, truly technical instruments, and technique properly speaking date from a certain point in history and act thereafter upon all historical and social evolution. Technique is only a modern thing. "*Machinery*, properly so-called, dates from the end of the eighteenth century."[5] A few lines farther on, Marx makes clear what he means by machines: "Simple tools; accumulation of tools; composite tools; setting in motion of a composite tool by a single hand engine, by man; setting in motion of these instruments by natural forces, machines; system of machines having one motor; system of machines having one automatic motor—this is the progress of machinery."[6] "Automation" alone is capable of transcending specialization for the sake of productive, universalized, but dealienated, labor. Marx firmly holds that this is how it will be. Vigorous as he is in his attacks upon the productive techniques of the present, he is just as strong in his praises for the productive techniques of the future. Those of the present imply bad *automatism*, a fragmentary and *atomized* activity; in a reified and *automatic* system, individuals can only be "*atomistic*." On the other hand, the technique of the future will have overcome automatization and atomism thanks to its integral development, permitting men the integral development of their essential forces, of their energy. The grand technique of future times will be the antidote that will cure us of the ills of underdeveloped, and therefore alienating, technique. Marx's hope in technique is unshakable, and it would be naïve to brand it as naïve. "What characterizes the division of labour in the automatic workshop is that labour has there completely lost its specialized character. But the moment every special development stops, the need for universality, the tendency towards an integral development of the individual begins to be felt. The automatic workshop wipes out specialists and craft-idiocy."[7] Proudhon, a man of much less genius and infinitely more scepticism than Marx, who represents for Marx that petty bourgeois mentality that sees everywhere both the "good side" and the "bad side,"[8] made the following note in the margin of his own copy of *The Poverty of Philosophy* next to the lines just quoted: "Fine! And how do you mean this integral development?" He never got an answer to his question, and at any rate the answer could not come simply on paper.

If the secret of the motor forces of history lies in technological development, one would have to be able not to think of productive technique as something derivative. It is technique that determines exchange, trade, and consumption. Industry *even produces needs* in order to have to satisfy them. There is a perpetual renewal, an unbroken rotation of the cycle—is it endless? Needs seek satisfaction, production serves needs but also generates them, consumption creates in turn new needs. What binds production to consumption and needs to satisfaction is more a matter of technique than of the pressure of "natural" needs. "Most often, needs arise directly from production or from a state of affairs based on production. World trade turns almost entirely round the needs, not of individual consumption, but of production."⁹ It lies in the essence of *modern technique* that production be a quasi-absolute creative force. "Production precedes consumption, supply compels demand."¹⁰ Marx placed a great deal of emphasis upon the dynamic element whose energy leads the present toward the future. The production of this future will come not only under the sign of communism but also in conjunction with the rhythm of planetary technique. The production of all kinds of products, "material" and "spiritual," globally regulates all the modalities of exchange and commerce, of distribution and consumption. "In general, the form of exchange of products corresponds to the form of production. Change the latter, and the former will change in consequence. Thus in the history of society we see that the mode of exchanging products is regulated by the mode of producing them."¹¹

Production, distribution, exchange, and consumption form a circuit activated by production. Still, production, material production—which holds first place in Marx's interest—even while determining and encompassing distribution, exchange, and consumption, is itself held in a perpetual movement of consumption; it consumes the means of production, utilizes them, and uses them up, just as it does with the products of production. Production and consumption are both regulated by the technological process, which is the creator and destroyer of every species of production. "The identity of production and consumption amounts to Spinoza's proposition: *Determinatio est negatio*."¹² In the technological process, the subject transcends itself in objectifying itself and in ceasing to be stationary, and the object is continually transcended by consumption and the production of new objects. When reification is overcome, this truth regarding the development of technique only becomes more dazzling. Subjectivism and objectivism will be simultaneously abolished to the enhancement of creative, productive activity.

Aiming as it does to satisfy the total needs of men, technique, the motor of world history, is engaged as well in an ever-new creation of needs that would be satisfied and of objects destined to be consumed (and therefore abolished) by subjects (subjects not as isolated, but as members of the community, as men who have therefore transcended their subjectivity). The greatness of Marx's thinking is to be sought in his grasp of the *movement of technique*, in the vision of absolute productivity, which, abolishing all economic, legal, political, moral, psychological, and ideological enclaves, far transcends the simple demand for the satisfaction of life needs. Understanding *man's self-alienation* as he does, Marx thinks that abolition of this alienation in its economic and technological roots will lead to the *conquest of the world by and for man*. Man is, of course, moved by his natural drives; yet technique, which, more than being the instrument for the satisfaction of needs, is the motor force of history, possesses a far stronger power. Technique commands the rhythm of the drives and sets the world in motion. The demand to abolish the existing state of things—the bourgeois, capitalist state, alienation and reification, private property and the division of labor, background worlds and pseudo-worlds of a celestial order—amounts to this: the total liberation of human, social productivity, so that productive technique, entirely "let go," will be responsible for practically solving all questions and all riddles in its becoming.

This Marxian conception of absolute productivity is not yet understood to its full extent and depth. Marx does not imagine the *physis* and the *cosmos* as an order one could not transgress, as a rhythm that dictates to us what we have to think and do. He does not see the World as a *Divine Creation*. He rejects any idea of Creation and looks to the hypothesis of *generatio aequivoca* for the one legitimate answer to this kind of question. Marx does not see the totality of the world from the perspective of the *modern Subject* to which *objects* correlate as objects of its representation and will. He sheds light upon another era, one no longer cosmocentric and Greek, theocentric and Christian, egocentric and modern: that is, the planetary epoch of a technique and a productivity that transcend every constituting subject and every fixed object and that hold to hardly any limits, natural or divine.

"Production is thus at the same time consumption, and consumption is at the same time production. Each is simultaneously its opposite. But an intermediary movement takes place between the two at the same time. Production leads to consumption, for which it provides the material; consumption without production would have no object. But consumption also leads to production by providing for its projects the subject for whom they are products. The product only attains its final consummation in consumption."[13] The product of production truly exists only from the moment when it ceases to exist, being abolished in and by consumption; its being calls for its annihilation. The product is distinguished from every natural or ideal object by the fact that it is really *produced*, that it only *becomes* an actual product and only asserts itself as a product in being consumed. To be *in reality* and not $\delta\upsilon\nu\alpha\mu\epsilon\iota$ ("potentially"), according to Marx's thinking, it must be absorbed by consumption, thus ceasing to be. "It is only consumption which, by destroying the product, gives it the finishing touch, for the product is a product, not because it is materialised activity, but only in so far as it is an object for the active subject."[14] The active subject in a nonsubjectivist sense is that subject endowed with objective essential forces and producing without stop nonreified objects for use by human subjects.

Marx tries to lay bare the essence of production. This essence, being a fluid phenomenon, does not manifest itself fully in the regime of capitalism, even though it may be the bourgeois, capitalist system that is the first in the history of the world to create the conditions for the prodigious development of technique. The bourgeoisie carried the development of productive forces to a certain point, even while keeping it within the limits of private property and capital. Modern technique remained held in chains. The bourgeoisie set off to conquer the world, but then stopped. "It has been the first to show what man's activity can bring about. It has accomplished wonders far surpassing Egyptian pyramids, Roman aqueducts and Gothic cathedrals; it has conducted expeditions that put in the shade all former Exoduses of nations and crusades."[15] The bourgeoisie has done many things, it has even introduced the reign of making, but its evolution took place only within the world of reification and alienation, and it pushed these inhuman powers to their ultimate limits. Based as it is on private ownership of the means of production, the reign of the bourgeoisie could only hold back the unlimited power of productivity, which by its very essence is capable—if there is no individual appropriation—of satisfying the entirety of human needs.

Now, it is only the proletariat that can lead the whole of society to social and socialistic society, where neither classes nor private ownership will have a place. Only the workers of the whole world can, for the first time in history, which has never yet been universal, realize the actual possibilities, the actual energies of universal, universalized technique. Only a mankind that sets no restraint upon production can forge ahead and, in a permanent revolution, constantly renew production, needs, the activity of objective subjects, and the objects of the social activity of men.

Marx begins from man as impelled by his natural drives, by his physical needs (to which false metaphysical needs come to be superadded, by way of sublimation). This "biologism," this "naturalism," in reality rests upon a technicism. More than by their natural, human drives, their physical and social needs, men are moved by what makes them produce their life, by what constitutes a veritable *producer instinct* transcending all naturalness and aiming for the complete transformation of nature by technique. All that alienates man was and is the result either of the nondevelopment of productive forces, as in ancient times, or the underdevelopment of technique, which in any case means the development of technique is limited by particular native circumstances that check its flight toward universality (as in the bourgeois, capitalist regime). Mythology, religion, politics, and metaphysics belong to the superstructure, for they are an idealization and a sublimation of the impotence of man and reflect on an ideological level the power of some men over others. They constitute a hindrance to the manifestation of the power of the human totality and the totality of men.

Marx goes so far as to talk of a drive, an *instinct for production* (*Trieb der Produktion*). What he says about it is to be taken literally, because it is consistent with the spirit of his undertaking. In his discussion of the circular movement, Marx links together into a single whole (*a*) the needs that drive men to production, which, through consumption, aims at the satisfaction of needs, and (*b*) the ever-recurring production of new needs and the propagation of need in and by consumption, so that production produces consumption and consumption produces production by creating the need for new production; in the context of this discussion Marx says this: "Consumption creates the impulse to produce (*Trieb der Produktion*), and also creates the object which acts as the determining purpose of the production."[16] It would be vain, then, to separate the terms and then ask the question, Does Marx's man begin from need or from production? if one keeps the terms separate. Man is set in motion by the instinct to produce, and the process of production-consumption-production creates and propagates need while aiming for its satisfaction, at the same time that it continually produces new objects. The cycle that binds production to consumption encircles the entire globe. It cannot be otherwise, and what alone is needed is that neither subject nor object be reified in this activity of making-to-be. Production, the creator of subject (man) and objects, must free itself from all that shackles it and makes it stiff and rigid, from all that holds it back and immobilizes it, from all that weighs it down with the burden of the past or suffocates it under an arrested present. Every "in-itself" will become thus a "for-us"; every given (*toute donnée*) will become a thing made (*un fait*), made by us and for us in an unlimited activity (and an infinite one?).

Marx knows perfectly well that "the starting point is of course the naturally determined factors; both subjective and objective. Tribes, races, etc."[17] He knows that productive action takes its point of departure in the natural givens that exist antecedent to it. Yet he does not consider these natural givens as the most essential element; the modification that technique imposes upon them is, in his eyes, much more important. For even these natural givens begin to exist only in and by historical action. In *The German Ideology*, this theme is touched upon in the following words: "Of course, we cannot here go either into the actual physical nature of man, or into the natural conditions in which man finds himself—geological, orohydrographical, climatic and so on. The writing of history must always set out from these natural bases and their modification in the course of history through the action of men."[18] What produces history, that is, the natural, human, and social life of man, is the production of goods and the production of want. Production "produces the object of consumption, the mode of consumption and the urge to consume (*Trieb der Konsumtion*)."[19] The identity between production and consumption, between "consumptive production" and "productive consumption," is not just formal; they are bound together by movement and this movement is productive. The productive subject objectivates and transcends himself in production, and the object is subjectivized in productive consumption. All the same, in order for this identity of production and consumption to appear in the full light of day, in order for their reciprocity and interdependence to become visible (for these relations are what make the two not just mutually complementary but mutually realizational and creative), one has to have a view of the totality of the process and not see the identity as a formal one. Identity is not undifferentiated and empty, whether taken in the economic order or in the logical. It contains the mutual action its constituent elements exercise upon each other. A vulgar or academic political economy and abstract and scholastic Hegelianism both flatten out this organic, dynamic unity. To make it transparent, one must understand that it is the expression and the framework of the relationship that bind individuals together. The basic social relation is a relationship of exploitation and can become a community relationship if the totality of production and products, of distribution and exchange, and of consumption and the reproduction of needs and goods becomes the charge of, and is set in movement by, the totality of human society. Marx makes his ideas about this identity clear in this way: "The conclusion which follows from this is, not that production, distribution, exchange and consumption are identical, but that they are links of a single ⟨totality⟩ (*Totalität*), different aspects of one unit. Production is the decisive phase, both with regard to the contradictory aspects of production and with regard to other phases. The process always starts afresh with production."[20] In every organic whole and in every reciprocal action, is there not always one dominant and determining power? "A distinct mode of production thus determines the specific mode of consumption, distribution, exchange and the *specific relations of these different phases to one another.*"[21] Understood in its totality, not in its one-sided, particular, and therefore specifically determined form, socialist production will be the determining power for socialist society on the move and will constitute its firmest footing.

Essentially, then, it is *modern production*, leading history toward its universalization and society toward its socialization, that captures Marx's whole attention and impassioned science. All the discussions about the universal import or lack of it in Marx's thought should be based on a keen understanding of the historical dimension of modern technique. World history has not existed from all time (even considering all historic time), and the world has never yet been world, that is, has not yet *become* through the activity of men. Marx is categorical on this score: "World history did not always exist; history as world history is a result."[22] Technique is precisely that power that has led to this result; it is technique that is entrusted with the task of radically universalizing history, which is the phenomenon of becoming as having-become, for in this it is a product. The bourgeoisie has set up the world market. By the abolition of the capitalistic laws of this market, by the abolition of that which turns beings and things into commodities, human history becomes fully world history and the world is transformed into a human world. The bourgeoisie, having forged the weapons for those who will put it to death, having produced the men who will be its undertakers, finds in modern workers both its negation and its accomplishing. The bourgeoisie inaugurates modern times, and the workers bring them to completion.

Not enough attention is given to what Marx thinks of class struggle as characterizing the last few centuries of history. The *Communist Manifesto*, nevertheless, rather explicitly states that class struggle characterizes *above all* recent centuries; it characterizes *principally* the last centuries of our era, the centuries of the emergence of modern technique—the very period that Marx chiefly attacks. The intensity and extent of class conflict coincide with modern technique, with the bourgeois, capitalist era. In the passage from the *Manifesto* that follows, because we find the word *centuries* used, must we not precisely understand *centuries* and not *millennial ages*? "But whatever form they [i.e., class antagonisms] may have taken, one fact is common to all past ages [*Jahrhunderten*—centuries], viz., the exploitation of one part of society by the other."[23]

Marx's schema essentially applies to the phase of history that is tending to become universal and world-wide. It is to this epoch of becoming for humanity that his language is giving voice. Does this mean that he is not concerned with or does not want to shed light at the same time upon the Oriental empires and the adventures of Asian peoples, the Greek city-states, the Roman Republic and the Roman Empire, as well as the fate of medieval and feudal forms of development? There is no doubt that Marx wants to include the whole of history. Upon the publication of his *Contribution to the Critique of Political Economy*, a number of objections were raised against him. Recall the brilliantly formulated thesis from the preface, according to which it is the specific mode of production—the mode of production of material life and the social relations that flow from it—that, as the basic economic structure, determines the legal and political superstructure and the forms of social consciousness that correspond to that specific condition. One critic, while admitting the truth of the thesis for the modern world, where material interests predominate, did not recognize its validity for the Middle Ages, when the Catholic religion held sway, nor for Athens and Rome, where politics was king. Replying to the objection, Marx wrote in *Capital*:

> In the first place it strikes one as an odd thing for any one to suppose that these well-worn phrases about the middle ages and the ancient world are unknown to anyone else. This much, however, is clear, that the middle ages could not live on Catholicism, nor the ancient world on politics. On the contrary, it is the mode in which they gained a livelihood that explains why here politics, and there Catholicism, played the chief part. For the rest, it requires but a slight acquaintance with the history of the Roman republic, for example, to be aware that its secret history is the history of its landed property. On the other hand, Don Quixote long ago paid the penalty for wrongly imagining that knight errantry was compatible with all economic forms of society.[24]

Marx in no way renounces the total validity of his schema or the universality of the method of historical materialism. He is not for all that, however, an orthodox Marxist. He does not lack at times a certain awareness, a certain irony. He recognizes that his vision and his schema, his method and categories are only fully valid, in the first place, for the historical conditions from which they draw their sustenance, in whose womb they develop. Marx acknowledges: "The example of labour strikingly demonstrates how even the most abstract categories, despite their validity in all epochs—precisely because they are abstractions—are equally a product of historical conditions even in the specific form of abstractions, and they retain their full validity only for and within the framework of these conditions."[25] The schema of historical materialism and materialist dialectic, though valid for the Oriental and Asian empires, the Greek and Roman cities and republics, and the medieval states, is nonetheless *fully valid* only for and within the economic, social, and historical conditions of which it is the product. The modern world, Western in origin but becoming universal, the bourgeois, capitalist regime, constituted the most favorable terrain for the development of the productive forces of technique. The bourgeoisie has not only produced the modern world, it has also engendered its negators; it has prepared the way for those who reject it. Its negators will be the protagonists of the socialist world, of the universe of universalized technique free of all particular appropriation, the universe of the human community that will have abolished all exploitation of man by man. The bourgeoisie has, in fact, prepared the future world. "The bourgeoisie has subjected the country to the rule of the towns. It has created enormous cities, has greatly increased the urban population as compared with the rural, and has thus rescued a considerable part of the population from the idiocy of rural life. Just as it has made the country dependent on the towns, so it has made barbarian and semi-barbarian countries dependent on the civilized ones, nations of peasants on nations of bourgeois, the East on the West."[26]

Is it not by the instigation of bourgeois, capitalist technique that the world has begun to be universalized, in that technique *levels down* differences and peculiarities? "Bourgeois society is the most advanced and complex historical organisation of production. The categories which express its relations, and an understanding of its structure, therefore, provide an insight into the structure and the relations of production of all formerly existing social formations the ruins and component elements of which were used in the creation of bourgeois society.... The anatomy of man is a key to the anatomy of the ape."[27] Marx is convinced that "bourgeois economy thus provides a key to the economy of antiquity."[28] Yet he does not admit that one can erase all differences, and he asks that one take vast economic insights with a grain of salt. Thus he voices a protest against the "insights" of (economistic) economists "who obliterate all historical differences and who see in all social phenomena only bourgeois phenomena."[29] Marx is not afraid to write: "Thus, although it is true that the categories of bourgeois economy are valid for all other social formations, this has to be taken *cum grano salis* . . ."[30] This grain of salt disappears in all the insight gained by post-Marxian Marxists.

The requirement of the grain of salt is inseparable from the need for self-criticism. For one social formation can understand those that preceded it only by making its own self-criticism, by becoming conscious of itself. To do this, it must judge itself not by its own ideology but by its real life. Likewise, the historical understanding of past social formations must not share the illusion that those times had regarding themselves. The *ideas* men of those periods had about their real practice must not be transformed into the actual force determining things, as viewed from the level of historical understanding. To take at its word everything any age says or thinks about itself is to be the dupe of its ideological illusions.[31] One must not imagine either that "later history" is the goal of "earlier history"; "what is designated with the words 'destiny,' 'goal,' 'germ,' or 'idea' of earlier history is nothing more than an abstraction formed from later history, from the active influence which earlier history exercises on later history."[32] Historical becoming, therefore, is not so one-sided as it may seem in the eyes of the naïve or the mystified, and the play of retroactive projections and two-way influences is in fact more complicated. No "world spirit" (*Weltgeist*) regulates the materials of history. The power that rules the present social disorder and holds individuals in subjection, while appearing to them as some alien, higher force, is the corollary to the universalization of history and to the "world market" (*Weltmarkt*).[33] The world market, this supreme power in world history that "hovers over the earth like the fate of the ancients, and with invisible hand allots fortune and misfortune to men, sets up empires and overthrows empires, causes nations to rise and to disappear,"[34] must be abolished for the sake of making history truly world history. Marx thinks that in this way economic forces will no longer hover over the earth. But is the world he envisions a world of open worldhood (*mondialité ouverte*)? Will the new world mean a new and radical *progress*?

"What is called historical evolution," Marx writes, "depends in general on the fact that the latest form regards earlier ones as stages in the development of itself and conceives them always in a one-sided manner, since only rarely and under quite special conditions is a society able to adopt a critical attitude towards itself."[35] From the moment when bourgeois society began to criticize itself, it revealed itself as capable of understanding Oriental, Asian, and medieval societies, even if in a one-sided way. Economy did not begin to exist simply when one began to talk of it. But *to talk about it* in its past and present one must be capable of exercising a minimum of self-criticism on the level of contemporary society and economy. The task of completing bourgeois self-criticism falls upon the proletariat, which pushes the critique of bourgeois, capitalist society to its ultimate consequences. Self-criticism is only the prelude to negation criticism. The negation criticism of the proletariat aims for the abolition of private property, the socialization of the means of production, and the freeing of the motor force of history, namely, technique. At the same time, the proletariat liberates itself and frees man and society as a whole. The completing of critique means, then, the abolition of all that reifies people and things, all that dehumanizes human life, all that stands as an obstacle to the full development of total productivity. Marx is absolutely convinced that all alienation will cease with the abolition of the private ownership of the means of production. Economic alienation, political alienation, existential alienation, and ideological alienation will be reduced to zero, since they all reduce to basic economic alienation and cannot but be abolished by its abolition.

Marx engaged in a lively polemic against all the ahistorical conceptions of vulgar political economy. He vigorously denounced crude economism. He rejected every claim of full value for economic categories beyond the world within which they were born. But all this did not keep him from being held in the grip of the economic. Besides, could he reflect upon the economic world without sacrificing to it? Thus the whole Marxian theory of total dealienation by the abolition of basic economic alienation carries the weight of the very thing it wished to lift off the back of mankind. With a foolish naïveté, Marx believes that the whole spectrum—still better, the whole pyramid—of alienations will be *ipso facto* eliminated with the abolition of its one basis. From time to time, at different periods in his life, he wondered about the "totally" open perspective that would be expected to open up with the abolition of private property. In September 1843, Marx wrote to his friend Ruge: "Especially *communism* thus is a dogmatic abstraction; I have in mind the actually existing communism . . . not some other imagined or possible one. . . . Dissolution of private property, therefore, is in no way identical with communism . . ."[36] At no time after that did Marx come out in this way with a statement about the nonidentity of the abolition of private property and communism. On the contrary, he always held to the thesis of their identity, namely, that "by the communist revolution and the abolition of private property which is identical with it . . . history becomes transformed into world history."[37] Or again: "*Communism* as the *positive* transcendence of *private property*, as *human ⟨self-alienation⟩*, and therefore as the real *appropriation of the human* essence by and for man."[38] Yet Marx could not escape the question whether one can reject a thing of power and remain unaffected by it in its rejection. He saw that communism, as a movement of negation of private property, continued to be infected by what it negated. Marx never did succeed in completely and radically distinguishing vulgar, crude communism from communism as the agency of total dealienation. He did not even come to see—and foresee—that the expropriation of the capitalists would not necessarily end the exploitation of man by man and that the simple suppression of the private ownership of the means of production did not coincide with the actual end of all expropriation. Marx was unable to recognize the will to power. He was not capable of foreseeing that the will for power and appropriation could reconstitute itself and reform after the abolition of private property. His passion for reduction prevented him from seeing that private ownership is not the first source and single basis for all exploitation and, still less, of all alienations. He firmly believed that an *exploitation of nature by men*, based on a technique freed from private ownership, would entail no *exploitation of men by men*.

Yet thinkers of genius, sometimes in a flash of light, do come to grasp things that do not substantiate their theses. Marx perceived that communism could not be defined *by reference to* the private property that it wished to abolish. Its proper truth is thus contaminated by what it opposes. Still, perhaps one should think that what subsists in it as alienation will be an alienation all the more keen in that men who experience it will be conscious of it. Communism is the negation that negates the negation of men, that is, private property. It is the negation of negation, and what it negates marks its positivity. Despite Marx's efforts to begin solely from the *positing* of man, the being par excellence, man as the totality of the human world, as the objective, productive subject that rests only upon his own foundation—despite all these efforts the Marxian perspective is strongly affected by everything that, in its own eyes, negates and reifies man. In a badly mutilated portion of the manuscript of Marx's 1844 writings, we can still read the following lines: "If we characterize *communism* itself because of its character as negation of the negation, as the appropriation of the human essence through the intermediary of the negation of private property . . . the real ⟨alienation⟩ of the life of man remains, and remains all the more, the more one is conscious of it as such, hence it may be accomplished solely by putting communism into operation."[39] And just after this touchy passage Marx says something we have already heard: that history will accomplish this movement, though a long and severe process it will be; that we must regard it as real progress to have gained beforehand a consciousness of both the limited character and the goal of this historical movement, a consciousness that reaches out beyond that real movement.

Communism, therefore, though regarded as overcoming all alienation, remains infected and affected by its contrary, and the negativity it contains will lead to its being itself transcended. In moments of his highest perceptiveness, Marx does succeed in thinking the problematic of communism and putting direct questions to it. He wants to set radical limits to the capitalism of socialism-communism, without seeing more closely those elements which, *coming from* the first, would pass into the second. From the perspective of socialistic and communistic economy and society, Marx could not and would not see the things that would in that economy and in that society generalize and collectivize capitalism, thus, so to speak, socializing bourgeois society. Marx did not see through to the very end the way in which socialism, communism, and practical materialism are the *heirs* of the bourgeoisie, capitalism, and positivism. He was not someone to whom it was given to understand that the so-called bourgeois forms of life are more fit to survive than the bourgeoisie itself. And so he refused to see in the new society a fundamental democratization and universalization of "bourgeois forms," a continuation of this strange, solid, errant "bourgeoisie"—with or without bourgeois man himself.

Bent on abolishing a separate economic life and the economic mode of the production of life as it has existed up to now, wanting to free men from economic forces, communism nevertheless defines itself by its essentially economic organization.⁴⁰ How could it be otherwise, if it is true that communism is a practical movement, pursuing practical ends with practical means? Is it not inevitable that it be so, since communism aims for the total liberation of technique and since technique is the spring at the very heart of the economy?

The Marxian grasp of economy is and remains ambiguous and polyvalent. Hurrying to see history advance, Marx does not spend time over what remains problematic. He has not very seriously considered the prospect that socialization of the means of production might not put an end to human exploitation. The expropriation of the expropriators and the abolition of private property and of the customary division of labor should lead man to a human and social appropriation of the totality of beings. There will, of course, be an *appropriation* in the new society, there will even be property, but the appropriation will not be of a possessive, reified kind, no more than property will be individual. Since all production is a process of appropriating nature by technique and since all consumption is an appropriation of the products of productive technique, no social formation would be able to do without appropriation and ownership (of which private ownership is but one form). For "it is tautological . . . to state that where no form of property exists there can be no production and hence no society either. Appropriation which appropriates nothing is a contradiction in terms."⁴¹ The *Communist Manifesto* also gives all the specification one could wish here: "The distinguishing feature of Communism is not the abolition of property generally, but the abolition of bourgeois property. . . . In this sense, the theory of the Communists may be summed up in the single sentence: Abolition of private property!"⁴² Would not socialistic society run the risk of becoming a kind of universal capitalist, even while doing more than making economy a state matter, even while actually socializing it? Are common and communal ownership and collective appropriation strictly incompatible with what would be called the regime of universal capitalism? Marx does not put the question that way, so his answer has a different orientation. "When therefore capital is converted into common property, into the property of all members of society, personal property is not thereby transformed into social property. It is only the social character of the property that is changed. It loses its class character."⁴³ Capital, which manifests itself as a "thing," a *res*, and holds dominion through being a fetishized power, results from a social relationship between people which remains masked. It "thingifies" (*chosifie*), it reifies all that it touches—and it touches everything—in that it converts everything into a commodity. The interhuman—and human—bonds upon which it is established do not in any way become transparent in the framework of capitalism. On the other hand—and as part of the same situation—to those it crushes capital appears to emanate from some mysterious social power that the members of society can take possession of by suppressing individual capitalists. When, after the abolition of capitalism, (capitalistic) capital is transformed into (socialistic) common property, the socialistic society should not thereby become a universal capitalist, for this social ownership would no longer exploit wage labor. Socialistic society in the mode of collective ownership would have radically broken the bondage of wage labor to capital. Wage labor as such, capital as such, and the production of a supplement to wage labor so that it too might be exploited would no longer exist.

One has to recognize that Marx never envisioned socialism and communism as the achievement of Paradise on earth. He forbade himself much peering into the future. Now and then he gives a few indications, and only that. Here is one of them:

> Let us now picture to ourselves, by way of change, a community of free individuals, carrying on their work with the means of production in common, in which the labour-power of all the different individuals is consciously applied as the combined labour-power of the community. All the characteristics of Robinson's labour are here repeated, but with this difference, that they are social, instead of individual. Everything produced by him was exclusively the result of his own personal labour, and therefore simply an object of use for himself. The total product of our community is a social product. One portion serves as fresh means of production and remains social. But another portion is consumed by the members as means of subsistence. A distribution of this portion amongst them is consequently necessary. The mode of this distribution will vary with the productive organisation of the community, and the degree of historical development attained by the producers.[44]

At any rate, collective ownership, in the framework of which all the members of the society would appropriate social products—according to their labor in the first phase of communism, according to their needs in fully achieved communism—would make it impossible to anyone whatsoever to subject the labor of someone else to his own control by appropriating it in the mode of expropriation. The instruments of production will be exploited in common, and all will appropriate the labor of all. One must not forget that the abolition of private property means as well the abolition of the division of labor (in its customary sense). The division of labor and private property are identical,[45] so that the abolition of one does not work without the joint abolition of the other. But we have already tried to see how it is that this abolition too, or in any case the prospect of this abolition, entails certain difficulties.

Achieving socialism-communism will not throw man back to a level that had long ago been superseded. It is in no way a question of reviving the world of man's immaturity, when he had "not yet severed the umbilical cord that unites him with his fellowmen in a primitive tribal community."[46] The conditions of immediacy and a quasi-natural transparency are not to be either rediscovered or recreated. They rested upon a nondevelopment of technique. Since that time, productive forces have undertaken their labor of conquest in a process of long and painful development; and this has culminated in the creation of technical conditions that permit the transcending of reification. Although the development of technique alienated man still more, it created the material possibilities for his liberation. The world and history can become the work of freely associated, socialized men, that is, a production "consciously regulated by them in accordance with a settled plan."[47] Without being a return to a state of affairs that preceded the introduction of capitalistic reification and the abstract rule of commodity production, socialistic society will have to be on constant watch against people taking on false masks in their lives and against human relationships becoming disguised as relationships between things. Before the bourgeois era, there was a time when, though men wore masks in their social life, "the social relations between individuals in the performance of their labour" clearly showed "as their own mutual personal relations . . . not disguised under the shape of social relations between the products of labour."[48] That took place on the base condition of a technique that was insufficiently developed. Capitalism put an end to this kind of relation by creating modern industry. In its turn, communism is to abolish capitalist reification by taking an endlessly developing technique under its conscious control according to the order of a total plan.

Will mankind thus pass from the reign of necessity to the reign of freedom? When the alienation of technique has been abolished (together with the alienation of the proletarians and the capitalists, for the possessor class as well as the exploited class represent the self-alienation of man), will human history overcome the power of necessity? Marx is in no hurry to quit this realm. In volume three of *Capital*, he writes:

> In fact, the realm of freedom actually begins only where labour which is determined by necessity and mundane considerations ceases; thus in the very nature of things it lies beyond the sphere of actual material production. Just as the savage must wrestle with Nature to satisfy his wants, to maintain and reproduce life, so must civilised man, and he must do so in all social formations and under all possible modes of production. With his development this realm of physical necessity expands as a result of his wants; but, at the same time, the forces of production which satisfy these wants also increase. Freedom in this field can only consist in socialised man, the associated producers, rationally regulating their interchange with Nature, bringing it under their common control, instead of being ruled by it as by the blind forces of Nature; and achieving this with the least expenditure of energy and under conditions most favourable to, and worthy of, their human nature. But it nonetheless still remains a realm of necessity. Beyond it begins that development of human energy which is an end in itself, the true realm of freedom, which, however, can blossom forth only with this realm of necessity as its basis. The shortening of the working-day is its basic prerequisite.[49]

Will socialism-communism, then, be able to overcome necessity and do so with necessity as its basis? Will it achieve the reign of "freedom"? Marx says it will. And thus, while declaring the essence of *reality, being, the world, totality,* and *human history* to lie in material production, he then talks about something lying "beyond the sphere of actual material production," a "beyond" that, while immanent to human history, is not thereby prevented from being *beyond material production*. Once material production is established in its omnipotence and once its achievement of the conquest of the world is everywhere under way, once technique has emerged victorious from its battle with nature, would the riddle of "reality," that is, of the being-in-becoming of the totality of the world, of the history of man and societies, be revealed in a horizon other than that of material production—and beyond it?

The Political Problematic

No less ambiguous and polyvalent than the economic problems are the political problems; in fact they are more so. Marx aims for the abolition of politics, politics being something that, in his eyes, was never too real a force. Sometimes he seems to concede a kind of almost essential reality to the State and to politics, to make them out as something more than simple form for a content that is real because it is economic—and here the forming action is also a deformation. He could, for example, see the State as having some basis. After Marx says that, in the contradiction between individual interests and the common interest, the "common" interest takes an independent form as the State and thus constitutes a kind of illusory community, he recognizes that this takes place "always based, however, on the real ties existing in every family and tribal conglomeration—such as flesh and blood, language, division of labour on a larger scale, and other interests—and especially, as we shall enlarge upon later, on the classes, already determined by the division of labour, which in every such mass of men separate out, and of which one dominates all the others."[50] Marx could also see that among nomadic populations and in the Middle Ages, "*horse and sword*, wherever these are the *true means of life*, are also acknowledged as the true political powers in life."[51]

Marx, "however," continues to blame all historiography for seeing in history only "the political actions of princes and States," being blind to the "real basis of history" upon which rose national and political formations. All kinds of historiography, whether pseudo-objective or subjective, of the French, English, or German sort, "in particular in each historical epoch have had to share *the illusion of that epoch*."[52] Politics has not been viewed as the action of giving form to real motives, as a power stirred by the impotence of actual practice to be brought to full human and social realization. Being a summation of the real, practical tendencies of society and the résumé of its struggles, politics, as a form of superstructure resting on the base of economy, has never yet been able to escape alienation, and it can achieve itself only in its abolition. "Just as *religion* is the table of contents of the theoretical struggles of mankind, the *political state* is that of the practical ones. The political state, therefore, expresses all social struggles, needs, and truths within its form *sub specie rei publicae*."[53] This table of contents, this résumé, however, has to be analyzed, for under the appearance of the sovereignty of man, the citizen, the sovereignty of private property deploys its forces. And as Marx wants to abolish economic life as a separate thing, private property, the division of labor, and the realm of commodities, he wishes equally well to abolish political life and the State, the bureaucratized civil service, and the realm of administrative organization. But do the State and politics allow of being abolished for the benefit of man's species being? Can man's own powers, which are by their very essence social forces, be organized, coordinated, and planned out without the mediation of political forms? With little hesitation in giving an answer to this serious problem, Marx replies in the affirmative.

In "On the Jewish Question," Marx earnestly called for the setting up of an identity between man's own powers and social forms. "Human emancipation will only be complete when the real, individual man has absorbed into himself the abstract citizen; when as an individual man, in his everyday life, in his work, and in his relationships, he has become a *species-being*; and when he has recognized and organized his own powers (*forces propres*) as social powers so that he no longer separates this social power from himself as political power."[54] Is not this handsome requirement abstract? Can individual man, in his empirical life, coincide, then, with human society as a whole? Can man's species-being overcome mediations, rise above its fragmentation, and win identity? With technique fully *planned out* and existing *on a planetary scale*, could the total society of total individuals do without state and political admininistration, do without any organization whatsoever of power, which is always power over things and beings?[55] Could the organs and organisms that intervene between the organization of needs and production and the organization of distribution and satisfaction be rendered superfluous?

To talk of a Marxian utopia in reference to the prospect of the abolition of politics and the State, of administration and bureaucracy, does not help us see the problem better. People have drawn attention to this anarchist accent in Marx's vision many times. Yet it is not certain that, from this point onward, we are even up to understanding how the integral development of total technique (more than of total men) could make politics as such useless. Technique is not incapable of absorbing politics. Nor is it impossible that the relations of both the private and public lives of men may actually enter a new phase. Perhaps institutions are so empty of living substance that they will be revealed as incapable of surviving, either as they are or in modified form. We should no longer be crude in the way we take hold of the specifically political problematic, as we are when we continue to operate with the concepts of nineteenth-century political theory. Technique is shaking up habitual political procedures more and more and is perhaps preparing to abolish politics as politics; it would in that event take over its tasks, without, however, necessarily opening up into the world that Marx envisions in terms a little too idyllic and harmonious.

Marx unquestionably did quite neglect the will to power and all the contradictions and sources of conflict that it implies. He believed that the abolition of economic alienation in general and capitalist alienation in particular would abolish as well all national and political superstructure, all specialized civil service, and all bureaucracy. He also saw the national States as absorbed in and by the universal history of mankind, which is a vision indisputably far more realistic than the preceding one. He was convinced that violent revolutions and wars would remain unknown in world socialistic society, a society without classes and without a State, that is, without any specific form of power. Marx drew his strength from this vision; we are rather inclined to see in it his weakness. Yet the distinction between strength and weakness should not be made artificially.

We would be within our rights to think that the abolition of economic life and political life, as Marx advocates it, would open up a new horizon to human activity, that of play. With labor in the customary sense of the term abolished as labor, would not the continuous and multiform productive activity of men then be of the order of play? With politics as such no longer necessary and power no longer asserting itself as a specific autonomous force, would not all that which manifests the cohesion of human undertakings be secured in a kind of play? The play of energies and forms would not resemble the known kinds of play; it would deploy its own "rules" without having to obey any kind of transcendent meaning whatever or any end that would be external to it. The whole economic and political history of mankind has been more or less of the order of tragedy; and when social forms and forces became exhausted, which is the same spectacle we ourselves are now witnesssing, it all underwent a second death in comedy. We saw Marx regard comedy as the last phase of any historic form, as the misshapen repetition of a drama whose real heroes were already dead. Marx thought that history followed this course so that mankind could blithely break with its past, and he demands this gladsome historical destiny, this burlesque end, for the political powers of the "modern *ancien régime*." The text that contains these remarks, Marx's "A Contribution to the Critique of Hegel's Philosophy of

Right, Introduction," puts it this way: "The modern *ancien régime* is nothing but the ⟨comedian⟩ of a world order whose real heroes are dead."[56] In indicating the possibility of a new horizon within which human social activity would manifest itself as play, we are thinking of a play that would no longer be tragic and dramatic, since the heroes of this kind of thing are dead, or comic, since comic parody would have undergone a second death, in total derision. This play would know how to transcend tragedy and comedy so that it might deploy its own forms and energies.

The political play that lays claim to Marx, the political movement that begins from him in order to establish societies and States that call themselves socialist, seems strangely remote from the aims of its founder.[57] Still, this movement, these societies, and these States do *in a certain way* realize the thought that lies at their origin; at any rate they actualize *a certain* dimension of this thinking (namely, its most massive one), while at the same time "betraying" an important axial element essential to its original intention. Marx the thinker many times stated, with regard to both philosophic and political systems, that the realization of their truth, which was always an errant thing, meant their abolition and their downfall and that every great victory was the prelude to a defeat. Would what he thus says of bourgeois democracy not be equally true, though in a different way, for the movement of socialist society? Marx gave all his energy to urging socialistic criticism to drive the representative system toward its final and fatal victory, so that in that victory, and by it, it would find its downfall. Thus he wrote to Ruge: "By elevating the representative system from its political context to a general context and by claiming the true significance that is due to it, he [the socialist critic] at the same time forces this party to go beyond itself, for its victory is simultaneously its loss."[58] For the moment, we can ask this question: By elevating the socialist system from its political form to a general form and by claiming the true significance that is due to it, do we not oblige it to go beyond itself, in the manner in which every consummated victory leads to a loss, which in turn lays the groundwork for a new conquest?

The Anthropological Problematic

The man that communism wishes to *institute* for the first time in the history of the world is species man (*Gattungswesen*), total man,

man as his own foundation, as producing both himself as man and the world. Though he has never yet existed, this man can and must exist, not according to some image of man, some ideal of man, but through constant transformation-working human activity. Though a totally developed *individual*, man will, in Marx's perspective, be totally *socialized*, since total society and total man will be one. Totality-minded (*totaliste*) rather than totalitarian, Marx's outlook understands this man as an organic portion of mankind; the individual being of every man integrates into the becoming of generations.

The man of times past had *not yet* come to maturity. Technique had not yet cut the umbilical cord that bound him to nature and to the primitive community. On the other hand, men in the different eras of extremely simple and utterly manifest production lay caught in the network of relations found in direct despotism and slavery. Then again, primitive or archaic man did not invest his essential, objective energies into productive forces that were suitable to self-development—self-development both in regard to those forces and in regard to man himself (and of course such development would be at the same time his alienation). As man gradually developed by developing his production, as he produced himself as man by developing the productivity of labor, he also further alienated himself, reified himself, and came to live in a fetishized world filled with product objects that remained alien to him. Bourgeois capitalist society completed this process, transforming men into things and falsely endowing things with "personal" properties. Men wear social masks and social relationships between people take the form of relationships between things—indeed they do more than "borrow" this form; they actually make themselves into things, reify themselves. The reign of commodity production makes everything a commodity, and *persona* and *res* become confused in world commodity exchange. The developed technique of this society, nevertheless, allows man to envision in a practical, technical way his total development and his full liberation, on the condition of the abolition of capital and the consequent freeing of the forces of technique as well as his own human and social energies. The high development of productive forces within the confines of the bourgeois regime and the consummation of alienation and reification form the conditions that allow man to gain reflective awareness of his power and to launch himself upon the conquest and appropriation of his being and of all that is and is made.

It is man who produces man, according to Marx. By producing

his life, man produces himself. Man owes his (human) being, his essence, and his existence to his productive labor alone. Man is created neither by God nor by Nature. As man, he has created himself. Marx's humanism is altogether radical. He recognizes no court of determination higher than that of human productivity. Production is an absolutely *thetic* power; in it resides first positing action. Production is as well the motor force of negativity, and it develops *antithetic* powers. Finally it is in production, and by production, that the supreme *synthesis* is worked.

The worker creates man, and developed technique creates the conditions for the global development of man's humanity. Man has always been alienated. The regime of private property and commodity production totally reifies him, but at the same time and for the first time in the course of nature's conversion into history (*dans le devenir-histoire de la nature*), it produces a man capable of total self-liberation. Human liberation, the abolition of reification, actual dealienation, means for Marx a *reintegration* and a *return*, a *reconquest* of man's being by man, though that mode of being has never yet deployed its power. The abolition of economic alienation and the consequent abolition of all alienations will permit man to appropriate his being, which, once again, has never yet existed and whose becoming led to the present state of things. And yet it is this state of things that makes possible the realization of man's energy-laden possibilities in human, social becoming.

Marx does not seem able to escape the fate that befalls every vision of History, though he makes an extraordinary effort not to yield to it. Do we not find in his vision a *point of departure*—namely, that of harmonious totality, of integral unity—which, though never having existed empirically, is to be regained on a higher level in a reconciliatory future, after the abolition of alienation? Do we not find there the idea of a *long and painful process* of progressive alienation that attains its culminating point in the present and that gives birth to the future? Do we not find a *future* that gives freedom to all the things that the process of becoming put into bondage and that institutes the man that shall accomplish reintegration, return, and reconquest? Without wanting to put Marx's thought into the form of a systematic schema, without wanting to cast it into molds that are foreign to it, without wanting to imprison it in a circle or in a schematic of circular movement, we still cannot keep from thinking that there is a path in Marx that leads the being of man, as something *posited*, through a process that negates man's humanity and thereby toward his reconciliation with his being and with the becoming of the world. Synthesis, the negation of negation—that is, the communism of the future, the abolition of alienation, the reconciliation of man with himself and with the world—negates the antithesis—that is, the realm of private property and reification, the bourgeois and capitalist present—and rejoins the thesis—that is, the first positing of the very being of man as total species man, who has nonetheless never existed in any past time whatsoever. It might be that no globally historical thought could be asserted without some elaboration of a vision of this kind. Nevertheless, Marx's vision is dynamic: for him what is at issue is conquering the world by and for man, who in this conquest as permanently established will achieve his own (re)conquest.

We do not learn a great deal from Marx himself about total species man, about this entity who only alienates himself by developing the productive technique that makes reconquest both possible and necessary for him, that permits him when fully developed to assert himself as total man. The totality of man lies in the future. Total man would no longer be alienated, would have overcome the cleavages within him. Though one must not understand total species man in a naïve and summary manner, though one ought to regard it as a total opening toward all that is and comes about by and for man in the historical becoming of time, Marx's species man does make one smile. Even more, we have the right to wonder if the Marxian demand for man as someone who will have overcome all his alienations—and all alienations—and will thereupon totally conquer himself and the world, is not itself ideological. We have the right also to think that this demand constitutes itself an "alienation," an "alienation" all the more abstract and illusory in that it pretends to put an end to all alienation.

Can man cease being "totally" broken and torn, ever unsatisfied, and deeply unhappy? Can he cease being moved by negativity? Can all the anthropological, human, and existential "alienations" come to be wholly abolished, except by the abolition of man himself?

The Marxian demand for total man is madly optimistic. It is a broad and generous demand, asking men to devote themselves — without protest—to a task that is indefinitely open-ended. Nevertheless, the species man Marx talks about wears a sad face, and we must recognize it: his language is impoverished; his thinking is tied to his praxis; his love and death are no longer a problem for him, being simply part of the reproductive process, something occurring within the series of human generations that follow one upon the other. "On the one hand . . . but on the other hand . . ." is the formula for the type of thinking Marx stigmatized as eclectic, mediocre, vacillating, and petty bourgeois. But is not the concept of man that Marx sketches "petty bourgeois"? Would not the end of the reign of tragic, noble individuals and great men, on the one side, and the emancipation of the workers, on the other, bring everything down to a universal averageness in a world-wide monochromatic gray?

Merely *saying* that man will be man on a species level in a total way simply because he is such by essence is hardly enough to enable him to assume that station, resplendent in the light of the being-in-becoming of open totality. The Marxian fanaticism for unity and totality is a poor success at disguising—in fact, it does not disguise at all—the sense in which it understands the unity of totality. Man named total is in fact the "practical materialist" (*der praktische Materialist*),[59] the "communist materialist" (*der kommunistische Materialist*),[60] man as deploying a polytechnical activity and indifferently carrying out divers and various labors. The being of "total" man is wholly poured out in technique and practice, sensuous, material, real, and actual activity. What, then, becomes of the other side? Is it only some "other side" that is to be abolished as belonging to spheres of superstructure and sublimation, to the clouds of ideology, as constituting a false mystical or mythical issue? Or are we witnessing in Marx a first grasp, still very much an approximation, of a new type of man, the man of technique? Is it impossible that a new human type is being created, one who would break through all customary metaphysical and psychological limits? Is it out of the question that metaphysics may cease having a place in man's nature?

We are not successful in our attempts to scrutinize the individuality and the subjectivity of the total individual, and the grounding basis for such an individual totally escapes us, supposing that one existed. What does this imply? That this "individual," this nonatomistic atom of a society that possesses atomic technique, will have surmounted subjectivity and individuality? Do the depths of subjectivity, which are not simply nor principally a matter of subjectivism, psychologism, individualism, and egoism, permit their being surmounted, if they are not at the same time "preserved" in *Aufhebung*? Can self-consciousness, struggling to become world-consciousness, be radically transcended? It might be possible. It may also be that all these questions, posed in this form, remain desperate philosophical questions, and thus are the sort of thing Marx wanted to abolish. Will personal history, once integrated into world history, no longer contain its own riddles, its own vicissitudes, and its own particular demons? Will it no longer be subject to questioning and anguish? Will the Sphinx who posed the riddle to Oedipus no longer ask crucial questions in the form of questions, so as not to have to receive answers? Marx does not answer questions like these. More than that, he does not ask them, at least not explicitly.

The Ideological Problematic

None of the questions that arise within the very movement of Marxian thought, not to mention the questions that it provokes, permit easy clarification. Moreover, we lack the horizon in which we might situate them. After the orgiastic experience of speculative thought and the intoxication of romanticism, Marx wants to remain the sober partisan of practice, to which the sole thing that would be added is the understanding of practice. In the course of finding his own way of thinking, Marx constantly rejects Hegel and romanticism, sometimes naming them, sometimes not. His conviction is unshakable: neither sublime ideas nor lofty feelings can effectively resolve the truly important problems, that is, the practical problems. The strength shown by speculative thought is the reflection of its actual impotence, and the development of the power of technique will make the ideological constructions of the philosophic mind powerless, indeed superfluous. "This ideal rising above the world is the ideological expression of the impotence of philosophers in face of the world. Practical life every day gives the lie to their ideological bragging."[61]

As long as it remained bogged down in its insoluble antinomies (Kant) or its speculative contradictions (Hegel), as long as it confused real presence and representation in the mind, speculative thinking hardly succeeded in surmounting the real antinomy, the true contradiction. The real world and abstract thought remained cut off from each other. As long as knowledge and consciousness turned their back upon real life, soared up above it, transcended it, or resolved it into thought, as long as idealistic and ideological constructions, logico-ontological systematizations, and enterprises of criticism kept being formed in the clouds of heaven, real life on earth continued on its way. The fundamental discord, which all the same expressed the complementarity of real alienation and ideological alienation, did not present a problem. Marx undertook to shake throughout its whole length and width the pretense of unity in a world that was neither united and one nor human in character, and he wanted to make its contradictions explode.

All that was proclaimed united is and remains torn and shredded. Conscious being and being simply put, thinking and being, theory and practice could be considered identical, but their unity or identity was always of a mystical kind. No one hit upon the thought "that there is a world in which *consciousness* and *being* are distinct; a world which continues to exist when I do away with its existence in thought, its existence as a category or as a standpoint; i.e., when I modify my own subjective consciousness without altering the objective reality in a really objective way; in other words, without altering my own *objective* reality and that of other men. Hence the speculative *mystic identity* of *being* and *thinking* is repeated in Criticism at the equally mystic identity of *practice* and *theory*."[62] The mystic identity that like a mask hides the real contradiction must be torn off, abolished, and transcended in order for true unity to be instituted.

The being-in-becoming of the totality of the world, as it exists and manifests itself through human activity, is incontestably seen as a whole by Marx; *nevertheless*, for him too totality has two aspects, one material and real, the other spiritual and ideal. This is the duality that must be abolished for the sake of unity. Yet, although nonmystical unity, the unity of conscious being and being, of thought and being, of theory and practice, of logos and praxis, may be proclaimed synthetic, total, dialectical, and unitary, although this unity may be assigned the task of bringing together the one-sided, individually different truths of each of these opposition powers (each of which is mistaken in taking itself for the total truth), it is nevertheless accomplished in the name of material, real being, under the stimulus of actual production, of material practice. Marx's answers to crucial questions too often lie on a level different from that of the question itself. In his parody of speculative thinking, Marx thinks he reaches speculative thought itself. He still thinks he transcends it, leaving Kant and Hegel far behind. But does he only jump over what he professes to transcend, and does he not also fall short of certain Kantian and Hegelian positions?

Marx does not show much concern for *reflecting* on the unity he proclaims, a unity which is unity only by the mere fact of full development on the one hand and abolition on the other; his aim lies in the simultaneous abolition of real differences and of the identification that he calls mystical or speculative. Whenever the question arises of identity and difference, of being and thought, he refers to the workers in Manchester and Lyons and to the existence of a (dealienated) mass of (also dealienated) men, as, for example, in *The Holy Family*:

> But these *massy*, communist workers, employed, for instance, in the Manchester or Lyons workshops, do not believe that *"pure thinking"* will be able to argue away their industrial masters and their own practical debasement. They are most painfully aware of the *difference* between *being* and *thinking*, between *consciousness* and *life*. They know that property, capital, money, wage-labor and the like are no ideal figments of the brain but very practical, very objective sources of their ⟨self-alienation⟩ and that they must be abolished in a practical, objective way for man to become man not only in *thinking*, in *consciousness*, but in massy *being*, in life.[63]

The unified world that Marx urges and foresees is like a vast workshop where the difference between employers and workers will be abolished, where everyone will have become both employer and worker. Thought would be at the service of practice—*ancilla technae?*—and the head would not point to worlds where the feet could not go. Philosophy would be absorbed in the first place by production, that is, by the techniques of material production, and then, secondarily, by the techniques of intellectual production; for Marx does not come up with a way of settling the lot of the latter.

Marx knew that "thinking and being are thus no doubt *distinct*, but at the same time they are in *unity* with each other."[64] The ancient νοεῖν ("thinking") and εἶναι ("being"), taken in the unity of αὐτό ("the same"), continue to raise problems.[65] Did Marx also recognize that in wanting to walk on one's feet one ran the risk of losing one's head? As for this "head," as for speculative spirit and "idealist" dialectic, "abstract" thought and consciousness as transcending real movement, Marx was not quite sure what to do with it. We note his embarrassment at a number of points on the path he follows. There is no getting around it—the thinking head is right there, and it continues to plague him. Can mankind cease thinking with its "head"? For the demand that intellectualism be transcended does not for all that resolve the question of the being and becoming of thought. Are the real, productive, practical, and technical works which the mass of mankind is called upon to accomplish and the endless project that humanity underway is asked to construct collectively as a mass—once labor as it was is abolished and once the division of labor and therefore the difference between manual labor with materials and intellectual labor in the mind are all abolished—are these a matter simply of walking with one's feet? Will mankind make use of the instruments of technique in order to modify and perfect its movement according to the dialectic that has been set on its feet? We have a number of times pointed out the difficulties that Marx experiences in regard to the question of thought and consciousness, theory and ideas—in a word, in regard to the "head." As much as he can, he reduces these "powers" to reality and practice, to activity that shows actual effects; yet they continue to be at work, a constant source of embarrassing problems. Occasionally Marx goes so far as to give them a privileged role, privileged in relation to the one played by the opposite agency that is supposed to vanquish them. The passage from *Capital* that follows should make us reflect on the questions raised by this head with its charge on initiative, this head upon which precisely the Hegelian dialectic walked and that had to be overturned in order to get it to walk—if not think—with its feet.

> We pre-suppose labour in a form that stamps it as exclusively human. A spider conducts operations that resemble those of a weaver, and a bee puts to shame many an architect in the construction of her cells. But what distinguishes the worst architect from the best of bees is this, that the architect raises his structure in imagination before he erects it in reality. At the end of every labour-process, we get a result that already existed in the imagination of the labourer at its commencement. He not only effects a change of form in the material on which he works, but he also realises a purpose of his own that gives the law to his modus operandi, and to which he must subordinate his will.[66]

Consequently, the operations of the head—operative ideas, the imagining of what is to be done, the knowledge from the beginning of the result to which practical activity is thereupon going to lead —play a determinant role. *Sometimes* this role is conceived by Marx in an "idealistic" way, and that is almost inevitable since he *usually* discredits the constructions of the head, the operations of ideas, imagination, knowledge, and consciousness, deeming them an alienated reflection of material practice and real technique. Rather than dialectical, this whole problematic remains equivocal and ambiguous. No dialectic helps us here to solve the problem, the problem, namely, of the dialectical ties that unite the so-called real dialectic to the one that is termed the dialectic of thought; and Marx himself says principally what the dialectic is not or must not be, not what it is and how it deploys its play.

Marx's "real humanism," his historico-dialectical and chiefly *practical* materialism, and his open-ended communism only reject traditional metaphysics and the economic, social, and political powers that correspond to it. But what the will wishes to abolish does not for all that instantly cease to exist; the movement of supersession carts along a lot of things that it would like to be done with, which is inevitable, since they are the basis for the movement of transcending. Marx, infinitely more than he thinks and wishes, remains tributary to a "metaphysics" that, even when set on its feet, continues to be at work. And this is what prevents him from gaining a clear view of the fate of ideologies and ideological alienation. Let us not too hastily draw a conclusion that Marx was unsuccessful in opening up the domain of a radically new reality, a new type of activity. Perhaps it belongs to the "essence" of this reality and this activity not to be able to resolve the questions and the problems that spring from thought. Marx wishes to fulfill the errant truth of philosophy, that is, of speculative, metaphysical thought; he wishes to bring it to fulfillment and eliminate it. Speaking of the war levied against *speculative metaphysics* and *all metaphysics*, he says: "It [metaphysics] will be defeated for ever by *materialism* which has now been perfected by the work of *speculation* itself and coincides with *humanism*."[67] The *labor* of speculation, all ideological, idealistic, and alienated as it may be, as a consequence led to practical materialism and real humanism, which were destined to put an end at once to the alienation of labor and ideological alienation, thanks to the institution of new labor as a total *but* practical thing. Philosophy is abolished in being realized; that is, it realizes its truth and abolishes its errance.

Philosophy *realizes* itself; it inscribes itself into *reality*; it forms one body with the *real*. Philosophy becomes *really* true reality, *genuine reality*—by ceasing to be what it had been since Heraclitus, according to Marx, by ceasing to be philosophy in order to become world. Real movement and practical activity will have as their task the continual actualization of the (dereified) "logic of fact" instead of being lost in the (alienated) "fact of logic."[68] Philosophic thought does away with the concrete determinations of all that is, for the sake of Being, the One, and Totality. It is time to do away with it itself by realizing it in the world. The philosophers have converted predicates into subjects and made both abstract. What must be done now is to smash the speculative proposition and its logico-dialectical deductions, to drive the copula out of the heaven of ideas. *Being* itself, the *is*, must cease to "be," in order to be made reality in the becoming of historical time.

Yet how aproblematical the real remains for Marx! It is a real that he does not demonstrate but that he begins from and then ends up reaching. For him, what is real, true, active, and actual is that which is real, true, active, and actual in the eyes of common sense, which is what it actually becomes for everyone after Marx. He saw what men did and would do. Rejecting the philosophic way of seeing things and claiming to transcend it, Marx posits this real through the negation of philosophy and negates philosophy in the name of the real. Does he thus fall short of the movement of philosophic thought, or does he transcend it in another direction? The time does not yet seem to have come in which this most serious and most decisive question might be *posed* before being "resolved." We are scarcely beginning to glimpse what Marx's gaze was directed toward. Posing this question becomes all the more difficult because Marx nonetheless spoke of something "beyond the sphere of actual material production,"[69] in which, however, philosophy *ought* to be realized in being abolished. The reign of technique follows upon the realm of ideas. Technique itself is the secret of material production, the motor force of real becoming. It must first realize itself, in depth and in extent. Will it in turn be transcended? Will there be something beyond technique, beyond, to be sure, global, universal technique? Let us leave the question open. We cannot do otherwise, at least for the moment.

The Problematic of Reconciliation as Conquest

In the realm of communistic reconciliation, naturalism, humanism, and socialism will realize their essence and reach fulfillment. "*Naturalism*" is the point of departure; it must therefore be fully realized in the final outcome. A being animated by natural drives and physical needs, man must accede to the real satisfaction of his drives and needs, through a perpetual renewal of needs as well as of the modes of their satisfaction. Marxian naturalism implies a kind of "biologism"; yet this naturalism and biologism are set in motion by the *instinct for production* that technically produces man and the human world. *Humanism* begins with man, man as himself the very root and foundation for himself; it posits the kind of being man has, indissolubly natural, human, and social. In the world of humanism achieved, everything, beings and things, will have become human, everything will come to stand transparently revealed in the practical activity of man objectifying his being in his works. Humanism, permitting the objective subjectivity of every man to deploy its energies, can be really achieved only in *socialism*. Socialism implies here a kind of "sociologism." The term *socialism* is, of course, imprecise. Sometimes it means for Marx a simple social organization that does not set up a truly new structure but just repairs the old edifice.[70] Taken in all strictness, however, *socialism* means the same as *communism*: socialism socializes and collectivizes all that is and is made; it institutes human society. Socialism, consequently, is perfected in communism. The latter presupposes a very high degree of development for technique together with the abolition of private property, then makes everything into common goods for men who will no longer be isolated individuals, members of a class, or citizens of a State but producers and consumers in a panhuman community, protagonists of a history that is ever in a process of becoming. With communism, the transformation of nature into history and history into world history reaches its culmination and, at the same time, truly begins. The essence of communism lies in the conquest of the earth and the universe by and for man as the one who set total technique into motion.

Naturalism, humanism, and socialism-communism are based on absolute productivity, practical activity, transformative praxis—in a word, on Technique. Technique is not reducible simply to machines and limited industrial production. It is the motor force of history, the power that transforms nature into history, the engine of movement for universal history. It is, therefore, at the same time the point of departure and the point of arrival (from which one sets off again). It is Technique that is the basis for naturalism, humanism, socialism, and communism. It is Technique that was underdeveloped in a primitive time, that was then alienated and itself brought about alienation, and that is going to be wholly liberated. The reign of the reconciliation (in conquest) of man with himself and with his world will be that of consummate technicism. While bringing about this full "return become conscious, and accomplished within the entire wealth of previous development,"[71] and while freely disposing of all the useful materials that it finds before it, that is, disposing of them following its own will, this technicism "consciously treats all natural premises as the creatures of hitherto existing men, strips them of their natural character and subjugates them to the power of the united individuals."[72] It seems that the distinction between what is "natural" and what is "artificial" is destined to disappear. Reattacking and reconquering all the acquisitions of past conquests, full technicism transforms them so profoundly that it strips them of their former character, maintaining them in a process of continuous actualization. All that passed for natural being was so artificially; on the other hand, the worst artifacts were accepted as natural. Liberated technique will, therefore, abolish at once "naturalness" and "artificiality," making everything utterly open to the sensuous activity of men and of the human mass.

The aim of naturalism-humanism-communism is not only the abolition of labor as it has been performed, of economic life as a separate thing, and of the very production of man (according to reified norms). Also to be abolished are the State, politics and the bureaucracy, morality and the family, and religion and the divers and various forms of ideology. In addition all alienation and all that strangeness in which the existence of men wallows must be transcended, as must "naturalness" and any "world" other than the one of sensuous activity. This series—this esemble, rather—of eliminations, abolitions, and transcendings is the proper action of communism, which will in turn one day be superseded. It is in this sense that we have spoken of the tenacious will to transcendence, the passion for annihilation on the part of the founder of Marxism. All that is, is declared nonexistent, being condemned to sink into nothingness.

All things that were regarded as suprasensuous are thus returned to their origin, namely, the sensuous. Being alienated products of alienated production, they must, after their annihilation, be taken up again by "total" praxis. But how can "suprasensuous" powers be taken up again after being abolished? What shall this taking up again consist of? Metaphysics altogether rests upon the distinction —and the difference—between the physical and that which is higher than the physical, between the sensuous and its ground, between the material and the spiritual, between things and ideas. The abolition of this distinction means the collapse of metaphysics. It becomes clear, then, that the kind of seeing found in metaphysics was a limited one and enclosed us in limits.

Does Marx in actuality abolish metaphysics? We have several times tried to address this question from a number of angles in order to circumscribe its central point. The essential thing is not to lose sight of it. Marx first of all overturns traditional Western metaphysics as elaborated by the Greeks, the Christians, and the moderns. He denies the primacy of the suprasensuous, of the metaphysical and the spiritual, and brings them back down to that from which they arise. He nevertheless gives a position of privilege to the sensuous, the historically physical, and the material. He consequently *reverses* metaphysics. Marx reverses the "upside down world" which is the "real one" so that true reality may be set up in a practical way in the world as dealienated and walking with its feet.[73] At the same time, Marx *fulfills* modern metaphysics, generalizing it after having reversed it. He does not succeed, it seems, in abolishing it, in eliminating it. Practice *and* the understanding of practice, real movement *and* the movement of consciousness as transcending real movement continue to form two orders despite all the assertions of unity. Moreover, the fact that real practice *takes up again* the ideological world indicates that that world is not in fact annihilated. Speaking of dealienated men, men therefore reconciled with themselves and with totality, Marx explains that they will "be brought into practical connection with the material (as well as intellectual) production of the whole world and be put in a position to acquire the capacity to enjoy this all-sided production of the whole earth (the creations of man)."[74] It seems, then, that there will be an "intellectual" (or "spiritual"—French *spirituel*) order of productive technique after the world of mind or spirit (*le monde spirituel*) is abolished. But the new intellectual (spiritual) dimension (*le nouveau spirituel*) cannot, however, even in the communist world, be of the order of material practice which, it is claimed, is one and all-encompassing. It will be *different*. We have several times tried to indicate the great difficulties in which Marx finds himself in regard to the final lot of "superstructural" powers. They keep springing up where they should no longer be. In mentioning this production of an *intellectual (spiritual) kind as well*, Marx puts it in parentheses;[75] yet for all that it does not cease to be active. The question is in no way settled by saying that these new intellectual (spiritual) powers will no longer be like the old ones. Their "intellectual" ("spiritual") character is exactly what raises the problem. As intellectual (spiritual) powers, as forces in thought and consciousness, they demonstrate by their simple existence—even if they should be powers of but second rank—that the unity is not total and unbroken. Will there be, therefore, no unitary foundation for the human enterprise of world conquest? Will that which transcends technique continue to subsist, despite all the efforts of technique to annex it? Will the world continue to bear within it a *double* aspect?

Marx was always a resolute opponent of dualism. Platonic dualism (which in its origins did not yet take leave of *physis*), Christian dualism (which Marx calls "real"), and the "abstract" dualism of modern times were all unsuccessful in truly overcoming the opposition between man's humanity, on the one hand, and what *negates* it, on the other. With modern times this opposition reaches the high point of its development, embracing everything in its own abstraction. "The abstract, reflected opposition belongs only to modern times. The Middle Ages was the real dualism; modern times is the abstract dualism."[76] Modern life, the epoch of civilization, Marx tells us farther on, "separates man's objective essence from him, taking it to be merely external and material. Man's content is not taken to be his true actuality."[77] The communist era will complete modern times and inaugurate at the same time a new epoch. Desiring and aiming for the conquest of the world in the name of man, will comunist practice be able to do away with dualism as such, as well as every kind of duality? Will the men of dealienated technique, being destined to realize the words of the Old Testament by assuming dominion over the earth in freedom from every sin and curse, have no knowledge of dualistic rents in the fabric of the whole? Setting off to conquer the earth—"the universal ⟨object⟩ of human labour"[78]—if not the universe, would objective subjects, free of every prohibition, be also free of any burden weighing upon their sensuous activity? In their struggle to appropriate the earth, the waters, and the air by making them human, will the conquerors of the planet *not know*—or not be *sensitive* to—any kind of corrosive action? They will indeed technically produce their life and will certainly reproduce (in one way or another). Will that be enough for them? Marx makes some room for "the material (as well as intellectual) production of the whole world"; but about the essential way in which intellectual (spiritual) production is to be manifested he is silent.

The Greek world lived and thought within the horizon of *Physis*, in the order of *Cosmos*, which held mastery over chaos. The era of *physis* was supplanted by the era of Creation. The Christian world found its support in the Creator who drew the world out of nothingness and preordained it to the Apocalypse. Then, at a certain point, a new kind of undertaking emerged: the passage to the era of production, to the era of man's creation by man, to the era of the fashioning of the world, to the universal appropriation of all (human) creations by all men. Man, in taking account of the fact that he is his own root or, rather, is becoming his own root, sets off on his way, by radical revolution, toward the total satisfaction of his root needs in a mobilization of full technique. The *root needs*, the *radical needs*, according to Marx's humanistic radicalism, are the *need for food*, the *need for a dwelling place*, and the *need for clothing*. Marx prudently adds "and many other things," leaving the door thus open for other needs to appear. He puts it this way: "But life involves before everything else eating and drinking, a habitation, clothing and many other things."[79] About these *many other things* Marx is not very explicit. Eager for totality, he always wants to begin from the totality of human needs, demanding their total satisfaction by an activity that is sensuous, practical, polytechnical, and universal. Since all needs and all the senses— "physical and mental [or spiritual] "[80]—have been and still are alienated, one has to make possible, by the abolition of "both aspects"[81] of alienation, the necessary satisfaction of *elementary needs*, those that are indispensable for life: food, drink, clothing, habitation. The *other needs* are much less essential, for they depend in some way upon the still obscure fate of the superstructure and spiritual powers.

Marx wants to return to man the dignity he has never yet had. He surely suspects that the simple, and complex, production of life and its reproduction will hardly suffice to make human life livable. But does he do more than just suspect it? He vehemently condemns the bourgeois and philistine conception of a life reduced to the actions animals must perform to keep from perishing and to perpetuate their species. "What they [the petty bourgeois] want, namely, to live and propagate themselves (and, as Goethe says, no one achieves more than this), the animal wants too,"[82] Marx writes to Ruge. But he wants precisely to do away with the world in which that is how it is. "Freedom, the feeling of man's dignity, will have to be awakened again in these men. Only this feeling, which disappeared from the world with the Greeks and with Christianity vanished into the blue mist of heaven, can again transform society into a community of men to achieve their highest purposes ..."[83] These *highest purposes* for the human community that will have overcome self-alienation and social fetishisms are not made very clear for us. Marx does not offer much help here. When reification is abolished and human beings entertain human relations with their peers and with things, what will spring up as the *highest purpose* in this new situation? And *over* what will this purpose rise higher? Marx *wants* man to recover his world, his humanity, his self-esteem, for, as we read in his "Contribution to the Critique of Hegel's 'Philosophy of Right,' Introduction," man "has either not yet gained himself or has lost himself again."[84] Marx leans toward the simultaneous and joint transcending of subject and objects. Both subject and objects are to be annihilated and transcended at one time and as such. There will then remain their bond, itself neither subjective nor objective, before which and by which they vanish and lose their individual specialness. That the subject is to be transcended—that we are almost ready to admit. At the same time, however, objects are to be transcended; technique transforms objects so profoundly that they cease to be objects. The ὄντα ("beings") and the φαινόμενα ("manifestings") of *physis* gave way to the *entia creata* of the Christians, and these in turn were followed by the objects for a subject in modern thinkers. Planetary technique following the order of a total plan aims for the transcending of subject and objects for the sake of a continuous process of unending production. The process of *production*—its rhythm and style—is far more important than the perishable things it produces.

The objective and, above all, *productive* subject, transcending in his activity both (individualistic) subjectivity and (reified) objectivity, is the "metaphysical" and "ontological" ground for technique. The liberation of technique which is here to be brought about leads to the dazzling development of productivity, the conquering power. Man's natural, human, and social being will therefore be coterminous with the being-in-becoming of technique, whose progress will lead human society to naturalism, humanism, and socialism, that is, to real, realistic, and positive humanism, to practical communism. Man will thus reintegrate his being, achieve a return to himself, thrust forward again and win himself back again, even though, because of the nontotal development of technique, he has never yet been what in the future he can be. The world of the totality of what is—and what can come about—is based upon action, praxis, production, and labor; and thus men will tirelessly, and with no purposefulness beyond technique, produce products destined to be consumed, that is, annihilated. The totality of being will coincide in this way with the totality of human productivity, which will be unto itself its own basis and purpose. What techniques can lay hold of constitutes not just the sum total of one order, the graspable, but Totality. Marx's thinking concerns the "sensuous" totality of being as it lends itself to being taken hold of and fashioned by technique. The *reduction* of the world to the *production* of technique implies the positing of technique at the ground of being and the motor of becoming. But does the World as open totality allow itself to be taken hold of by technology? How will it respond to this enormous *provocation*?

Perhaps it is time to begin understanding Marx's thought as *Technology*, provided we take the term in its full amplitude and true depth. Technology would constitute the center of Marxian thinking, its intention and its axial nerve. Technology holds the keys to the world; it is by technological becoming that man produces himself as man, that nature becomes history, and that history is transformed into universal world history. It is technology that builds the bridges between the past, the present, and the future, that forms the rhythm of historical time, of the coming-to-be of man's conquests. In technology, too, and principally in technology, is the secret of the twofold aspect of the world and the double side of alienation whereby the spiritual world reflects and sublimates the insufficiencies of the material world. Finally, it is technology that holds the secret of the bonds that untie *theory* and *practice, thought* and *action, logos* and *technē*. For it is a certain conception of technique that gives rise to the very distinction between (theoretical) thinking and (practical) activity, *logos* being measured by the standard of a *praxis* posited as different and thinking being discredited from the first as *theoretical* activity, thereupon losing all decisive importance—all of which means that all thought will in the future be technical in essence. The full development of technique, that is, the flowering of *technology*, will absorb thoughts and ideas, that is, the development of *ideology*. Ideologists will have no further voice in deliberations; they will have become nonessential and inoperative.[85]

Marx, the "last philosopher" and he who has thought through the question of technique, offers us his thinking in all its richness and all its poverty. He sets us to thinking about all that which, thanks to him, emerges as a problem and a question. The richness of his thought implies questions he had not thought of, that present themselves to us and impose themselves upon our minds. It is through Marx's thought that what he has not reflected on stands revealed in its wealth (and poverty). It is in the light of what a thinker has seen that we can ourselves detect the lacunae and breaks in his thinking and in what remains for us to bring into view. The more someone's thinking is rich and profound, so much the more does it contain questions that are not answered and remain open, questions that did not exist before as such. Marx has begun thoughtful reflection on technique. It lies upon us now to think about and to experience technique as a planetary conquering force.

AFTERWORD

Before undertaking to criticize Marx and before attempting to move beyond him, one has to understand what he says. To achieve dialog with his thinking and to bring his thought into confrontation with world historical reality presuppose a long meditation upon and a meticulous attention given to all that is and is made. For reality is not so easily divorced from idea, or theory from practice.

Marx introduces us to the movement of the negativity that runs through universal history, that shakes it to its foundations, and that continues unabated within Marx's own work. Marx wants man to transcend radically his alienation by launching himself into the conquest of the world by means of a totally freed technique. What philosophy has been pursuing in thought since Heraclitus is to be realized in and through practice. Yet all realization at the same time means a loss, and the riddle of thought remains unsolved. What will the world be that is utterly accessible to human activity and that, in an unlimited way, of course, is reduced to the totality of human productivity and technique? What will this totality be that is essentially practical rather than total? Once the duality between what is *material* and what was spiritual is abolished, for the sake of basic, practical activity, what will become of freedom? Marx writes: "The realm of freedom begins only where labour which is determined by necessity and mundane considerations ceases; thus in the very nature of things it lies beyond the sphere of actual material production."[1] After the abolition of the metaphysical world for the sake of the historical "physical" world, after the abolition of the two aspects of alienation for the sake of a unitary, global productivity, what will be the meaning of the intellectual (or spiritual) production of which Marx speaks? For he tells us regarding the men of fully achieved universal history that they shall "be brought into practical connection with the material (as well as intellectual) production of the whole world and be put in a position to acquire the capacity to enjoy this all-sided production of the whole earth (the creations of man)."[2]

The greatest questions still remain open and cannot receive a univocal solution. Marx from time to time knew how to give a problematic character to his own perspectives. He advocates the world of planetary technique according to the order of a total plan, free of all exploitation, alienation, and bureaucracy. The zealous enemy of ideology and utopia, he keeps urging a world that is one and global, made by and for man. But this world continues to have implications that put it in question. And the unresolvable questions do not seem to belong only to the first stages in the realization of socialism and communism.

Socialism and communism, the *heirs* of the bourgeoisie and the capitalism they would reject, remain affected, as the negation of negation, by what they reject, namely, the world of private property. Socialism-communism, the movement of appropriation, cannot be wholly stranger to that from which it issues and does not mean the transcending of all alienation. It is Marx himself who introduces the questioning of his vision, thus making communism, "the solved riddle of history,"[3] itself a riddle. He presents to us for our meditation and our experience that which it is our lot to think about and to live.

These things that still await our understanding we can find in the *Manuscripts of 1844*:

> The transcendence of ⟨self-alienation⟩ follows the same course as ⟨self-alienation⟩.[4]

> Communism is the position as the negation of the negation, and is hence the *actual* phase necessary for the next stage of historical development in the process of human emancipation and rehabilitation. *Communism* is the necessary pattern and the dynamic principle of the immediate future, but communism as such is not the goal of human development—which goal is the structure of human society.[5]

> It takes *actual* communist action to abolish actual private property. History will come to it; and this movement, which in *theory* we already know to be a self-transcending movement, will constitute *in actual fact* a very severe and protracted process. But we must regard it as a real advance to have gained beforehand a consciousness of the limited

character as well as the goal of this historical movement —and a consciousness which reaches out beyond it.[6]

And, in *The German Ideology*, Marx exposes the powerful mainspring for forward movement, powerful but essentially economic, practical, and technical, and therefore not total.

> Communism differs from all previous movements in that it overturns the basis for all earlier relations of production and intercourse, and for the first time consciously treats all natural premises as the creatures of hitherto existing men, strips them of their natural character and subjugates them to the power of the united individuals. Its organisation is, therefore, essentially economic.[7]

> Communism is a highly practical movement, pursuing practical aims by practical means . . .[8]

Thus Marx opens himself to the future that lies at hand, which through its negativity will in turn generate what more distantly is yet to come.

NOTES

Translator's Introduction

1. See the preface, pp. 3–5, and the last paragraph of the conclusion.
2. On basic life needs, see the beginning of chap. 6 and pp. 331–332.
3. See the beginnings of chap. 4, chap. 6, and chap. 12. On defining human being according to Marx, see also Joseph J. O'Malley, "History and Man's 'Nature' in Marx," *Review of Politics* 28 (October, 1966):508–527, reprinted in *Marx's Socialism*, ed. Schlomo Avineri (New York: Lieber-Atherton, 1973), pp. 80–100.
4. Karl Marx and Frederick Engels, *The German Ideology*, ed. and trans. S. Ryazanskaya (Moscow: Progress Publishers, 1964), p. 39.
5. See the beginning of chap. 4.
6. In the end, though this is not discussed in Axelos's book, the species-proper characteristics of human being result from the appreciation and articulation of *meaning as such* rather than from only causal involvements with things. In other words, objects *made* are objects *cultural*, i.e., mediated and ordered explicitly by *meanings*, and thus able to be historical.
7. See in particular chap. 7. Two recent books devoted to an analysis and discussion of the concept of alienation in Marx are István Mészáros, *Marx's Theory of Alienation*, 3d ed. (London: Merlin Press, 1972), and Bertell Ollman, *Alienation: Marx's Concept of Man in a Capitalist Society* (Cambridge: At the University Press, 1971).
8. See chaps. 12 and 13 and the sections entitled "The Economic Problematic" and "The Problematic of Reconciliation as Conquest" in the conclusion; also chap. 4.
9. Up to this point I have not used the words in accord with the conventions that I begin to explain from here on. From now on, however, I shall keep to the distinction in usage introduced in the discussion now beginning.
10. "En pleine époque de la mise à mort d'un monde par la technique conquérante et planétaire, n'aurions-nous pas besoin d'une *technologie* qui commencerait à penser tout ce dont la technique se saisit, et la technique elle-même?" (Kostas Axelos, "Thèses sur Marx–V," in *Vers la pensée planétaire*, Arguments 21 [Paris: Editions de Minuit, 1970], p. 175).
11. "La technique, s'appelant elle aussi *technē*, consiste dans la production

et transformation de tout ce qui est.... La technique doit être comprise au sens le plus global et radical: elle embrasse toutes les techniques effectives et imaginaires; de la grammaire à l'érotisme, de l'écriture jusqu'à la fabrication de rêves.... C'est la technique qui est le moteur de l'époque planétaire, surdéterminant tous les secteurs culturels et les autres" (Kostas Axelos, *Arguments d'une recherche*, Arguments 40 [Paris: Editions de Minuit, 1969], p. 174). In this passage Axelos broadens the notion of technique beyond the productive making of objects, to which it is largely restricted in the Marx book. He thus shows in this other context a greater semblance to Jacques Ellul's notion of technique than in the present work. See the next section of this introduction.

12. A remark made to me in a personal conversation in the summer of 1973.

13. P. 331.

14. Axelos himself, in his own thinking, does not agree with Marx that technique is the base determinant for all human activities. Rather he distinguishes technique as one primary agency *among others* in the general scheme of the world. (See note 82 of this introduction.) But technique *now* does *in fact* have the dominant role, owing to the way historical contingencies have played it up in human development. The modern situation, then, is much the same for both, only the overall perspective for interpreting it is different. See Kostas Axelos, "Introduction à la pensée planétaire," in *Vers la pensée planétaire*, pp. 13–49.

15. Cf. pp. 294–295.

16. Cf. pp. 245, 325–326, and *passim*.

17. Cf. p. 296.

18. Cf. pp. 297, 300, and *passim*.

19. It is in the context of an emphasis on *modern* technique that Axelos uses the terms *technology* and *technological* most deliberately. Cf. pp. 293 and 294 ff. His usage is in a few cases inconsistent with the distinction suggested here. French admits a looser usage of *la technologie* much like the English. See note 23 of this introduction.

20. Paris: Armand Colin, 1954; English translation by John Wilkinson under the title *The Technological Society* (New York: Vintage Books, 1964).

21. Ellul, *Technological Society*, p. 19.

22. Ibid., p. xxv.

23. See ibid., p. xxvi. This meaning for the word *technology* is found in common usage in English. In contrast, *technique* in English means not some *generic* phenomenon of which *technology* would be one form or one component (as with Ellul) but rather a *particular* operation of *particular* means serving some *particular* purpose, as, e.g., "metallurgical techniques," or "the technique of spinning wool into thread." This usage of the two terms is the one found, e.g., in the encyclopedic English work on the history of technical performances, *A History of Technology*, ed. Charles Singer, E. J. Holmyard, A. R. Hall, and Trevor I. Williams, 5 vols. (New York: Oxford University Press, 1954–1963).

24. Paris: Presses Universitaires de France, 1962–: vol. 1, *Les origines de la civilisation technique* (1962); vol. 2, *Les premières étapes du machinisme* (1965); vol. 3, *L'expansion du machinisme* (1968). An English translation by Eileen Hennessy of the first two volumes has appeared under the title *A History of Technology and Invention: Progress through the Ages* (New York: Crown Publishers, 1969). Maurice Daumas is curator of the French national

museum of techniques and technologies, the Musée du Conservatoire National des Arts et Métiers.

25. Daumas, *History of Technology*, 2:11.

26. Daumas, *Histoire générale des techniques*, 2:xvi—xvii (my translation). See the English translation, *History of Technology*, 2:7—9, where Daumas's conceptual point has been blurred by the shifting choice of words used to translate *la technique*. It remains to be noted that Daumas later on suggests that his term *la technologie* means the same phenomenon designated in English by the word *engineering* (*Histoire générale des techniques*, 3:xxiii).

27. Daumas, *History of Technology*, 2:7—9, and *Histoire générale des techniques*, 3:xviii—xx.

28. See the excellent treatment of this same idea of a new order of activity, fusing science and technique, by Hans Jonas, "The Practical Uses of Theory," in his *The Phenomenon of Life: Toward a Philosophical Biology* (New York: Delta Books, 1968), pp. 188—210.

29. Note 11 points out that Axelos in the Marx book uses *technique* in a narrower sense than he does in his other writings, where a broader sense suggests a much greater similarity to Ellul. However, *technique* in this broader sense comes to share in the all-inclusiveness and universality achieved by *technique* in the narrower sense of the industrial making of objects (Axelos's primary usage of the word in the Marx book, where such making is dealt with as a base-level determinant for social development) because it too, in these other domains, adopts science and scientific methods to achieve specific objectives. In other words, when the making of objects becomes technology, other domains come to be governed by a basically similar technoscientific order. Such, at least, seems to be the argument one would have to develop in order to make clear the similarities that subsist despite differences of emphasis and usage in the different authors.

30. One can see here a divergence between Axelos and Marx. For Axelos, in his own thinking, the power relationship need not be, as it is for Marx, basic and ultimate for either the stance of knowledge or that of technical action. This is the important of his thoughts on play.

31. On the meaning of this word, see pp. xxvi—xxviii.

32. See pp. 54, 55, 78, and throughout.

33. See "The Economic Problematic" in the conclusion. It is a debatable point whether for Marx nature could ever be thoroughly humanized. It very much depends upon whether one believes that Marx's thinking on the humanizing of nature in the *1844 Manuscripts* changes in his later works. Axelos does not think such a change occurs, others do. See Alfred Schmidt, *The Concept of Nature in Marx* (London: NLB, 1971).

34. "What Marx opposes as alienation is not mediation in general but a set of *second order* mediations (PRIVATE PROPERTY—EXCHANGE—DIVISION OF LABOUR), a 'mediation of the mediation,' i.e., a *historically specific* mediation of the *ontologically fundamental* self-mediation of man with nature. . . . Labour (Productive activity) is the one and only absolute factor in the whole complex: LABOUR—DIVISION OF LABOUR—PRIVATE PROPERTY—EXCHANGE. (Absolute because the human mode of existence is inconceivable without the transformation of nature accomplished by productive activity.) Consequently any attempt at overcoming alienation must define itself in relation to this absolute as opposed to its manifestation in an alienated form.

... If 'productive activity' is not differentiated into its radically different aspects, if the ontologically absolute factor is not distinguished from the historically specific form, if, that is, activity is conceived—because of the absolutization of a particular form of activity—as a homogeneous entity, the question of an actual (practical) transcendence of alienation cannot possibly arise" (Mészáros, *Marx's Theory of Alienation*, p. 79). Axelos himself does not make this distinction altogether clear, although he certainly seems to imply agreement with Mészáros's point. See pp. 55–56, 126–129, and 245 as examples.

35. That alienation can in fact be overcome *entirely* or only *in part*, while human history continues, is one of the issues taken up in Mészáros, *Marx's Theory of Alienation*, pp. 247–250.

36. See the beginning of chap. 12.

37. "Objectification under conditions when labour becomes external to man assumes the form of an alien power that confronts man in a hostile way. This external power, private property, is 'the product, the result, the necessary consequence, of *alienated labour*, of the external relation of the worker to nature and to himself.' Thus if the result of this kind of objectification is the production of a hostile power, then man cannot 'contemplate himself in a world that he has created,' in reality, but, subjected to an external power and deprived of the sense of his own activity, he invents an unreal world, he subjects himself to it, and thus he restricts even further his own freedom" (Mészáros, *Marx's Theory of Alienation*, p. 159). The quoted material in this citation is from Karl Marx, *The Economic and Philosophic Manuscripts of 1844*, trans. Martin Milligan, ed. Dirk J. Struik (New York: International Publishers, 1964), pp. 117 and 114.

38. See the quote from Marx on p. 227. Indeed, for Marx, the historically developing condition of alienation *had* to be passed through in order for human being to be realized. As Mészáros points out: "Alienation, reification, and their alienated reflections are therefore socio-historically *necessary* forms of expression of a fundamental ontological relationship" (*Marx's Theory of Alienation*, p. 112).

39. "It goes without saying that a form, or *some* form of *externalization*— i.e., objectification itself—is as absolute a condition of development as activity itself: a non-externalized, non-objectified activity is a non-activity. In this sense some kind of mediation of the absolute ontological condition of man's interchange with nature is an equally absolute necessity. The question is, however, whether this mediation is in *agreement* with the objective ontological character of productive activity as the fundamental condition of human existence or *alien* to it, as in the case of capitalistic second order alienation" (ibid., p. 91). Axelos indicates much the same point in chap. 7.

40. Mészáros, *Marx's Theory of Alienation*, p. 36.

41. Ibid., p. 45 (Mészáros's italics).

42. See pp. 296–297, 298–299, 325–326. On this, Axelos reflects the insights of Heidegger into the nature of technology. See the latter's "Die Frage nach der Technik" and "Wissenschaft und Besinnung" in *Vorträge und Aufsätze* (Pfullingen: Neske, 1967). An English translation by William Lovitt of these essays and others of Heidegger's on technology has appeared as *The Question Concerning Technology and Other Essays* (New York: Harper Torchbooks, 1976).

43. Ellul, *Technological Society*, p. xxv.
44. Ibid., p. xxix.
45. Ibid., p. 128 (Ellul's italics).
46. Ibid., "The Characterology of Technique," chap. 2, pp. 61–147.
47. Ibid., p. 131.
48. Hans Jonas, "Seventeenth Century and After: The Meaning of the Scientific and Technological Revolution," in *Philosophical Essays: From Ancient Creed to Technological Man* (Englewood Cliffs: Prentice-Hall, 1974), pp. 45–80. The passage quoted is from p. 47.
49. Jonas, "Practical Uses of Theory," p. 198. Others have also made this point, and earlier than Jonas, most notably Max Scheler, Herbert Marcuse, and Max Horkheimer. See William Leiss, *The Domination of Nature* (Boston: Beacon Press, 1974), for a representation of their various analyses.
50. Jonas, "Practical Uses of Theory," p. 202.
51. For a study of this situation in the context of a broader perspective, i.e., one that sees it as expressing a historically conditioned, general social development, see Leiss, *Domination of Nature*. Leiss presents a position that, in attempting to clarify the meaning of "mastery of nature," is in effect critical of some of the central contentions in Axelos's interpretation of Marx.
52. See chaps. 11 and 14 and "The Ideological Problematic" and "The Problematic of Reconciliation as Conquest" in the conclusion.
53. Marx seems to *retain* science as empirical investigation even after alienation is superseded, precisely because the empirical study of natural structure and properties is concordant with technique as position-in-opposition, which is itself retained as basic. This is one of the points on which Axelos, then, finds Marx's explicit thought insufficiently radical. See chaps. 11 and 14.
54. See, e.g., pp. 155–158. Though this is Axelos's basic criticism of Marx, he makes critical suggestions of a variety of kinds throughout the book.
55. The set of eight questions and their replies was published under the title "Vers une pensée planétaire" in Axelos, *Arguments d'une recherche*, pp. 168–170. The passage quoted is from p. 168 (my translation).
56. From a vita provided by Axelos.
57. Axelos is founder and editor of the series, Arguments, for Les Editions de Minuit. A number of excellent books by various authors have appeared in this collection, both as original works and as translations. (Understandably, Axelos's own books appear in this series.) On the history and purpose of the series, see Axelos's remarks in *Arguments d'une recherche*, pp. 160–167.
58. Axelos, "Vers une pensée planétaire," pp. 169–170 (my translation).
59. On the meaning of this word, *errance*, see pp. xxvi ff.
60. Arguments 8 (Paris: Editions de Minuit, 1962). Axelos presented this work as the complementary thesis for satisfying the requirements for the French *doctorat ès lettres*.
61. The Marx book formed the principal thesis submitted for the *doctorat ès lettres*. It was first published by Minuit in 1961.
62. Arguments 21 (Paris: Editions de Minuit, 1964).
63. For a discussion of Heidegger that shows his importance, in tandem with Marx, for Axelos's thinking, see Kostas Axelos, "Marx und Heidegger,"

in *Einführung in ein künftiges Denken: Uber Marx und Heidegger* (Tübingen: Max Niemeyer, 1966), pp. 3—42; also the round-table discussion "Karl Marx et Heidegger," in Axelos, *Arguments d'une recherche*, pp. 93—107.

64. Some of these Heideggerian influences are apparent, e.g., in Axelos's method of "dialog" with a thinker from the past (here Marx), in his way of reviewing the history of Western philosophy, in his use of Greek words as philosophically primordial, and last, in certain neologisms noted here and elsewhere.

65. On translating this term, see Axelos's introduction, note 2.

66. Here Axelos follows Heidegger's thinking to a great extent, particularly as the latter develops it in *Der Satz vom Grund* (Pfullingen: Neske, 1957). See Axelos, *Vers la pensée planétaire*, pp. 21—22.

67. Axelos, *Arguments d'une recherche*, p. 170 (my translation).

68. See note 82 in this introduction.

69. The term is, of course, Heidegger's, and Axelos deems it best expressive of this copenetration of truth and error despite the fact that he finds earlier explicit recognition of this sort of thing in such cardinal thinkers as Heraclitus and Hegel. See *Vers la pensée planétaire*, pp. 185—186.

70. On the multiple meanings Axelos sees in the word *planetary*, depending in great part on a play of words, see ibid., p. 46.

71. From Axelos, "Thèses sur Marx—X," in ibid., pp. 176—177 (my translation).

72. Ibid., pp. 47—48 (my translation). It should be noted that there are, in fact, two ways of taking *errance* in Axelos. One is negative, and according to it *errance* means the inherent mistakenness of any position that, following the exigency to seek unequivocal explanation of the unambiguously fundamental, makes specific assertions about the truth of things. The other is positive, and according to it *errance* means the inherent ambivalence of the "play" of reality, which, in being recognized and made the rule of one's own activities, is thus taken seriously, as Axelos implies should be done, p. 12. The shift between the positive and the negative senses is evident, e.g., on pp. 292—293. (Cf. p. 297.)

73. Axelos, *Arguments d'une recherche*, p. 199 (my translation). The parallel between this statement and the last sentence of the conclusion of the present book should be noted.

74. Ibid., p. 174 (my translation).

75. Axelos, *Vers la pensée planétaire*, p. 17 (my translation).

76. Ibid. (my translation).

77. Ibid., p. 18 (my translation).

78. See Axelos, *Einführung in ein künftiges Denken*, pp. 27—42.

79. In "Introduction à la pensée planétaire," Axelos goes into detail regarding (1) "the great powers," i.e., "myth, religion, poetry, art, politics, philosophic thinking, science, and technique"; and (2) "the elementary forces" served by the great powers, namely, "language, labor, love, fight, play" ("play" in the sense of *games*, not in the sense of Axelos's all-encompassing *Play*).

80. See Axelos, "Art, technique et technologie planétaires," in *Arguments d'une recherche*, pp. 173—176.

81. See Axelos, "L'interlude," the last piece of *Vers la pensée planétaire*, pp. 321—328. This is one of the only two portions of Axelos's work that,

to my knowledge, have appeared previously in English. Together with pp. 20—24 of *Vers la pensée planétaire*, "L'interlude" has appeared under the title "Planetary Interlude" in *Game, Play, Literature*, ed. Jacques Ehrmann (Boston: Beacon Press, 1971), pp. 6—18.

82. Nicholas Lobkowicz, *Theory and Practice: History of a Concept from Aristotle to Marx*, International Studies of the Committee on International Relations (Notre Dame: University of Notre Dame Press, 1967), p. 354.

83. Some passages in the present book where these distinctions are indicated are the following: pp. 55, 56 note 6, chap. 7 throughout, 125 note 6, 219—220, 295, 298 ff., 307—308. In making the distinctions in the way I have, I have also relied on Mészáros, *Marx's Theory of Alienation*, esp. pp. 34—35, 87—92, 159, 245—250, and 14 note 3.

Introduction. On Understanding Marx's Thought

1. See the author's essays, "Pourquoi étudions-nous les présocratiques? " and "Le Logos fondateur de la dialectique" in *Vers la pensée planétaire*, pp. 67—92.

2. *Errance* in English is an odd neologism but one justified here by the distinctive usage Axelos makes of the French word *errance*. In French literary usage, *errer* has a double sense: "to be in error, to go astray," and "to wander, to journey." English retains the second only in the adjectival form, *errant*, of the verb *err*. The French, then, has a native linguistic resource for rendering a notion peculiar to the writings of Martin Heidegger, namely, that of a congenital mistakenness in Western thinking, even while—and precisely because—it searches wanderingly for a path of unmitigated, basic truth: *die Irre*. Axelos adopts the term *errance* with this complex meaning throughout the present study of Marx, as has been explained in my introduction. For a discussion of the Englishing of this notion of Heidegger's, see William J. Richardson, *Heidegger: Through Phenomenology to Thought*, Phaenomenologica 13 (The Hague: Martinus Nijhoff, 1963), pp. 223—227. —R. B.

3. Freud, too, will undertake a work of liberation from the curse attached to original sin, but for him it will be love, not productive labor, that is to be thus set free. To the Marxian objective, the emancipation of productive labor, corresponds the Freudian aim, the emancipation of love—both by the reflective consciousness of oppressive prohibitions.

4. I have tried to open discussion of "Marxist philosophy," a question that has been too much neglected in its most problematic aspects, in an article that was followed by answers to objections and questions that were raised. See Arguments 4, 1957, "Y a-t-il une philosophie marxiste?" and 5, 1957, "Marxisme 'ouverte' ou 'en marche'?." See also 7, 1958, "Thèses sur Marx." These texts are brought together in my *Vers la pensée planétaire*.

5. It should be mentioned that the French translation of these manuscripts which Axelos uses here (Karl Marx, *Economie politique et philosophie*, trans. J. Molitor, Ouevres philosophiques [Paris: Costes, 1927—1938], 6 [1937]: 9—135) omits the whole set of writings comprising the first manuscript. As a result, Axelos makes no references to any of the essays that make up that first manuscript, the most famous of which is the one that has been given

the title "Alienated (or Estranged) Labor." On the physical composition and disposition of contents in the original manuscript sheets, see "Publisher's Note," in Marx, *Manuscripts of 1844*, pp. 5—6.—R. B.

Part I. From Hegel to Marx

1. This expression, and the notion behind it, is taken directly from Heidegger. The way Axelos himself adopts it comes out in the course of chap. 1, but one can find Heidegger's own ideas on the matter in Martin Heidegger, "The Onto-Theo-Logical Constitution of Metaphysics," in *Identity and Difference*, trans. Joan Stambaugh (New York: Harper and Row, 1969), pp. 42—74. Briefly, Heidegger's point in regard to the course of Western metaphysics is that the intellectual search for the structure of being—ontology—by way of a rational articulation of that structure as fully grounded—logic—has generally been the quest for some ultimate and supreme Being—theology. Thus, Western metaphysics has had an *onto-theo-logical* constitution, culminating in Hegel.—R. B.

1. From Absolute Knowledge to Total Praxis

1. The writings of the young Hegel (*Die Phänomenologie des Geistes* [1807], *Wissenschaft der Logik*, 1 and 2 [1812 and 1816], *Encyklopädie der philosophischen Wissenschaften* [1817], *Grundlinien der Philosophie der Rechts* [1821], as well as *Vorlesungen über die Philosophie der Religion, Vorlesungen über die Aesthetik, Vorlesungen über die Geschichte der Philosophie,* and *Vorlesungen über die Philosophie der Geschichte*—sets of university lectures published after his death [1831] contain the methodic elaboration and systematic exposition of his thought. In the movement of his dialectical thinking, a thinking that is at the same time indissolubly logical, ontological, metaphysical, systematic, encyclopedic, and historical and in the doctrine thus constituted, Hegel wished to grasp in thought and express in language the logos of the total movement of universal and concrete reality.
2. G. W. F. Hegel, *Philosophy of Mind*, Pt. 3 of the *Encyclopedia of the Philosophical Sciences* (1830), trans. William Wallace, together with the *Zusätze* in Boumann's text (1845), trans. A. V. Miller (Oxford: At the Clarendon Press, 1971), p. 224 (no. 465).
3. G. W. F. Hegel, *The Phenomenology of Mind*, trans. J. B. Baillie (New York: Harper Torchbooks, 1967), p. 70.
4. One should not understand Hegel's role as the last philosopher in a dogmatic way. With Hegel, philosophy achieves its supreme destiny, whereupon it continues to lead an existence as an activity, and, more than that, becomes universal. It is to Heidegger's credit that he at least poses the question, when he asks: "Where must we seek the completion of modern philosophy? In Hegel, or not until the later philosophy of Schelling? And how about Marx and Nietzsche? Do they already step out of the course of modern philosophy? If not, how can we determine their place?" (Martin Heidegger, *What Is Philosophy?* trans. William Kluback and Jean T. Wilde [New Haven: College and University Press, 1962], p. 89). Heidegger tries too to make us attentive to Marx's voice: "Absolute metaphysics belongs with its inversions via Marx and

Nietzsche to the history of the truth of Being. Whatever stems from it cannot be affected or done away with by refutation." But it is not easy to open productive dialog with Marx and Marxism. "Because Marx, in discovering this alienation, reaches into an essential dimension of history, the Marxist view of history excels all other history. Because, however, neither Husserl nor, as far as I can see, Sartre recognizes the essentially historical character of Being, neither phenomenology nor existentialism can penetrate that dimension within which alone a productive discussion with Marxism is possible." Heidegger goes even further: "One can take various positions in regard to the theories (and arguments) of communism, but from the point of view of the history of Being, it is indisputable that in it an elementary experience has been made manifest of what is world-historical. He who takes 'communism' only as a 'party' or as '*Weltanschauung*,' is thinking just as narrowly as those who by the term 'Americanism' mean—and what is more in a depreciatory way—a particular mode of life" (Martin Heidegger, "Letter on Humanism," trans. Edgar Lohner, in *Contemporary European Thought*, pp. 206, 209—210, vol. 3 of *Philosophy in the Twentieth Century*, ed. William Barrett and Henry D. Aiken, 3 vols. [New York: Harper and Row, 1971]). What Heidegger says does not mean that he has himself accomplished what he proposes as a task.

5. Karl Marx and Frederick Engels, *The German Ideology*, ed. and trans. S. Ryazanskaya (Moscow: Progress Publishers, 1964), p. 206. The passage to which Marx refers is found in Hegel, *Phenomenology*, pp. 94—95.

6. Hegel, *Phenomenology*, p. 95.

7. Karl Marx and Frederick Engels, *The Holy Family or Critique of Critical Critique*, trans. R. Dixon (Moscow: Foreign Languages Publishing House, 1956), p. 160.

8. Ibid., pp. 253—254. [The word in angle brackets indicates a change of the original English rendering of Marx's term in order to maintain a conformity with Axelos's usage. See chap. 2, n. 6, and my introduction.—R. B.]

9. Ibid., p. 254.

10. Ibid., p. 106.

11. Hegel, *Phenomenology*, p. 805.

12. This and the following quotations are from ibid., pp. 807—808. [The italicizing in these quotations has been changed to accord with that in Axelos's French text, which, incidentally, is more faithful to that in Hegel's original German. The bracketed word is an insertion by Axelos.—R. B.

13. These two lines are a deformed quotation of the last two verses of the poem by Schiller entitled "Freundschaft."

14. Marx and Engels, *German Ideology*, p. 29.

15. Hegel, *Phenomenology*, p. 800. Baillie has *notion* where others would put *concept*. The latter term is the one normally used here to translate Hegel's term *Begriff* as Axelos treats it in French. See Walter Kaufmann, *Hegel: A Reinterpretation* (Garden City: Anchor Books, 1965), p. 145, for some discussion of the matter.

16. Marx and Engels, *German Ideology*, p. 27.

17. Ibid., p. 63.

18. Marx, *Manuscripts of 1844*, p. 64.

19. Ibid.

20. Ibid.

21. Ibid., p. 172.

22. Ibid.

23. Marx, "Theses on Feuerbach," in *German Ideology*, pp. 646–647.
24. Ibid., p. 647.
25. Marx and Engels, *German Ideology*, p. 47.
26. Hegel, *Phenomenology*, p. 366.
27. "... die Vernunft sich *alle Dingheit*, auch die *rein gegenständliche selbst* ist; sie ist aber dies *im Begriffe*, oder der Begriff nur ist ihre Wahrheit ..." (Hegel, *Phänomenologie des Geistes*, ed. Johannes Hoffmeister [Hamburg: Felix Meiner, 1952], p. 254. [My translation.—R. B.] See Hegel, *Phenomenology*, p. 371.
28. Karl Marx, *Capital: A Critique of Political Economy*, ed. Frederick Engels, 3 vols. (New York: International Publishers, 1967). The quotation is from vol. 1, p. 9, a reprint of the 1887 translation by Samuel Moore and Edward Aveling.
29. Marx, "Theses on Feuerbach," p. 645.
30. Italicizing by the author.
31. Marx, "Theses on Feuerbach," p. 647.

Part II. Economic and Social Alienation

2. Labor, the Division of Labor, and the Workers

1. Marx and Engels, *German Ideology*, p. 31.
2. Marx, *Manuscripts of 1844*, pp. 136–137.
3. Ibid., p. 182.
4. Ibid., p. 177. [Angle brackets indicate a change of the original English rendering of Marx's term in order to conform with Axelos's usage. See n. 6 below and my introduction. Here, as elsewhere, brackets indicate German words Axelos inserted for clarification.—R. B.]
5. Ibid.
6. Ibid., pp. 187–188. [Once again the words in angle brackets indicate the terms one should adopt in order to conform to Axelos's usage.—R. B.] One has to distinguish carefully and understand the specific difference between the terms for *expression, externalization*, and *alienation* that Marx uses in joining and opposing these notions. *Aussern* means "to express," "to manifest," while *Ausserung* is "expression," "manifestation." *Entäussern* means "*to externalize by relinquishing*" and *Entäusserung* is "externalization," "relinquishment." *Veräussern* means "to alienate," and *Veräusserung* is "alienation." *Entfremdung* also means "to alienate," and *Entfremdung* is "alienation," "alienness."
7. Ibid., p. 159.
8. Marx and Engels, *German Ideology*, p. 64.
9. Ibid., p. 46.
10. Ibid., p. 45.
11. Marx, *Manuscripts of 1844*, p. 129.
12. "Division of labour and private property are, moreover, identical expressions: in the one the same thing is affirmed with reference to activity as is affirmed in the other with reference to the product of the activity" (Marx and Engels, *German Ideology*, p. 44).
13. Marx, *Manuscripts of 1844*, p. 163.
14. It is essentially in Marx's economic writings (*A Contribution to the Critique of Political Economy* and *Capital*) that he will give his still unequaled analysis of

use-value, exchange-value, and surplus-value and of the relationship between commodities, prices, money, and capital. In résumé, one might recall here that labor, which is the source of all wealth, though not all by itself but in concert with nature, constitutes economic and social values in the form of objectified, realized, materialized, and crystallized labor. Values are distinguished into *use-values* and *exchange-values*. The first are the means of existence, the objects of human needs, things useful or enjoyable. These do not enter as such into the world of commodities. "To be a use-value is evidently a necessary prerequisite of the commodity, but it is immaterial to the use-value whether it is a commodity. Use-value as such . . . lies outside the sphere of investigation of political economy" (Karl Marx, *A Contribution to the Critique of Political Economy*, trans. S. W. Ryazanskaya, ed. Maurice Dobb [New York: International Publishers, 1970], p. 28). Exchange-value, on the other hand, "seems at first to be a quantitative relation, the proportion in which use-values are exchanged for one another . . . Quite irrespective . . . of their natural form of existence, and without regard to the specific character of the needs they satisfy as use-values, commodities in definite quantities are congruent, they take one another's place in the exchange process, are regarded as equivalents, and despite their motley appearance have a common denominator" (ibid., p. 4). Marx makes reference also to Aristotle (*Politics*, 1:9, 1257a), who distinguishes οἰκεία χρῆσις ("proper use") from ἀλλαγή ("barter"). Exchange-values constitute the economic value of commodities that result from crystallized (and alienated) social labor (Marx, *Critique of Political Economy*, p. 27).

The worker sells not his *labor* to live but his *labor-force*, and *surplus-value* results from the buying of the worker's labor-force by the capitalist. For the salary for labor does not represent the value (or the price) of the labor furnished; it is a *disguised* form of the value (or the price) of the labor-force. The worker works for the capitalist gratuitously a certain length of time, and this *surplus-labor* is the source of the *surplus-value* for the capitalist. "Therefore, the value of labour-power, and the value which that labour-power creates in the labour-process, are two entirely different magnitudes; and this difference of the two values was what the capitalist had in view, when he was purchasing the labour-power. . . . The seller of labour-power, like the seller of any other commodity, realises its exchange-value, and parts with its use-value" (Marx, *Capital*, 1:193). The worker is thus paid for only a part of his labor; the lengthening of the workday and the intensifying of productivity increase his gratuitous labor. "If we now compare the two processes of producing value and of creating surplus-value, we see that the latter is nothing but the continuation of the former beyond a definite point. If on the one hand the process be not carried beyond the point, where the value paid by the capitalist for the labour-power is replaced by an exact equivalent, it is simply a process of producing value; if, on the other hand, it be continued beyond that point, it becomes a process of creating surplus-value" (ibid., p. 195).

15. Author's italics. Here it is a matter of tasks that are *practical* and concrete, technical and real. Marx is little interested in great questions and problems that do not involve a practical solution.

16. Marx, *Critique of Political Economy*, pp. 20–21.

17. Ibid., pp. 21–22.

18. Engels will add: "that is, all *written* history." It is Engels who, with his *Origin of the Family, Private Property, and the State* (published in 1884,

therefore after Marx's death) gives to *Marxism* its dialectical schema of universal history: *thesis*—primitive communism, a society without private property and without classes; *antithesis* (negation)—societies divided into classes, especially Greco-Roman, slave societies, medieval and feudal societies, modern and bourgeois-capitalist societies; *synthesis* (negation of the negation)—socialism-communism, which is the passage from "pre-history," primitive *and* civilized, to true history. We do not find in Marx *this* vision of a primitive communism with entirely communal property, a kind of golden age and lost paradise which will be once again realized at the end of the dialectical process of universal history but on a conscious and higher level. Marx begins directly with historical man as someone who through social labor sets himself over against nature and alienates himself in and by labor, the division of labor, and his status as "laborer." But men can transcend that alienation and wholly dealienate themselves economically and socially. Nevertheless, Marx does admit a primitive sort of collective property, a form of primitive community from which private property and private right issued. "History has shown, on the contrary, that common property (*e.g.*, among the Indians, Slavs, ancient Celts, etc.) is the original form, and in the shape of communal property it plays a significant role for a long time" (Karl Marx, "Introduction to a Critique of Political Economy," in *Critique of Political Economy*, Appendix, pp. 192—193.) Marx admits that work in common under its still natural form is at the origin of the history of all civilized peoples, without this state of things being for all that a harmonious one and man truly man. In the *Critique of Political Economy*, he has a footnote consideration that he himself later quotes in a footnote in the second edition of *Capital* (1:77). He says: "At present an absurdly biased view is widely held, namely that *primitive* communal property is a specifically Slavonic, or even an exclusively Russian, phenomenon. It is an early form which can be found among Romans, Teutons and Celts, and of which a whole collection of diverse patterns (though sometimes only remnants survive) is still in existence in India. A careful study of Asiatic, particularly Indian, forms of communal property would indicate that the disintegration of different forms of primitive communal ownership gives rise to diverse forms of property. For instance, various prototypes of Roman and Germanic private property can be traced back to certain forms of Indian communal property" (*Critique of Political Economy*, p. 33).

Maximiliam Rubel wants to show, with quotations as his supporting evidence, that Marx thought that future society would be a rebirth of the archaic rural commune in a higher form. "Recalling the works of Henry Morgan, he [Marx] expresses his conviction that the rural commune 'answers to the historical current' of his time, and that the 'new system' toward which modern society is tending 'will be a rebirth in a higher form of an archaic social type' " (*Karl Marx: Essai de biographie intellectuelle* [Paris: Marcel Rivière, 1957], p. 433). [My translation—R. B.] Rubel quotes this phrase of Marx's without much reflection on it. He does not collate it with the ensemble of Marx's thought as aiming before all else at an original solution to the riddle of history, one made possible only by the amazing development of technique; for this is what, for the first time, allows the conquest of the socialized world by socialist man.

19. Karl Marx and Frederick Engels, "Manifesto of the Communist

Party," in *Birth of the Communist Manifesto*, ed. Dirk Struik (New York: International Publishers, 1971), p. 89.

20. Ibid., p. 90.

3. Private Property, Capital, and Money

1. Marx, *Manuscripts of 1844*, p. 136.
2. Marx and Engels, *German Ideology*, p. 33.
3. Ibid., pp. 35–36.
4. Marx, *Manuscripts of 1844*, p. 126.
5. Marx and Engels, *German Ideology*, p. 65.
6. Marx, *Manuscripts of 1844*, p. 132.
7. William Shakespeare, *Timon of Athens*, act 4, sc. 3.
8. Goethe, *Faust*, p. 1, trans. Philip Wayne (Hammondsworth: Penguin, 1949), p. 91; quoted in Marx, *Manuscripts of 1844*, p. 166.
9. Marx, *Manuscripts of 1844*, p. 168. In *Capital* (1:132), Marx refers also to a passage in Sophocles's *Antigone* (lines 295–301), in which the corrupting power of money is denounced.
10. Kathleen Freeman, *Ancilla to the Pre-Socratic Philosophers*, a complete translation of the Fragments in Diels, *Fragmente der Vorsokratiker* (Oxford: Basil Blackwell, 1948), p. 31. [It should be recalled that the present book on Marx forms a trilogy, *Le déploiement de l'errance*, of which Axelos's book on Heraclitus is the first volume: *Héraclite et la philosophie* (Paris: Editions de Minuit, 1962).–R. B.]
11. See Marx, *Capital*, 1:105. Marx knew very well Hegel's interpretations of Heraclitus (in Hegel's published lectures on the history of philosophy), as well as the huge monograph by Ferdinand Lassalle, *Die Philosophie Herakleitus des Dunklen von Ephesos*, published in Berlin in 1858 in two volumes. In a letter to Engels of February 1, 1858, Marx criticizes Lassalle for repeating in all their particulars Hegel's thoughts on Heraclitus. See Eduard Bernstein, "Vorbemerkung," in *Die Philosophie Herakleitus des Dunklen von Ephesos*, ed. Eduard Bernstein (Berlin: Paul Cassirer, 1920), 1:7–8.

4. The Machine, Industry, and Technicist Civilization

1. Marx and Engels, *German Ideology*, p. 39.
2. Marx, *Capital*, 1:179.
3. Marx and Engels, *German Ideology*, p. 39.
4. Ibid., p. 41.
5. Marx, *Manuscripts of 1844*, p. 142.
6. Marx and Engels, *German Ideology*, pp. 75–76.
7. Ibid., p. 76.
8. Marx, *Manuscripts of 1844*, p. 148.
9. Ibid., p. 153.
10. "In our days everything seems pregnant with its contrary. Machinery, gifted with the wonderful power of shortening and fructifying human labour, we behold starving and overworking it. The new-fangled sources of wealth, by some strange weird spell, are turned into sources of want. The victories of art

seem bought by loss of character. At the same pace that mankind masters nature, man seems to become enslaved to other men or to his own infamy. Even the pure light of science seems unable to shine out but on the dark background of ignorance. All our invention and progress seem to result in endowing material forces with intellectual life, and in stultifying human life into a material force. This antagonism between modern industry and science on the one hand, modern misery and dissolution on the other hand; this antagonism between the productive powers, and the social relations of our epoch is a fact, palpable, overwhelming, and not to be controverted. Some parties may wail over it; others may wish to get rid of modern arts, in order to get rid of modern conflicts. Or they may imagine that so signal a progress in industry wants to be completed by as signal a regress in politics. On our part, we do not mistake the shape of the shrewd spirit that continues to mark all these contradictions. We know that to work well the new-fangled forces of society, they only want to be mastered by new-fangled men—and such are the working men" (Karl Marx, "Speech at the Anniversary of the *People's Paper* [April 14, 1856]," in *Selected Works in Two Volumes*, by Karl Marx and Frederick Engels [Moscow: Foreign Languages Publishing House, 1955], 1:359—360). Marx is a man whose thinking centers on the question of technique. In making it his purpose that man move beyond self-alienation by the conquest of the world, he does not mean, by his observation that "the victories of art seem bought by loss of character," that he hopes for men to be delivered from modern technique. Marx's intention here is that technique be delivered from all that renders it inhuman, thus setting free both its own powers and those of human beings. [The word *art* in these quotations obviously means not one of the fine arts but craft, ingenuity, artifice—in short, technique. And, in fact, the French translation of this passage has rendered the word by *technique*, which in turn relates to the generally accepted translation of τέχνη by "art" in Greek philosophy, particularly Aristotle.—R.B.]

11. Marx, *Critique of Political Economy*, Appendix, pp. 209—210.
12. Ibid., p. 210.
13. Ibid.

Part III. Political Alienation

5. Civil Society and the State

1. Marx and Engels, *German Ideology*, pp. 47—48.
2. Ibid., p. 48.
3. Ibid., p. 91.
4. Ibid., p. 78.
5. G. W. F. Hegel, *Philosophy of Right*, trans. T. M. Knox (Oxford: Oxford University Press, 1967).
6. Karl Marx, "A Contribution to the Critique of Hegel's 'Philosophy of Right': Introduction," in *Critique of Hegel's "Philosophy of Right,"* ed. Joseph O'Malley, trans. Annette Jolin and Joseph O'Malley (Cambridge: At the University Press, 1970), pp. 129—142.
7. Marx, *Hegel's "Philosophy of Right,"* pp. 1—127.
8. Marx, *Manuscripts of 1844*, p. 184.
9. Ibid., pp. 185—186.

10. Ibid., pp. 184–185.
11. Marx, *Hegel's "Philosophy of Right,"* p. 45.
12. Marx is relentless when he launches an attack against established political life, bureaucratic administration, and state apparatus. It is a critique that ought as well be applied to the political realities that lay claim to Marx and Marxism. Marx makes his point with striking expressions: "Political life in the modern sense is the Scholasticism of popular life" (ibid., p. 32.) All through his critique of Hegel's political philosophy, Marx is pleased to salute his adversary every time the latter exposes administrative routine and typical civil service procedurism, mechanisms that have only the "horizon of a restricted sphere" (ibid., p. 54).
13. Marx and Engels, *German Ideology*, pp. 36–37.
14. Marx, *Hegel's "Philosophy of Right,"* p. 30.
15. Ibid., pp. 31–32.
16. Ibid., p. 32.
17. Ibid., pp. 29–30.
18. Ibid., p. 31.
19. Ibid., p. 46.
20. Ibid., pp. 46–47.
21. Ibid., p. 47.
22. Ibid., p. 48.
23. Marx and Engels, *German Ideology*, p. 45.
24. See ibid.
25. It goes without saying that in the mind of the founder of Marxism communist principles are to be given application in modalities that are concrete and particular. In the 1872 preface to the *Communist Manifesto*, Marx and Engels stress that "the practical application of the principles will depend ... everywhere and at all times on the historical conditions for the time being existing ... " ("Preface to the German edition of 1872," in *Birth of the Communist Manifesto*, p. 129). What we ourselves wish to underline is the *political* orientation that was already taken in the communist movement while Marx was still living. As the fierce adversary of Hegel's political philosophy and of politics in general, he could in no way avoid giving the emancipation of the workers a role that was political right from the first. Already in 1844 Marx wrote, in the Paris journal *Vorwärts*: "*Revolution* in general—the *overthrow* of the existing ruling power and the *dissolution* of the old conditions—is a *political act*. Without *revolution*, however, *socialism* cannot come about. It requires this *political act* so far as it needs *overthrow* and *dissolution*. But where its *organizing activity* begins, where its *own aim* and *spirit* emerge, there socialism throws the *political* hull away" ("Critical Notes on 'The King of Prussia and Social Reform' " [1944], in *Writings of the Young Marx on Philosophy and Society*, ed. and trans. Lloyd D. Easton and Kurt H. Guddat [Garden City: Anchor Books, 1967], p. 357).
26. V. I. Lenin, "What Is To Be Done? Burning Questions of Our Movement," in *Collected Works* (Moscow: Progress Publishers, 1964), 5:347–529 (originally published by Lenin in 1902).
27. It is remarkable to see Lenin calling attention to the way Marxian truth is forgotten: "And it is on this particularly striking point, perhaps the most important as far as the problem of the state is concerned, that the ideas of Marx have been most completely ignored! In popular commentaries, the thing done is to keep silent about it as if it were a piece of old-fashioned

'naïveté' just as Christians, after their religion had been given the status of a state religion, 'forgot' the 'naïveté' of primitive Christianity with its democratic revolutionary spirit" (V. I. Lenin, "The State and Revolution," in *Collected Works*, 25:420).

28. Plato, "The Republic," trans. Paul Shorey, in *The Collected Dialogues*, ed. Edith Hamilton and Huntington Cairns, Bollingen Series 71 (New York: Pantheon Books, 1961), p. 712 (473a).

29. Marx, *Capital*, 1:82.

30. Aristotle, "Metaphysics," trans. W. D. Ross, in *The Basic Works of Aristotle*, ed. Richard McKeon (New York: Random House, 1941), p. 712 (993b, 20).

Part IV. Human Alienation

6. The Relationship of the Sexes and the Family

1. Karl Marx, *The Poverty of Philosophy* (New York: International Publishers, 1963), p. 147.

2. See the beginning of chap. 2, "Labor, the Division of Labor, and the Workers," and of chap. 4, "The Machine, Industry, and Technicist Civilization."

3. The German word, *Verhältnis*, which Marx uses in this passage, means at the same time "condition" and "relation." Conditions are for Marx relations that both bind men and set them in opposition to nature and among themselves.

4. Marx and Engels, *German Ideology*, p. 40.

5. Ibid., pp. 40–41.

6. It is Engels who took up the task of tracing—in schematic fashion, as it turns out—the history of the family, in his *Origin of the Family, Private Property, and the State*, which appeared in 1884 (after the death of Marx). It is Engels, too, who contributed to the creation of the "dualist" theory of history, which Marxists, by the way, do not accept. In the preface to the first edition of *Origin*, Engels says: "According to the materialistic conception, the determining factor in history is, in the last resort, the production and reproduction of immediate life. But this itself is of a twofold character. On the one hand, the production of the means of subsistence, of food, clothing and shelter and the tools requisite therefore; on the other, the production of human beings themselves, the propagation of the species. The social institutions under which men of a definite historical epoch and of a definite country live are conditioned by both kinds of production; by the stage of development of labour, on the one hand, and of the family, on the other" (Friedrich Engels, "Preface to the First Edition 1884," in *The Origin of the Family, Private Property and the State*, in Marx and Engels, *Selected Works in Two Volumes*, 2:170–171). Engels, friend, collaborator, and popularizer of Marx, shows his acquaintance with the famous work of Bachofen entitled *Das Mutterrecht* (1861), and he relies on the two books of L. H. Morgan, *Systems of Consanguinity and Affinity* (1871) and, especially, *Ancient Society or Researches in the Lines of Human Progress from Savagery through Barbarism to Civilization* (1877), when he gives

the following outline of the evolution of the family: *primitive promiscuity*, consisting of untrammeled sexual commerce whose nature is not very precisely given by Engels, was followed by the *consanguine* family, in which parents and children are prohibited from mutual sexual commerce. This in turn was followed by the Punaluan family, in which relationship between brothers and sisters was forbidden, and then the *pairing family*, based on the union of couples, though the union was still of a rather elastic form. These are the principal forms of evolution of the family in the course of savage and barbarous prehistory, of which examples are still found in contemporary primitive peoples. With the *patriarchal family*, which marks the great historical defeat of the female sex, that is, the overturning of maternal right and the loss for woman of her domestic governing role, we begin the domain of (written) history. Finally, the last form of this narrowing process is the *monogamous family*, based on male domination and being the type of relation that, from the Middle Ages on, says Engels, if not to say beginning with Christianity, implies *individual sexual love*. Monogamy and individual sexual love, even while owing their origin to determinate property conditions, seem to constitute the highest type of relationship; and so, even in a postcapitalist future, they will be maintained. But what will disappear will be the predominance of the male and the indissolubility of the tie, that is, the difficulties that stand in the way of divorce. *Full equality of the two sexes* and *individual, shared sexual love* (as long as it lasts) will be then fully realized after capitalism is overturned, even if one cannot see what forms will follow in the evolution of the ties between the two sexes. The mutual inclination of man and woman will itself find adequate forms for its realization after the disappearance of all economic pressure.

7. Marx and Engels, *German Ideology*, pp. 42–43.
8. Ibid., p. 44.
9. Marx, *Manuscripts of 1844*, p. 133.
10. Ibid., p. 169.
11. Ibid. [Inserted comments by Axelos.–R. B.]
12. See Marx and Engels, *German Ideology*, p. 40, the passage given in a note at the bottom of the page.
13. Marx, *Manuscripts of 1844*, p. 133.
14. Ibid.
15. Marx, *Hegel's "Philosophy of Right,"* p. 40.
16. Marx, *Manuscripts of 1844*, p. 134.
17. In December of 1836, when Marx was eighteen, he gave to his "dear and ever-beloved Jenny von Westphalen" two exercise books full of lyric poems entitled *Buch der Liebe* [Book of Love].
18. Marx and Engels, *Holy Family*, pp. 31–32.
19. Ibid., p. 33.
20. Marx and Engels, "Manifesto of the Communist Party," pp. 108–109.
21. Marx and Engels, *German Ideology*, pp. 192–193.

7. Being, Making, and Having

1. *Faire* in French has a wide range of meanings, the basic sense of which is a transitive doing, or "making," in contrast to *agir*, which is an *in*transitive doing, or "acting." In this chapter *faire* is translated "making."–R. B.

2. When we find the word *Gegenstand* in Marx's German text, as here, it should be understood in general in its original meaning: *Gegen-stand*, "that which is over against and facing," "that which constitutes the other pole of a dialectical bond."

3. The word *sense* (French *sens*; German *Sinn*) used here, means in German, as in French, both "organs of sensation" and "meaning." [This obviously holds also in English.—R. B.]

4. Marx, "Economic and Philosophic Manuscripts (1844)," in Easton and Guddat, *Writings of the Young Marx*, pp. 307–308, 309, 310. The translation used in the edition usually adopted here (Karl Marx, *The Economic and Philosophic Manuscripts of 1844*, ed. Dirk J. Struik, trans. Martin Milligan [New York: International Publishers, 1964], pp. 139, 141) differs significantly from the French translation Axelos quotes. The rendering in Easton and Guddat conforms more to the latter and hence has been given.—R. B.

5. " . . . the so-called spiritual sense," Marx writes ("Economic and Philosophic Manuscripts [1844]," Easton and Guddat, p. 309).

6. One ought to keep in mind here once again this whole play of forces, powers, and impotencies that is revealed through the being of man and the being of things. Human being is a being that *is manifested* and *manifests* (*äussern*); and its essential activity is this *manifestation* (*Äusserung*). Nevertheless, this constitutes an *externality* (*Äusserlichkeit*) that, more than that, is a *giving up* (*Entäusserung*), an *alienation* (*Veräusserung, Entfremdung*). "Private property alienates the individuality not only of people but also of things" (Marx and Engels, *German Ideology*, p. 248. [On these distinctions see also Axelos's note 6 in chap. 2.—R. B.]

7. Marx, *Manuscripts of 1844*, p. 180.
8. Ibid., pp. 180–181.
9. Ibid., p. 140.
10. Ibid., p. 156.
11. Ibid., p. 150 [Inserted comment by Axelos.—R. B.]
12. Ibid.
13. Ibid., p. 151.
14. Ibid., p. 129.
15. Ibid., p. 138.

16. Marx calls the *drives* that activate natural human needs *Triebe*, which is the same word Freud uses to designate what we usually translated "instincts." For one, as for the other, man is first defined—and this is a metaphysical kind of definition—as a being that is animal, natural, and human and that is driven by its instincts toward the satisfaction of its needs in terms of sensuous objects (but not only in terms of materiality), all this being accomplished in the society of men. Yet the needs and desires remain unsatisfied and frustrated until a new order comes about. Needs themselves are nowhere arrested but increase to infinity. Marx and Freud require a full recognition and full satisfaction of the instincts (drives), without asking whether the dialectic thus conceived implies a term. The study of Marx and Freud—and of Marxism and psychoanalysis—is still to be made. There is a convergence particularly between the younger Marx and the older Freud, the Freud of *The Future of an Illusion* and *Civilization and Its Discontents*. (See "Le 'mythe médical' au XXe siècle" and "Freud analyste de l'homme," in Axelos, *Vers la pensée planétaire*, pp. 226–242, 243–272.) But neither Marx nor Freud said or suspected what someone

else, coming between them, was able to discern and dared to define as *will to power*, as will to infinite will. For it is the will to power that takes hold of men after the death of God and propels the human subject—not the subjective subject but the objective subject-man—toward the ever-unsatisfied and ever-increasing domination over the totality of "objects," all in a nihilistic world that has precipitated and dissolved Being into an endless and pointless becoming. It is of course Nietzsche that we are talking about.

17. Marx, *Manuscripts of 1844*, p. 181.
18. Ibid., p. 182. [The phrase, "i.e., to be real," omitted by Struik, has been added here to make the text conform both to the French version Axelos uses and to the original German text.—R. B.]
19. Ibid., p. 171.
20. Ibid., pp. 178—179.
21. Ibid., p. 138.
22. Marx, "Theses on Feuerbach," p. 645.
23. Ibid.
24. Marx, *Manuscripts of 1844*, p. 188. [The original text reads as follows: "... darum wird die Aufhebung der Entäusserung zu eine Bestätigung der Entäusserung, oder für Hegel ist jene Bewegung des *Selbsterzeugens*, des *Selbstvergegenständlichens* als *Selbstentäusserung und Selbstentfremdung* die *absolute* und darum die letzte, sich selbst bezweckende und in sich beruhigte, bei ihrem Wesen angelangte *menschliche Lebensäusserung*" (Karl Marx, "Ökonomisch-philosophische Manuskripte [1844]," in *Marx/Engels Gesamtausgabe* [*MEGA*], 1932; reprint ed. [Glashütten im Taunus: Verlag Detlev Auvermann, 1970], Abt. I, Bd. 3, p. 167). Bracketed words are the terms to be adopted to conform to Axelos's usage, from which, however, he departs in indicating that the first two instances of the word *Entäusserung* in the quotation means "alienation" here more than "externalization." See chap. 2, n. 6.—R. B.]
25. Marx, "Theses on Feuerbach, III," p. 646.
26. Marx, *Manuscripts of 1844*, pp. 151—152. [Inserted remark by Axelos.—R. B.]
27. Ibid., p. 148.
28. Ibid., pp. 155—156.

Part V. Ideological Alienation

8. Thought and Consciousness: True and Real?

1. "... finden wir, dass der Mensch auch 'Bewusstsein' hat. Aber auch dies nicht von vornherein, als 'reines' Bewusstsein" (Karl Marx and Friedrich Engels, *Die Deutsche Ideologie*, ed. V. Adoratskij, in *MEGA*, Abt. I, Bd. 5, p. 19). [My translation, conforming to Axelos's French version and differing slightly from the English translation generally quoted here: Marx and Engels, *German Ideology*, p. 41.—R. B.]
2. "Die Menschen haben Geschichte, weil sie ihr Leben *produzieren* müssen, und zwar auf *bestimmte* ... Weise: dies ist durch ihre physische Organisation gegeben, ebenso wie ihr Bewusstsein" (Karl Marx, *Die Frühschriften*, ed. Siegfried Landshut [Stuttgart: Alfred Kröner, 1968]). The

reading in an earlier edition of Landshut (1932) is the basis for the French translation Axelos uses. The text in *MEGA*, Abt. I, Bd. 5, p. 19, has the sentence punctuated differently, with some effect on the meaning. The standard English translation interprets the ambiguous last phrase of Marx's words in one determined way, whereas Axelos leaves it open to ambiguity (see Marx and Engels, *German Ideology*, p. 41).—R. B.

3. Marx and Engels, *German Ideology*, pp. 41–42.
4. Ibid., p. 42.
5. Ibid.
6. Marx, *Manuscripts of 1844*, p. 137.
7. Marx and Engels, *German Ideology*, p. 645.
8. Ibid., p. 42.
9. Ibid., p. 43.
10. Axelos's version replaces Marx's expression (*Organisation*) here with Engels's term (*Verhältnis*) (French *attitude*). See *MEGA*, Abt. I, Bd. 5, p. 569, for the critical text of this quote. The phrase in brackets is an explanation by Axelos.—R. B.
11. Marx and Engels, *German Ideology*, p. 37. [Axelos does not mention that this paragraph, which includes the first sentence of the next passage cited, was crossed out by Marx in the original. See *MEGA*, Abt. I, Bd. 5, p. 569.—R. B.]
12. This sentence is part of the crossed-out material referred to in the previous note. Its translation is modified here to conform to Axelos's reading of the text: *ideas* instead of *imagination, on its head* instead of *upside-down*. The German words are inserted by Axelos for the sake of precision.—R. B.
13. Marx and Engels, *German Ideology*, p. 37.
14. Ibid. [German words inserted by Axelos.—R.B.]
15. An alternate reading gives *Sublimate*. [The critical text has *Sublimate* (*MEGA*, Abt. I, Bd. 5, p. 16), while Landshut gives *Supplemente* (*Frühschriften*, p. 349). The English translation has *sublimates*, following *MEGA*, but this has been changed here to *supplements* to conform to Axelos's version.—R. B.]
16. Marx and Engels, *German Ideology*, pp. 37–38. [Inserted comment by Axelos.—R. B.]
17. So it is with language. All through *The German Ideology*, Marx speaks in ironic polemic against the tendency to make the words of a language say something through etymological analysis, amphibology, synonymy, and play on words. Language, as a historical, social product, falls under the blows of alienation and scarcely discloses any particular intrinsic truth. "For the bourgeois it is so much the easier to prove on the basis of his language, the identity of commercial and individual, or even universal, human relations, since this language itself is a product of the bourgeoisie, and therefore in actuality as in language the relations of buying and selling have been made the basis of all others. For example, *propriété*—property (*Eigentum*) and feature (*Eigenschaft*); property-possession (*Eigentum*) and peculiarity (*Eigentümlichkeit*); "*eigen*" ("one's own")—in the commercial and in the individual sense; *valeur*, value, *Wert*; commerce, *Verkehr*; *échange*, exchange, *Austausch*, etc., all of which are used both for commercial relations and for features and mutual relations of individuals as such" (Marx and Engels, *German Ideology*, p. 249). One has to know how to listen to and read that of which words and propositions are the paraphrase, for "the verbal masquerade only has meaning when it is the unconscious or deliberate expression of an actual masquerade" (ibid., p. 449).

18. *Ancilla to the Pre-Socratic Philosophers*, Fragment 50, p. 28. [In accord with Axelos's way of reading Heraclitus, the Greek *Logos* has been retained here instead of following the rendering in the English translation quoted, which has "Law" for *Logos*.—R. B.]

19. Ibid., Fragment 3, p. 42.

20. Marx, *Manuscripts of 1844*, p. 138.

21. See the letters Engels wrote to J. Bloch (September 21, 1890), H. Starkenburg (January 25, 1894), Conrad Schmidt (October 27, 1890), and F. Mehring (July 14, 1893). Engels recognizes that the production and reproduction of real life is the determining factor of history *only as what is ultimately decisive*. He says also that he and Marx partially bear the responsibility for the exaggeration of the economic dimension; for, preoccupied as they were with bringing out something their opponents denied, they did not always have the time, the place, or the occasion to do justice to other factors in the reciprocal action and reaction (*Wechselwirkung*) of all the forces involved. "Political, juridical, philosophical, religious, literary, artistic, etc., development is based on economic development. But all these react upon one another and also upon the economic base. It is not that the economic position is the *cause and alone active*, while everything else only has a passive effect. There is, rather, interaction on the basis of the economic necessity, which *ultimately* always asserts itself" (letter of Engels to Starkenburg, January 25, 1894, in *Selected Correspondence*, by Karl Marx and Frederick Engels, ed. V. Adoratsky, trans. Dona Torr [New York: International Publishers, 1942], p. 517). Yet, even though he says this, Engels continues to think that "ideological" powers are not worth very much: "The history of science is the history of the gradual clearing away of this nonsense [ideological stupidity resulting from economic underdevelopment] or of its replacement by fresh but already less absurd nonsense" (letter of Engels to Schmidt, October 27, 1890, in *Selected Correspondence*, pp. 482—483). [Explanatory insertion by Axelos.—R. B.]

9. Religion and Ideas

1. Marx, "Critique of Hegel's 'Philosophy of Right': Introduction," pp. 129—142.

2. Ibid., p. 131.

3. Ibid.

4. Karl Marx, "Theses on Feuerbach," pp. 646—647. The same idea is expressed as well in *Capital*: "The religious reflex of the real world can, in any case, only then finally vanish, when the practical relations of every-day life offer to man none but perfectly intelligible and reasonable relations with regard to his fellowmen and to Nature. The life-process of society, which is based on the process of material production, does not strip off its mystical veil until it is treated as production by freely associated men, and is consciously regulated by them in accordance with a settled plan. This, however, demands for society a certain material ground-work or set of conditions of existence which in their turn are the spontaneous product of a long and painful process of development" (Marx, *Capital*, 1:79—80).

5. Marx is the son of Jewish parents who "converted" to Protestantism. His father, son of a rabbi, was baptized before Karl was even born. His mother was descended also from a whole line of rabbis, and after the death of her parents

she entered the Evangelical Church. But Marx's parents, especially his father, were liberal Protestants, without deep religious faith. Karl Marx was baptized when he was six years old.

6. Karl Marx, "On the Jewish Question," in *Karl Marx, Early Writings*, ed. and trans. T. B. Bottomore (New York: McGraw-Hill, 1964), p. 34.

7. Ibid., pp. 37–38.

8. See also the first and eighth theses on Feuerbach, already cited.

9. Marx, "On the Jewish Question," p. 38.

10. Ibid., p. 39.

11. Ibid., p. 40.

12. In Axelos's own thinking, there are certain "things of great power" (*les grandes puissances*) that do act positively, that do variously perform a discovering, ordering, and transforming action upon the world: myth and religion, poetry and art, thought of a philosophical sort, science and technique. Unlike Marx, Axelos sees no one of them as basis for the others; indeed, all are modes of "deployment" of the one encompassing movement, play or, more specifically, the Play of the World (*le Jeu du Monde*). See Axelos, "Introduction à la pensée planétaire," pp. 11–49.–R. B.

13. See chap. 5, pp. 100–101.

14. This paragraph was crossed out by Marx in the original manuscript of *The German Ideology* and does not appear in the standard English translations. Axelos does not take note of its removal by Marx and quotes it as an integral part of the text. See Marx and Engels, *Die Deutsche Ideologie, MEGA*, Abt. I, Bd. 5, pp. 567–568.–R. B.

15. Marx and Engels, *German Ideology*, p. 43.

16. The terms *thoughts* (*Gedanken*) and *ideas* (*Ideen*) are, so to speak, used synonymously by Marx. The term *ideas*, however, is used in a more pejorative sense.

17. Marx and Engels, *German Ideology*, p. 60.

18. Marx and Engels, "Manifesto of the Communist Party," p. 99.

19. Lenin, "What Is To Be Done?" pp. 375–376.

10. Art and Poetry

1. Parts of a "confession-game" played by Marx, quoted by N. Rjasanoff, "Marx' Bekenntnisse," *Die Neue Zeit*, 1913, pp. 856–857. Reprinted also in Karl Marx and Frederick Engels, *Literature and Art: Selections from Their Writings* (New York: International Publishers, 1947), p. 145.

2. Marx and Engels, *German Ideology*, p. 657. [Axelos quotes this text without special mention, following Landshut. See Marx, *Die Frühschriften*, p. 412. *MEGA*, and the English translation, place it in an appendix. It figures in a set of what appears to be outline notes for further work, written on the last two pages of section 1, "Feuerbach," of *The German Ideology*.–R. B.]

3. Attacking Max Stirner, the apostle of oneness, in a passage dealing with Raphael, Leonardo da Vinci, and Titian, Marx writes: "Sancho [Stirner] imagines that Raphael produced his pictures independently of the division of labour that existed in Rome at the time. If he were to compare Raphael with Leonard da Vinci and Titian, he would know how greatly Raphael's works of art depended on the flourishing of Rome at that time, which occur-

red under Florentine influence, while the works of Leonardo depended on the state of things in Florence, and the works of Titian, at a later period, depended on the totally different development of Venice. Raphael as much as any other artist was determined by the technical advances in art made before him, by the organisation of society and the division of labour in his locality, and, finally, by the division of labour in all the countries with which his locality had intercourse" (Marx and Engels, *German Ideology*, p. 430).

4. Marx, *Manuscripts of 1844*, p. 136.

5. Ibid. [Parenthetic remark by Axelos.—R. B.]

6. The sole source for this early poem of Marx's seems to be Franz Mehring, ed., *Aus dem literarischen Nachlass von Karl Marx, Friedrich Engels und Ferdinand Lassalle, I—Gesammelte Schriften von Karl Marx und Friedrich Engels von März 1841 bis März 1844* (Stuttgart: Verlag J. H. W. Dietz, 1902), p. 28. My translation.—R. B.

7. Karl Marx, "Scorpion und Felix, Humoristischer Roman," *Werke und Schriften bis Anfang 1844*, ed. D. Rjazanov, in *MEGA*, Abt. I, Bd. 1, hbbd. 2, p. 83.

8. On Marx's poems, as well as his other poetic works, see his own comments in the beautiful letter he wrote to his father, November 10, 1837 (*Writings of the Young Marx*, pp. 40–50).

9. Karl Marx to Ferdinand Freiligrath, February 29, 1860, in Karl Marx and Friedrich Engels, *Werke* (Berlin: Dietz, 1972), 30:490. [My translation.—R. B.]

10. Karl Marx, "Aus den Vorarbeiten zur Geschichte der epikureischen, stoischen und skeptischen Philosophie," in *Werke und Schriften bis Anfang 1844, MEGA*, Abt. I, Bd. 1, hbbd. 1, p. 99. [My translation.—R. B.]

11. Karl Marx, "Differenz der demokritischen und epikureischen Naturphilosophie," in ibid., p. 30. [My translation.—R. B.]

12. Ibid., p. 40. [My translation.—R. B.]

13. Marx quotes this line from Lucretius, *De rerum natura*, 3:882 ("Differenz der demokritischen und epikureischen Naturphilosophie," *MEGA*, Abt. I, Bd. 1, hbbd. 1, p. 41), and Axelos takes it as typifying Marx's early interpretation of the passing of Greek civilization. My translation.—R. B.

14. This "introduction" is customarily found as an appendix to Marx's *A Contribution to the Critique of Political Economy*. Though he was working on it at the same time, Marx himself did not include it in that publication. It belongs more properly with the group of manuscripts Marx never did publish and that are now put together as the *Grundrisse der Kritik der politischen Ökonomie* (1939–1941; reprinted [Berlin: Dietz Verlag, 1953]. English translation by Martin Nicolaus, *Grundrisse: Foundations of the Critique of Political Economy* [New York: Random House, 1973]). Axelos always refers to it as the appendix to *Critique of Political Economy*, and so an English translation with the same placement was chosen to reflect this. See the following note.—R. B.

15. Marx, *Critique of Political Economy*, Appendix, p. 215.

16. Marx, *Critique of Political Economy*, Appendix, pp. 215–216. [Axelos's reading of the last sentence is simpler and more correct than the English translation quoted. The change is indicated.—R. B.]

17. This is where the *London Times* is printed.

18. Marx, *Critique of Political Economy*, Appendix, p. 216.

19. Axelos's italics.
20. Marx, *Critique of Political Economy*, Appendix, pp. 216–217.
21. Ibid., p. 217.
22. See Marx, *Capital*, 1:365.
23. *Franz von Sickingen* is a historical tragedy by Lassalle that deals with the uprising of the knights against the princes in the autumn of 1522, that is, two years before the Peasants' Revolt.
24. Karl Marx to Ferdinand Lassalle, April 19, 1859, in Marx and Engels, *Werke*, 29:592.
25. Marx and Engels, "Manifesto of the Communist Party," p. 92.
26. Ibid., p. 93.
27. Marx, *Manuscripts of 1844*, p. 142.
28. Ibid.
29. Ibid., p. 186.
30. Karl Marx, "Debatten über Pressfreiheit und Publikation der Landständischen Verhandlungen" (*Rheinische Zeitung*, May 19, 1842, no. 139), *Werke und Schriften bis Anfang 1844*, *MEGA*, Abt. I, Bd. 1, hbbd. 1, pp. 222–223.
31. Ibid., p. 223.
32. Marx, *Critique of Political Economy*, Appendix, p. 197.
33. Karl Marx, "The Eighteenth Brumaire of Louis Bonaparte," in Marx and Engels, *Selected Works in Two Volumes*, 1:247.
34. Marx, "Critique of Hegel's 'Philosophy of Right': Introduction," p. 134.
35. Marx and Engels, *German Ideology*, pp. 431–432. Marx makes clear, on the page preceding the passage just quoted, that one has to carefully distinguish "directly productive labour, which has to be organised" in communist society, and "labour which is not directly productive," by which each should be able to express himself freely.
36. G. W. F. Hegel, *Esthétique*, trans. S. Jankélévitch, 2 vols. (Paris: Aubier, 1944), 1:24. [Here and in the next quotation the French translation Axelos uses differs considerably from the German text (G. W. F. Hegel, *Asthetik* [Frankfurt: Europäische Verlagsanstalt, 1955], 1:21–22). The version given here is a translation of the French translation, inasmuch as that is the one Axelos is reading. The standard English translation follows the German more closely (G. W. F. Hegel, *The Philosophy of Fine Art*, trans. F. P. B. Osmaston [London: G. Bell, 1920], 1:12–13).–R. B.]
37. Hegel, *Esthétique*, 1:30.
38. Hegel, *Philosophy of Fine Art*, 1:13. [Italics in the first sentence are by Axelos.–R. B.]
39. Marx, *Critique of Political Economy*, Appendix, pp. 216–217.
40. Marx is strangely near to and yet distant from his contemporary, Rimbaud. Rimbaud also proclaims the death of language, poetry, and art, our absence from the world and the absence of the world; and at the same time he leaves himself open to a bright future, eager to hail "the birth of a new labor, and a new wisdom." But the language and vision of this "last poet" are essentially *poetic*. Beyond any established world and any revolutionary project, he sees something that need not "really exist" in order to have being. He speaks of what goes beyond "action, this dear point of the world," and says: "Eternal [i.e., omnitemporal] art would have its functions, since poets are citizens. Poetry will not lend its rhythm to action; it *will be in advance*

These poets will exist!" (Arthur Rimbaud, letter to Paul Demeny, May 15, 1871, in *Complete Works, Selected Letters*, trans. Wallace Fowlie [Chicago: University of Chicago Press, 1966], p. 309). Rimbaud is making an appeal to the new worker-poets, who were extremely bad. See the author's study, "Rimbaud et la poésie du monde planétaire," in *Vers la pensée planétaire*, pp. 139–171.

41. Marx, *Manuscripts of 1844*, p. 140.
42. Ibid., p. 141.
43. We have here the opening of Axelos's study of Marx in the direction of his own thinking, along the axis of what he names as Play. See below, pp. 257–258, 313–314. See also Kostas Axelos, *Le jeu du monde* (Paris: Editions de Minuit, 1969).—R. B.

11. Philosophy (Metaphysics) and the Sciences

1. Marx, *Manuscripts of 1844*, p. 174.
2. Marx and Engels, *German Ideology*, p. 255.
3. Marx, *Manuscripts of 1844*, p. 190.
4. Ibid., p. 175.
5. Marx and Engels, *German Ideology*, p. 57.
6. Ibid., p. 58.
7. Ibid. [The modification in the English translation follows the reading by C. J. Arthur, ed., *The German Ideology* (New York: International Publishers, 1970), p. 63.—R. B.]
8. Marx and Engels, *German Ideology*, p. 57.
9. Ibid., p. 38.
10. Marx, *Manuscripts of 1844*, p. 174.
11. Ibid., p. 171.
12. Ibid., pp. 153–154.
13. Marx, *Hegel's "Philosophy of Right,"* p. 89.
14. Marx, "Critique of Hegel's 'Philosophy of Right': Introduction," p. 132.
15. "Die Negation der seitherigen Philosophie, der Philosophie als Philosophie" (Karl Marx, "Zur Kritik der Hegelschen Rechtsphilosophie, Einleitung," in *Werke und Schriften bis Anfang 1844, MEGA*, Abt. I, Bd. 1, hbbd. 2, p. 613). The translation Axelos uses emphasizes the precise temporal meaning of *seitherig* out of its derivation more strongly than the standard English rendering, "previous." See "Critique of Hegel's 'Philosophy of Right': Introduction," p. 136.—R. B.
16. Marx attacks the whole philosophic tradition, pre-Hegelian, Hegelian, and post-Hegelian, and all philosophers and thinkers; whether they be writers and thinkers or professors of philosophy makes little difference. Regarding the latter, Marx has no great difficulty showing that they manage their "thinking" like a business in the manner of professionals. "The alteration of consciousness divorced from actual conditions—which philosophers pursue as a profession, i.e., as a *trade*—is itself a product of existing conditions and inseparable from them" (Marx and Engels, *German Ideology*, p. 414).
17. Marx, "Critique of Hegel's 'Philosophy of Right': Introduction," p. 135.
18. Marx, *Manuscripts of 1844*, p. 177.
19. Ibid., p. 173.
20. Marx, "Economic and Philosophic Manuscripts (1844)," in Easton and

Guddat, *Writings of the Young Marx*, p. 314. [The translation usually used here makes this passage quite obscure. See *Manuscripts of 1844*, p. 145.—R. B.]
    21. Marx, *Manuscripts of 1844*, p. 183.
    22. Ibid., p. 185.
    23. Ibid., p. 189.
    24. Hegel, *Philosophy of Mind*, p. 224 (no. 465).
    25. Marx, *Manuscripts of 1844*, p. 193.
    26. Hegel, *Phenomenology of Mind*, p. 801. [Axelos quotes this text from the French translation of Marx's *Manuscripts of 1844*. It is, as he indicates, a direct quotation by Marx from the *Phenomenology of Mind*, in, and ending, a few pages of unnumbered manuscript that comprise a fourth grouping, as the editor of the *MEGA* edition explains (*MEGA*, Abt. I, Bd. 3, p. 106). It is omitted in the available English translations of Marx's 1844 manuscripts.—R. B.]
    27. Marx and Engels, *Holy Family*, p. 30. [Engels is indicated as having written this passage.—R. B.]
    28. Marx, *Capital*, 1:19—20. [The emendation in the first line of the quotation follows both Marx's own German and Axelos's French version. Without it, the original English translation is misleading. The comment in brackets is by Axelos.—R. B.]
    29. Easton and Guddat, *Writings of the Young Marx*, p. 122.
    30. Marx, *Critique of Hegel's "Philosophy of Right,"* p. 89.
    31. Ibid.
    32. Ibid., p. 91.
    33. Marx cannot be more explicit in the formulation of this central idea: science is a product of industry. "Even this 'pure' natural science is provided with an aim, as with its material, only through trade and industry, through the sensuous activity of man" (*German Ideology*, p. 58). This is quite important. For even if *modern technique* appears as a product, as a result of science, it is in truth the activating spring and the secret of scientific development.
    34. Marx, *Manuscripts of 1844*, p. 152.
    35. An ironical allusion to Stirner's *Der Einzige und sein Eigentum* [The unique (i.e., the solitary individual) and his own] (Leipzig: Otto Wigand, 1845).
    36. Marx and Engels, *German Ideology*, p. 431.
    37. Marx, *Manuscripts of 1844*, p. 143.
    38. Ibid.
    39. Ibid., pp. 142—143.
    40. Ibid., p. 142.
    41. Marx and Engels, *German Ideology*, p. 658. (See n. 2 to chap. 10.)
    42. Marx, *Manuscripts of 1844*, p. 136. [Insertion by Axelos.—R. B.]
    43. Hegel, *Philosophy of Right*, p. 13.
    44. Marx and Engels, *Holy Family*, pp. 115—116.
    45. Marx and Engels, *German Ideology*, p. 647.

Part VI. The Prospect of Reconciliation as Conquest

12. The Premises for Transcending Alienation

    1. Marx, *Manuscripts of 1844*, p. 136.
    2. Ibid., p. 143.

3. Ibid., p. 137.
4. Marx's expression is "das Durchsichselbstsein der Natur und des Menschen . . . " (Karl Marx, "Ökonomisch-philosophische Manuskripte aus dem Jahre 1844," *MEGA*, Abt. I, Bd. 3, p. 124).
5. Marx, *Manuscripts of 1844*, p. 144.
6. Ibid., p. 130.
7. Ibid., p. 182.
8. Ibid., p. 181.
9. Ibid., p. 175.
10. Ibid., p. 135.
11. Ibid.
12. Ibid., p. 163.
13. Ibid., p. 139.
14. Ibid., p. 142.
15. Ibid., p. 143. [Explanatory insertion by the editor of the English edition. —R. B.]
16. Ibid., p. 145.
17. Marx and Engels, *German Ideology*, p. 647.
18. Ibid., p. 31.
19. Ibid., p. 57.
20. Ibid., p. 41.
21. Marx, *Manuscripts of 1844*, p. 182.
22. Ibid., p. 132. [Marx's words are: "Die Aufhebung der Selbstentfremdung macht denselben Weg wie die Selbstentfremdung" (*MEGA*, Abt. I, Bd. 3, p. 111).—R. B.]
23. Axelos has *suppression* in the French since that is the rendering of the German *Aufhebung* given in the French translation he uses. One should remember that the Hegelian term, *Aufhebung*, adopted here by Marx, means not just simple elimination, but rather doing away with X by fulfillment of the inherent possibilities of X as a conflict situation through transcending its conflict character into a higher positive integration. So the English in Struik has "transcendence." At other times the emphasis is more strongly upon the negative sense of *Aufhebung* and so the term is translated "abolition" or "annulment" (Axelos: *suppression*). Needless to say, how *Aufhebung* should be rendered in particular instances is a debatable point (see below, chap. 13, n. 8).—R. B.

13. Communism: Naturalism, Humanism, and Socialism

1. Marx, "Critique of Hegel's 'Philosophy of Right': Introduction," p. 133.
2. Ibid., p. 137.
3. Ibid.
4. Ibid.
5. Ibid.
6. Ibid., p. 142.
7. Ibid.
8. See chap. 12, n. 23, on Axelos's use of *suppression* in French for *Aufhebung* in German. Here, however, the term seems to be taken with the emphasis on the negative rather than positive aspect of a transcending/fulfilling action. Thus Axelos's *suppression* and O'Malley's "abolition," instead of a more comprehensive *dépasse-*

ment, or *transcendance* ("transcendence"). In addition *suppression* might also have a nondialectically negative meaning, in a matter that does not necessarily imply a positive Hegelian-Marxian movement. I have accordingly chosen to use "suppression" in instances where Axelos (using *suppression*) seems to be implying a doing away with by coercive dictate or coercive action, and "abolition" where there is elimination by more dialectically necessitated and integrated measures. Needless to say, the line between the two may not always be easily drawn, and coercion may be dialectically necessitated. At the same time it may occasionally be best to use the neutral "elimination" instead of either of the other renderings. —R. B.

9. Marx, *Manuscripts of 1844*, p. 154. Immediately after this sentence, Marx adds specific examples of dominant powers as forms of alienation: in Germany, *self-consciousness*, in France, *political equality*, in England, *real, material practical need* that has no other standard but itself.

10. Ibid., p. 133. [In Struik, following *MEGA* (Abt. I, Bd. 3, p. 112), the last sentence of the passage quoted here comes before the others given. Axelos's French version has inverted the order, according to Landshut's reading (Marx, *Die Frühschriften*, p. 233).—R. B.]

11. The expression is Marx's: "ganz rohen und gedanklosen Kommunismus" ("Ökonomisch-philosophische Manuskripte aus dem Jahre 1844," *MEGA*, Abt. I, Bd. 3, p. 112). See *Manuscripts of 1844*, p. 133.

12. Marx, *Manuscripts of 1844*, p. 133.

13. In the section devoted to the alienation of the relationship between the two sexes and the alienation of love, we had occasion to speak also of the kind of communism that opposes marriage in favor of the simple community of women.

14. Marx, *Manuscripts of 1844*, pp. 133–134.

15. Ibid., p. 134.

16. Marx, "On the Jewish Question," p. 15.

17. Ibid., p. 11.

18. Ibid., p. 16.

19. Ibid., p. 24.

20. Marx uses the French expression.

21. Marx, "On the Jewish Question," p. 31.

22. Marx, *Manuscripts of 1844*, p. 136.

23. Ibid., p. 126.

24. Ibid., pp. 145–146. [Parenthetic insertion by Axelos.—R. B.]

25. Ibid., p. 187.

26. Axelos: *suppression*. See this chap. n. 8.

27. Marx, "Hegel's 'Philosophy of Right': Introduction," p. 131.

28. Marx, *Poverty of Philosophy*, pp. 125–126. [Parenthetic comments by Axelos.—R. B.]

29. Marx and Engels, *German Ideology*, p. 56. [Parenthetic comment by Axelos. —R. B.]

30. Ibid., p. 87. "Seine Einrichtung ist daher wesentlich ökonomisch . . . " (Marx and Engels, *Die Deutsche Ideologie*, *MEGA*, Abt. I, Bd. 5, p. 60). [Parenthet phrase by Axelos.—R. B.]

31. "Dieser Kommunismus ist als vollendeter Naturalismus = Humanismus, als vollendeter Humanismus = Naturalismus . . . " (Marx, "Ökonomisch-philosophische Manuskripte aus dem Jahre 1844," *MEGA*, Abt. I, Bd. 3, p. 114).

32. Marx, *Manuscripts of 1844*, p. 135. [Parenthetic comment by Axelos. —R. B.]
33. Ibid., p. 137.
34. Ibid., pp. 137–138.
35. Ibid., pp. 138–139. [The English translation quoted here introduces a sentence division between the two phrases placed within angle brackets here. This is in accord with *MEGA*, Abt. I, Bd. 3, p. 118. Axelos's version is based on Landshut's reading (*Frühschriften*, p. 240), which, instead of introducing a sentence division between the two phrases, begins a second independent clause after them. The English translation has been modified in conformity with this latter version.—R. B.]
36. Ibid., pp. 143–144.
37. Ibid., p. 181.
38. Ibid., p. 158.
39. Sometimes Marx contrasts *idealism* with materialism, and sometimes *spiritualism* (or intellectualism—*spiritualisme*), and he does it quite summarily, without getting into the intricacies of logico-ontological problems. While wanting to transcend together (theoretical) materialism and idealism-spiritualism (these being also and even more theoretical and abstract) in the interest of a higher unity that would contain within itself their one-sided truths, and while not being a materialist ontologically speaking, since for Marx the material world is the product of human labor, he still defines himself, and the man of reconciliation, as a *practical materialist*, as a *communistic materialist*, as an *active, real humanist*. Practical, communistic, active, real materialism means this: all that is and becomes exists only as materiel and materials for human, social productivity.
40. " ... Man sieht, wie die Lösung der *theoretischen* Gegensätze selbst *nur* auf eine *praktische* Art ... möglich ist ... " (Marx, "Ökonomisch-philosophische Manuskripte aus dem Jahre 1844," *MEGA*, Abt. I, Bd. 3, p. 121). Marx himself underlines the words *only* and *practical*.
41. Marx, *Manuscripts of 1844*, pp. 141–142. [The phrase in angle brackets has been modified to accord with Axelos's reading of the text. Parenthetic comment by Axelos.—R. B.]
42. Ibid., p. 154.
43. Ibid., p. 183. [The French translation Axelos quotes (which is here represented by way of a distortion of the syntactic arrangement of the English version) does not quite follow the German. The entire sentence, which deals chiefly and directly with *Hegel's* notion of objectification and not Marx's own, runs as follows: "As we have already seen, the appropriation of what is estranged and objective, or the annulling of objectivity in the form of *estrangement* (which has to advance from indifferent foreignness to real, antagonistic estrangement), means equally or even primarily for Hegel that it is *objectivity* which is to be annulled, because it is not the *determinate* character of the object, but rather its *objective* character that is offensive and constitutes estrangement for self-consciousness."—R. B.]
44. Axelos quotes Landshut's reading here (*Frühschriften*, p. 243).—R. B.
45. Marx, *Manuscripts of 1844*, p. 141. [The expression in angle brackets is a modification of the English translation for the sake of clearer accord with Axelos's point in quoting the text.—R. B.]
46. Here, as elsewhere, especially in this chapter (see pp. 242 and 243),

Axelos seems to be referring to a distinction drawn from classical French linguistics: something sensuous (*le sensible*), on the one hand, that, on the other, operates as a sign and carrier (*le signifiant*) for the meaning signified, which is a spiritual element grasped by the mind. The distinction obviously implies a dualism, with the sensuous element having, consequently, two roles: (1) as itself, in some kind of material, mechanical, or behavioristic function, and (2) as semiotically instrumental to mental operations. See Ferdinand de Saussure, *Course in General Linguistics*, ed. Charles Bally, Albert Sechehaye, and Albert Riedlinger, trans. Wade Baskin (New York: Philosophical Library, 1959), pt. 1, "General Principles," chap. 1, "Nature of the Linguistic Sign," pp. 65–70.—R. B.

47. In the *Manuscripts* from 1844, Marx lays the foundation for his thinking as beginning with radical alienation and ending with total, communistic reconciliation. Deeming this foundation established, he no longer returns to it. The first part of *The German Ideology* (1845–1846), which he left also in manuscript state, tries to make clear what communism is in a more practical way. Both these works remained unpublished until 1932. The philosophical Marxism of the first thirty years of the twentieth century (Plekhanov, Lenin, Lukács, Korsch, etc.) was unaware of these basic writings, and that has had and still has an important effect historically and philosophically.

48. Marx and Engels, *German Ideology*, p. 46.

49. Marx, "Hegel's 'Philosophy of Right': Introduction," p. 142. [The italics are not in the English translation. Axelos's version has them, in accord with Marx's original German, and so they have been added here.—R. B.]

50. Marx and Engels, *German Ideology*, pp. 46–47.

51. This is the way Marx saw things "in his day," as one would say. The revolution, of course, began as *thought out* in Europe, but it did not begin to become reality in Europe, and especially not in Germany, as Marx had anticipated. Nor did it come about at once, simultaneously in many countries. This was and is a real historical fact heavy with consequences. Lenin characterized the epoch that came after Marx's as imperialist, imperialism being based on monopolistic concentration and the desire to take over the market in underdeveloped countries; and he saw that, owing to inequality in capitalistic development, revolution could be set in motion where the links in the imperialist chain were weakest, that is, in those countries that were economically backward and overexploited. It thus became easier to strike down capitalism at its most vulnerable point; for, in the handful of States that were especially rich and powerful and were exploiting the whole world to ensure their own excessive profit, one stratum of the proletariat, namely, the worker aristocracy, was corrupted by the bourgeoisie and broke off from the proletariat, making socialist revolution difficult. (See V. I. Lenin, "Imperialism, the Highest Stage of Capitalism," in *Collected Works*, 22:185–304.)

So, indeed, the weakest links in the imperialist chain, those countries that are technically underdeveloped and economically overexploited and, therefore, essentially agricultural and non-Western, with a relatively poorly developed state and military apparatus, can become the countries where revolution is launched. This revolution will begin as a bourgeois, democratic revolution whose aim is to liquidate those surviving elements of feudalism that are still strong, but then it will follow a path through the dictatorship of the proletariat (and its allies) toward socialism and communism. These countries will attain modern technique and become industrialized, thanks to this revolution. And this is what actually began to happen in Russia and then in China.

52. Marx and Engels, "Manifesto of the Communist Party," p. 125.
53. Marx and Engels, *German Ideology*, p. 47.
54. Marx and Engels, *Holy Family*, p. 52.
55. Ibid., pp. 52–54.
56. Ibid., p. 83.
57. Ibid., pp. 50–51.
58. Ibid., p. 62.
59. Ibid., pp. 85–86.
60. Ibid., p. 86.
61. Ibid., p. 85.
62. Marx and Engels, "Manifesto of the Communist Party," p. 99. See the end of chap. 9, "Religion and Ideas."
63. Marx and Engels, *German Ideology*, p. 51.
64. Ibid., p. 50.
65. Ibid., p. 62. [Parenthetic insertion by Axelos.–R. B.]
66. Ibid., p. 87. (See the quote on p. 238 of this chap.)
67. Ibid., p. 84.
68. Ibid., p. 47. [The sentence quoted here is not freestanding in Marx. The first comma in the quote has been added in accord with Marx's original text, which Axelos's version follows here.–R. B.] Marx's words at the end of this sentence read: " . . . wieder in ihre Gewalt Bekommen" (*MEGA*, Abt. I, Bd. 5, p. 25); yet "men get . . . under their control again" what never was under their actual control.
69. Ibid., p. 49.
70. Ibid., p. 91.
71. Ibid., p. 67.
72. Axelos reads Marx's expression *Abstreifung aller Naturwüchtigkeit* rather differently from the English translators. The standard rendering of the expression *Naturwüchtigkeit* is "natural limitations" (*German Ideology*, p. 84). Axelos sees it as meaning more radically the quality by which something is of nature, natural. In other words, he sees an explicit ontological sense of heavy import in what Marx is saying.–R. B.
73. Marx and Engels, *German Ideology*, p. 65.
74. Ibid., pp. 44–45.
75. Marx, *Capital*, 1:178.
76. Marx and Engels, *German Ideology*, p. 45.
77. Marx to Kugelmann, London, April 12, 1871, in Marx and Engels, *Selected Correspondence*, p. 309.
78. Karl Marx, "Critique of the Gotha Program," in Marx and Engels, *Selected Works in Two Volumes*, 2:32–33.
79. Marx details the nature of his contribution to the problem of social classes in a letter written from London to Joseph Weydemeyer, March 5, 1852: "And now as to myself, no credit is due to me for discovering the existence of classes in modern society nor yet the struggle between them. Long before me bourgeois historians had described the historical development of this class struggle and bourgeois economists the economic anatomy of the classes. What I did that was new was to prove: (1) that the *existence of classes* is only bound up with *particular, historic phases in the development of production*; (2) that

the class struggle necessarily leads to the *dictatorship of the proletariat*; (3) that this dictatorship itself only constitutes the transition to the *abolition of all classes* and to a *classless society*" (Marx and Engels, *Selected Correspondence*, p. 57).

80. Marx, "Critique of the Gotha Program," 2:23.
81. Ibid., p. 24.
82. Marx, *Poverty of Philosophy*, p. 174.
83. Marx and Engels, "Manifesto of the Communist Party," p. 111.
84. Ibid., p. 112.
85. Marx and Engels, *Selected Correspondence*, pp. 336–337.
86. In "The State and Revolution, the Marxist Theory of the State and the Tasks of the Proletarian in the Revolution" (*Collected Works*, 15:381–491), Lenin wanted to "*re-establish* what Marx [and especially Engels] really taught on the subject of the state" (p. 386). He sets the tasks for the proletarian revolution in regard to the State as follows: destroy the old bureaucratic, military apparatus, *smash* the old bureaucratic, military machine of the existing State. The abolition of the bourgeois State is, on the basis of "proletarian centralism" (p. 430) and "the socialist reorganization of the State," the transitional revolutionary form indispensable to the State. The new State abolishes the standing army and replaces it with an armed population. It likewise abolishes parliamentarianism and bureaucracy; it establishes a general eligibility for all civil servants and at the same time makes them all subject to dismissal, at any moment and without exception, with all their salaries reduced to the level of the workers' pay. All public functions are transformed into administrative functions, and thus they cease to be, properly speaking, political. The new State advances toward its withering away; "at a certain stage of this process, the state which is withering away may be called a non-political state" (p. 438). Institutions themselves will cease to exist. "For, in order to abolish the state, it is necessary to convert the functions of the civil service into the simple operations of control and accounting that are within the scope and abilities of the vast majority of the population . . . " (p. 452). The process of withering away will be long and harsh, and it will actually take place only in integral communism. Marxists and anarchists, therefore, have a common *goal*: the classless society free of any State. They are, however, radically distinct as to the *means* advocated to get there. Lenin wrote all this in the fall of 1917, on the eve of the revolution whose purpose was to achieve the planned organization of Russian economy and society. The soviet State has since covered some ground along its way. It should be emphasized that Lenin gave no indications about the proletariat *party* that, from being revolutionary, would transform itself one day into the party in power, the State party.
87. Marx, *Manuscripts of 1844*, p. 140.
88. Marx and Engels, *German Ideology*, pp. 83–84.
89. Marx to Kugelmann, London, April 17, 1871, in Marx and Engels, *Selected Correspondence*, pp. 310–311. This letter follows the one to Kugelmann previously quoted.
90. Marx and Engels, *German Ideology*, pp. 86–87.
91. Ibid., p. 87.
92. Ibid., p. 49. [The italics in the parenthetic phrase are Axelos's. The rendering of that phrase in the English translation has been modified to accord with Axelos's reading here, which more closely follows Marx's actual wording. See

Conclusion, note 75.—R. B.]
93. Marx, "Critique of the Gotha Program," 2:23.
94. Marx, *Capital*, 1:79.
95. Marx, "Critique of the Gotha Program," 2:24.
96. Marx and Engels, *German Ideology*, p. 47.
97. Ibid., p. 231. [Marx is here referring to, and rejecting as insufficient, Stirner's representation of communism as being primarily based upon a theoretical definition, namely, the specifically social definition of the essence of man. —R. B.]
98. Marx, *Manuscripts of 1844*, p. 146. [Struik remarks on the uncertainty in interpreting the point Marx is making in this last sentence of this particular manuscript portion.—R. B.] The German wording is as follows: "Der *Kommunismus* is die notwendige Gestalt und das energische Prinzip der nächsten Zukunft, aber der Kommunismus ist nicht als solcher das Ziel der menschlichen Entwicklung—die Gestalt der menschlichen Gesellschaft—"(*MEGA*, Abt. I, Bd. 3, p. 129).
99. Marx, *Manuscripts of 1844*, p. 154. The German text of this crucial passage is as follows: "Die Geschichte wird sie bringen und jene Bewegung, die wir in *Gedanken* schon als eine sich selbst aufhebende wissen, wird in der Wirklichkeit einen sehr rauhen und weitläufigen Prozess durchmachen. Als einen wirklichen Fortschritt müssen wir es aber betrachten, dass wir von vornherein sowohl von der Beschränktheit als dem Ziel der geschichtlichen Bewegung, und ein sie überbietendes Bewusstsein erworben haben" (*MEGA*, Abt. I, Bd. 3, pp. 134–135).
100. Marx, *Manuscripts of 1844*, p. 146.
101. Marx and Engels, *German Ideology*, p. 89.
102. Marx, "Zur Kritik der Hegelschen Rechtsphilosophie, Einleitung," *MEGA*, Abt. I, Bd. 1, hbbd. 1, p. 616. [My translation—R. B.] See "Hegel's 'Philosophy of Right': Introduction," p. 138.

14. The Abolishing of Philosophy by Its Realization

1. Marx, "Differenz der demokritischen und epikureischen Naturphilosophie," *MEGA*, Abt. I, Bd. 1, hbbd. 1, pp. 64–65. [My translation.—R. B.]
2. Ibid., p. 64. [My translation.—R. B.]
3. Ibid. [My translation.—R. B.]
4. Marx, "Zur Kritik der Hegelschen Rechtsphilosophie, Einleitung," in *MEGA*, Abt. I, Bd. 1, hbbd. 1, p. 613. [Axelos emphasizes the dual meaning of the Hegelian term *Aufheben* here, which is not done in the English translation of this passage by O'Malley and Jolin (*Critique of Hegel's "Philosophy of Right,"* p. 136). My translation.—R. B.]
5. Marx, "Critique of Hegel's 'Philosophy of Right': Introduction," pp. 141–142.
6. Karl Marx, "The Leading Article in No. 179 of the *Kölnische Zeitung*," in Easton and Guddat, *Writings of the Young Marx*, p. 129.
7. Ibid., pp. 214–215.
8. Ibid., p. 215.
9. Marx, *Critique of Hegel's "Philosophy of Right,"* p. 45.
10. See Marx and Engels, *German Ideology*, p. 41.
11. Ibid., p. 56.
12. Ibid., p. 87.

13. Marx, *Manuscripts of 1844*, p. 180.
14. Ibid., p. 165.
15. Ibid.
16. Axelos's expression, which may seem odd, is *l'être ontologique*, and it is plainly modeled upon Marx's term in the very next quote here from Marx's *Manuscripts: das ontologische Wesen*. There the translator, Milligan, has "ontological essence." Struik explains in a note (*Manuscripts of 1844*, pp. 59–60) the many senses *Wesen* can have, clarifying the seeming inconsistency in English. Here Marx and Axelos mean the expression in the sense of that core structure that gives a particular phenomenon its specific place in being precisely as actual being.—R. B.
17. Marx, *Manuscripts of 1844*, p. 165.
18. See n. 16.
19. Marx, *Manuscripts of 1844*, p. 165.
20. Ibid., pp. 142–143.
21. Ibid., p. 143.
22. Ibid.
23. Ibid.
24. Ibid.
25. Ibid., p. 138.
26. Ibid., p. 143.
27. Ibid., p. 135.
28. Marx and Engels, *German Ideology*, p. 28. [The text quoted is part of a passage Marx crossed out in the original manuscript. The English translation has been modified to accord with the stronger sense of the German *bedingen*, as read in the French translation Axelos uses.—R. B.]
29. Ibid., p. 647.
30. Ibid., p. 646.
31. Ibid., p. 38.
32. Ibid.
33. Ibid., p. 41.
34. Marx, *Critique of Political Economy*, p. 21.
35. Freeman, *Ancilla to the Pre-Socratic Philosophers*, Fragment 52, p. 28.
36. Ibid., Fragment 93, p. 31.
37. *Met.*, Z, 1, 1028*b*, in *The Basic Works of Aristotle*, pp. 783–784. [Translation slightly modified in accord with the author's following note.—R. B.] The translation of οὐσία by "beingness" (*étantité*) rather than by "essence" (*essence*) or, what is worse, by "substance" (*substance*) retains the participial form. See translators' notes to "Qu'est-ce que la philosophie?" by Martin Heidegger, trans. Kostas Axelos and Jean Beaufret, in Heidegger, *Questions II* (Paris: Gallimard, 1968), pp. 39–40.
38. Marx and Engels, *German Ideology*, p. 58.
39. Ibid., p. 59.
40. The very problem of thought and language, i.e., of the Logos rather than of Logic, is held in suspense in Marx's thought. Does the transcending of philosophy open the way to a new type of *thought*? Marx does not speak of *thought* of this kind. As for language, he tells us that it will be taken under the absolute control of the individuals in the communist society. Does this mean that it will be entirely subjected to planning and brought down to one level, that it will become a technical instrument? "As a matter of course, the individuals at some

time will take completely under their control this product [i.e., language] of the species as well" (Marx and Engels, *German Ideology*, p. 469).
41. Ibid., p. 49. [Translation slightly modified. See chap. 13, n. 92.—R. B.]

Conclusion. Open Questions

1. The young Lukács, writing *History and Class Consciousness* in 1923, sets off to war against the dialectic of nature. For him, materialism is essentially historical, and materialist dialectic aims for *totality* as the whole of experience, the coherent system of all the facts known and produced by historical, social subjects, objective subjects, object/subjects; for facts are *facts*. For Lukács, the "total character of dialectical method" lies in its "knowledge of what is real in historical becoming" (Georg Lukács, *History and Class Consciousness*, trans. Rodney Livingstone [Cambridge: MIT Press, 1971], pp. 10—15). Lukács writes: "Only when the core of existence stands revealed as a social process can existence be seen as the product, albeit the hitherto unconscious product, of human activity. This activity will be seen in its turn as the element crucial for the transformation of existence. Man finds himself confronted by purely natural relations or social forms mystified into natural. They appear to be fixed, complete and immutable entities which can be manipulated and even comprehended, but never overthrown. But also this situation creates the possibility of praxis in the individual consciousness" (ibid., pp. 18—19). Lukács, however, does not go deeper into what he calls the real. Need one recall here that this work was condemned by the Communist International (1924) and by social-democratic orthodoxy, i.e., by Zinoviev as well as Kautsky, and that the author himself repudiated it? Need one as well mention that all this agreement about the *idealist* tendency of the work explains little in regard to what this idealism is or is not? [Axelos is cotranslator, with Jacqueline Bois, of Lukács's book into French: *Histoire et conscience de classe* (Paris: Editions de Minuit, 1960).—R. B.]
2. Axelos's point depends on the phrasing of the French here: "Par un immense effort de réduction, il ramène—en le rétrécissant—tout ce qui est et se fait (et *est* ce qui *est fait*) à la production."—R. B.
3. Marx, *Poverty of Philosophy*, p. 133.
4. Ibid., p. 109.
5. Ibid., p. 138.
6. Ibid.
7. Ibid., p. 144.
8. *The Poverty of Philosophy* is Marx's reply to Proudhon's *Système des contradictions économiques ou Philosophie de la misère*. Rejecting Proudhon's position, he sets off to battle against the notion of constantly distinguishing two sides to something, one good and one bad, one worthwhile and full of advantages, the other worthless and full of disadvantages. He charges Proudhon with conceiving contradiction as formed by a combination of a good and a bad side and with wanting to keep the good side while eliminating the bad (Marx, *Poverty of Philosophy*, pp. 111—113). In this way no decision can really be taken. The wholesale freeing of the motor agency of development and the elimination of what stands as an obstacle to it are made impossible. Marx is not afraid to be one-sided; does he not work for the total and brilliant flowering of productive forces, the releasing of technique from any restraints? Proudhon in his eyes is

a coward. "The petty bourgeois is composed of On The One Hand and On The Other Hand. This is so in his economic interests and *therefore* in his politics, in his scientific, religious and artistic views. It is so in his morals, in everything. He is a living contradiction. If, like Proudhon, he is in addition a gifted man, he will soon learn to play with his own contradictions and develop them according to circumstances into striking, ostentatious, now scandalous or now brilliant paradoxes" (ibid., p. 202). Inasmuch as Marx's confidence lay in negativity, the motor force of history, his position is that "it is always the bad side that in the end triumphs over the good side. It is the bad side that produces the movement which makes history, by providing a struggle" (ibid., p. 121). Does this man whose thinking is a reflection on technique follow this idea to its ultimate consequences? Did the bad side, the productive agency of historical movement, always remain before his mind? For it would seem he could not escape the temptation to want to neutralize it.

9. Ibid., p. 42.
10. Ibid., p. 68.
11. Ibid., p. 78.
12. Marx, "Critique of Political Economy," Appendix, p. 195.
13. Ibid., p. 196.
14. Ibid.
15. Marx and Engels, "Manifesto of the Communist Party," p. 92.
16. Karl Marx, *Grundrisse der Kritik der politischen Okonomie* (1939–1941; reprinted [Berlin: Dietz Verlag, 1953]), p. 25. [My translation.–R. B.] See Marx, *Critique of Political Economy*, Appendix, pp. 196–197.
17. Marx, *Critique of Political Economy*, Appendix, p. 215.
18. Marx and Engels, *German Ideology*, p. 31.
19. Marx, *Critique of Political Economy*, Appendix, p. 197.
20. Ibid., pp. 204–205. [I have changed "whole" in the translation to "totality" simply because of the extent to which Axelos uses the latter word, emphasizing it here as well by inserting the original German term Marx employs.–R. B.]
21. Ibid., p. 205.
22. Ibid., p. 215.
23. Marx and Engels, "Manifesto of the Communist Party," p. 110. [Axelos is obviously urging us to read the German more literally than is usually done, as for example here in the English translation.–R. B.]
24. Marx, *Capital*, 1:82.
25. Marx, *Critique of Political Economy*, Appendix, p. 210.
26. Marx and Engels, "Manifesto of the Communist Party," pp. 93–94.
27. Marx, *Critique of Political Economy*, Appendix, pp. 210–211.
28. Ibid., p. 211.
29. Ibid. In *Capital*, Marx once again insists upon this limited validity for the categories of bourgeois economics. "They are forms of thought expressing with social validity the conditions and relations of a definite, historically determined mode of production, viz., the production of commodities" (1:76).
30. Marx, *Critique of Political Economy*, Appendix, p. 211.
31. See Marx and Engels, *German Ideology*, pp. 51, 52–53.
32. Ibid., p. 59.
33. See ibid., pp. 48–49.
34. Ibid., p. 47.

35. Marx, *Critique of Political Economy*, Appendix, p. 211.
36. Easton and Guddat, *Writings of the Young Marx*, pp. 212–213.
37. Marx and Engels, *German Ideology*, p. 49.
38. Marx, *Manuscripts of 1844*, p. 135.
39. Ibid., p. 154.
40. See Marx and Engels, *German Ideology*, p. 87.
41. Marx, *Critique of Political Economy*, Appendix, p. 193.
42. Marx and Engels, "Manifesto of the Communist Party," p. 104.
43. Ibid., p. 105.
44. Marx, *Capital*, 1:78–79.
45. Marx and Engels, *German Ideology*, p. 44.
46. Marx, *Capital*, 1:79.
47. Ibid., p. 80.
48. Ibid., p. 77.
49. Ibid., 3:820.
50. Marx and Engels, *German Ideology*, p. 45.
51. Marx, *Manuscripts of 1844*, p. 155.
52. Marx and Engels, *German Ideology*, p. 51.
53. Marx to Ruge, September 1843, in Easton and Guddat, *Writings of the Young Marx*, p. 213.
54. Marx, "On the Jewish Question," p. 31.
55. On these topics, see the author's essay, "La politique planétaire," in *Vers la pensée planétaire*, pp. 297–318.
56. Marx, *Hegel's "Philosophy of Right,"* p. 134. [The word in angle brackets has been modified to conform to Axelos's reading of Marx's term here.—R. B.]
57. In his reflections Marx never carried right to the end, in the dimension of the future, the lessons taught him by the revolutions of the past whose destiny he had learned to analyze. He did not see the kind of fate that befalls every revolution (even a permanent one) when it is transformed into an existing state of things (even if still fluid); he did not stop to consider the difference between movement and forward impetus, on the one hand, and an established regime, on the other. For him, the truth of innovative drive and the action of technical organization could, for the first time in the history of the world, become one thing.
58. Marx to Ruge, September 1843, in Easton and Guddat, *Writings of the Young Marx*, p. 214.
59. Marx and Engels, *German Ideology*, p. 56.
60. Ibid., p. 59.
61. Ibid., p. 414.
62. Marx and Engels, *Holy Family*, pp. 254–255.
63. Ibid., p. 73. The antagonism between individual and mass is destined to be transcended in the mass existence of individuals and in the individualization of members of the mass. The depth of historical action is inseparable from its breadth. "With the thoroughness of the historical action the size of the mass whose action it is will therefore increase" (ibid., p. 110). No more will there be a certain few individuals, the "elect," who are bearers of spirit and who, as creative and active spirit, stand over against the rest of mankind considered as the mindless mass, as mere matter, as a supply of simple material.
64. Marx, *Manuscripts of 1844*, p. 138.
65. See Heidegger, "Principle of Identity," in *Identity and Difference*, pp. 23–41.

66. Marx, *Capital*, 1:178.
67. Marx and Engels, *Holy Family*, p. 168.
68. "The philosophical task is not the embodiment of thought in determinate political realities [and in practical realities in general], but the evaporation of these realities in abstract thought. The philosophical movement is not the logic of fact but the fact of logic" (Marx, *Hegel's "Philosophy of Right,"* p. 18). [The parenthetic insertion is by Axelos.—R. B.]
69. See this chapter, p. 310.
70. See "Principles of Communism," by Frederick Engels (October 1847), in *Birth of the Communist Manifesto*, p. 187.
71. Marx, *Manuscripts of 1844*, p. 135.
72. Marx and Engels, *German Ideology*, pp. 86–87.
73. For until alienation is overcome, "the upside down world is the real world," Marx writes to Ruge ("Ein Briefwechsel von 1843," in *MEGA*, Abt. I, Bd. 1, hbbd. 1, p. 563). See Easton and Guddat, *Writings of the Young Marx*, p. 208.
74. Marx and Engels, *German Ideology*, p. 49. [The modification in the English wording here, indicated for the first time in chap. 13, n. 92, is explained in n. 75 immediately below.—R. B.]
75. Marx's wording in the German reads almost as if he were making a concession in regard to the existence of this other-than-material order of production: "... mit der Produktion (auch mit der geistigen) der ganzen Welt in praktische Beziehung ... " (*MEGA*, Abt. I, Bd. 5, p. 26). At least this is how Axelos reads it, following the more literal French translation. The English translation as unmodified has flattened out the qualifying phrase, eliminating its parenthetic character together with this more explicit problematic overtone.—R. B.
76. Marx, *Hegel's "Philosophy of Right,"* p. 32.
77. Ibid., p. 82.
78. Marx, *Capital*, 1:178. [In the English, the angle brackets indicate a modification of the translation, "object" for "subject," to conform to Axelos's French version, which follows the German more accurately.—R. B.]
79. Marx and Engels, *German Ideology*, p. 39.
80. Marx, *Manuscripts of 1844*, p. 139.
81. Ibid., p. 136.
82. Marx to Ruge, May 1843, in Easton and Guddat, *Writings of the Young Marx*, p. 206.
83. Ibid.
84. Marx, "Hegel's 'Philosophy of Right': Introduction," p. 131.
85. It is Marx himself who sets us straight here. A whole century of Marxists and anti-Marxists has failed to understand this spring powering Marxian thinking. Yet Marx himself names it: technique. "Darwin has interested us in the history of Nature's Technology, *i.e.*, in the formation of the organs of plants and animals, which organs serve as instruments of production for sustaining life. Does not the history of the productive organs of man, of organs that are the material basis of all social organisation, deserve equal attention? And would not such a history be easier to compile, since, as Vico says, human history differs from natural history in this, that we have made the former, but not the latter? Technology (*die Technologie*) discloses man's mode of dealing with Nature, the process of production by which he sustains his life, and thereby also lays bare the mode of formation of his social relations, and of the mental conceptions that flow from them" (*Capital*, 1:372, n. 3).

Afterword

1. Marx, *Capital*, 3:820.
2. Marx and Engels, *German Ideology*, p. 49. [Translation modified as explained above, chap. 13, n. 92, and conclusion, n. 75.—R. B.]
3. Marx, *Manuscripts of 1844*, p. 135.
4. Ibid., p. 132.
5. Ibid., p. 146.
6. Ibid., p. 154.
7. Marx and Engels, *German Ideology*, pp. 86—87.
8. Ibid., p. 231.

INDEX

Absence, 273, 286, 288
Absolute, defined as mind (Hegel), 205
Abstract categories, validity and limits of, 302
Abstraction
　characterized, 166
　scientific, 211
Abstractness, characteristic of the modern, 101
Accidents, 263–264
Adam and Eve, 17–18
Administration and civil service, 97–98, 104. *See also* Bureaucracy
Adultery, 116
Aeschylus, 175, 178
Aestheticism, 179
Aesthetics, 192–193
Agriculture, 219
Aisthesis (ἀίσθησις—"sense reception/perception"), 193
Aletheia (ἀλήθεια—"truth"), 190, 192, 246
Alienation
　defined, xxxii
　double aspect, 176, 213, 236, 272, 329
　　causes of, 194, 331
　economic, 53–66
　　abolition of, not alone sufficient, 230, 232, 238
　　root of all alienation, 213, 230
　essential? 129–130
　general horizon for Marx, 41–42
　human
　　at core of all alienations, 132, 141
　　duration of. *See* Man, alienated from the beginning
　ideological, 151–152, 166–173
　　double character, 170
　　religion first form of, 159
　in externalization, 166, 219–220
　most evolved form, 131
　necessity of, xiii, xx n. 38
　nonalienation not questioned, 106
　of art
　　and alienation of technique, 176–177, 185, 186
　　consummated in capitalism, 176, 185
　　consummated in realistic art, 185
　of technology, xii–xiv, xix–xx, xxi
　of the bourgeois, 60
　philosophic, 195–214, 271–288
　　as critique of German philosophy, 201
　political, 89–109
　　abolition of, not alone sufficient, 232
　religious, 159–173
　　continued by Hegel, 203
　　elimination of, not alone

sufficient, 234
scientific. *See* Science, alienation of
technicist, 84
transcendence of
    as *return*, 217, 220–221, 227
    follows same path as, 227–228, 334
    possible? xix n. 35
    real premises, 267
    reflective grasp, 220
    universalized in West, 185
Anarchy, in Marx, 93, 105, 312
Anatomy, of man and ape, 302–303
Anti-Marxism, 153
Anxiety, bourgeois, 134
Appropriation
    conditions for, 263
    contrasted with expropriation, 127
    explained, 125, 223, 242
    expropriative, 309
    nonpossessive, 125, 231, 239, 307
    ontological, 278
    universal, 254, 329
Archē, 153, 199
Architect, and spider, 322
Aristotle, 59 n. 14, 180, 246, 273
    on the metaphysical question, 8, 286
    on truth and action, 109
Art
    ambivalent value in Marx, 187
    and technique, 181–183, 185
    as production, 177, 181, 182, 187–188
    Christian, 183–185
    communist generalization of, 189–190
    Greek, 180–183, 184
    oriental and Asian, Marx's opinion of, 180
    realistic, 184–185
    reflects technical impotence, 186, 192
    romantic, 184

    sector of division of labor, 189
    separations in, 186
    supersession of, 176–177, 186–190, 193–194
    transformed into commodity, 185
    twofold character of, 177–178
    writings as end in themselves, 186
Atheism, place in communism, 234
Aufhebung ("transcendence/abolition"), 96, 201, 228 n. 23, 230, 230 n. 8, 254, 271, 276, 318
Automation, 294
Automatism, bad, 294
Axelos, Kostas, intellectual biography, xxv–xxvi

Balzac, Honoré de, 178, 185
Barbarism, in technicist civilization, 83–84
Bauer, Bruno, 39, 161, 199
Beautiful, the idea of, 190
Bebel, August, 262
Being, to coincide with productivity, 330
Being and thinking, mystic identity of, 320. *See also* Thought, and reality
Being and thought, identity in Hegel, 29–30
Being-becoming of totality, definition, 246
Biologism in Marx's thought, 281, 297, 324
Boredom, 196
Bourgeoisie
    an estate, not a class, 93
    inaugurates modern times, 30
Bureaucracy
    Marx's criticism of, 89–109 *passim*, 103–104
    rejected by Marx, 258–259, 262, 312

Capital, 67–75
    and *res*, 307–308
    defined, 71

Capitalism
  last stage of class struggle, 71
  modern, accommodating to workers, 72
  risk of, for communism, 307–309
  universalized in crude communism, 231–232
Capitalist, also alienated, 60
Categorical imperative, Marx's, 229
Cause and effect, "dialecticized," 157
Cervantes, Miguel de, 178
Christianity
  contribution to love, 121
  relation with Judaism, 164
  religion par excellence, 101, 165
  responsible for democracy? 100–101
Circle
  of being, making, and having, 129
  of Marx's thought, 276
Citizen
  abstract, 96
  distinction between, and man, 105, 232–233, 312
  figure of human alienation, 233
  split between citizen and man to be overcome, 105
City, artificial, 83
City-state, 100
Civilization
  division from nature, 57
  technicist, 82–85
    opposition to nature in, 79
Civil service, 97–98, 104
Civil society, 89–91, 225
Class
  defined, 60
  historical nature of, 259 n. 79
  revolutionary
    ambiguous, 251, 253, 261
    character and role, 249–250
  receives revolutionary ideas, 172
  struggle
    as recent phenomenon, 300–301
    endemic to the State, 106
    not alone explanatory, 92, 93
Comedy. *See* Tragedy, and comedy
Commerce, paradigm for sexual relationship, 120
Commodity, and money, 75
Common sense, 324
Communism
  affected by property, 278, 305–308, 334
  aimed toward future, 242, 292
  alienation not wholly transcended, 334
  anarchy, 105
  and conquest of universe, 325
  and Western society, 248
  and Western technique, 247
  as humanism, 235, 239
  as naturalism, 239
  as universalizing capitalism, 117
  basis, the economic, 238, 254, 306–307, 335
  classless, Stateless society, 106
  crude and thoughtless variety, 117, 119, 122, 230–231, 245
  Marx's difficulty with, 305–308
  essentially practical, 267, 335
  foundations of, 105
  heir to capitalism, 306, 334
  limited forms of, 234, 237
  makes all men workers, 233, 234
  neither fate nor free will, 267
  not simply suppression of private property, 232, 235, 305

380 Index

political, 232–233
primitive, 64 n. 18, 241
simultaneously everywhere, 248
solved riddle of history, 234, 239
to abolish *bourgeois* property, 307
to be itself transcended, 268, 291, 292, 314, 334
Communists, theoreticians of proletariat, 236
Communist society, generalized human *ego*, 49
Community, tribal, 309
Community of women, 119, 233, 234. *See also* Woman, communalization of
Concepts, character and role, 167–173
Conceptuality and art (Hegel), 191
Confession, 274
Conquest as founding, 220
Consciousness, 166–173, 195–214, 319–332
   Bewusstsein, 152, 154, 166
      dualism with *Sein* ("being"), 276, 320
   conducts practice, 143, 149, 206
   coproduces human development, 147, 171–173
   determined by life, 154
   development of, 147
   distinguishes man from animal, 153
   expresses practice, 143, 151, 154–155
   herd-, 147, 150, 154, 159
   ideological, 148
   revolutionary
      ambiguous dialectical situation, 252–253
      among nonproletarians, 252
      comes "from outside," 171–173, 249
      role in revolution, 170–173, 282
   social product, 145, 147
   supplementary cofactor for history? 146, 153, 154, 155
   term interchangeable with *thought* and *knowledge*, 145
   the unhappy, 138
   transcends real action, 49, 158, 171–173, 268–269, 281–282, 285, 292, 306, 333–334
   translates practice, 143, 150, 157, 166, 167, 207
   true, and thought described, 148, 158
   true, has never existed, 148, 152, 157
   See also Sense-consciousness; Thought
Contrary and contradictory, distinction absent in Marx, 206, 208
Contribution to the Critique of Political Economy, offers summary text, 61–63
Copula, 324
Cosmos, 197, 296, 328
Crafts, and labor, 256
Creation, 8, 246, 274, 296
   privileged Western idea, 63
   rejected by Marx, 55, 218, 296
Criticism
   critical, 161–162
   Marx's critique of, 161–162
   nature of, for Marx, 229, 304
   role of, 273–274
Critique, completion of, 304

Daumas, Maurice, xvii, xxiii
Dealienation
   primary condition for, 238
   real appropriation of the human essence, 221
Death, in terms of species being, 241–242
Degradation, in alienated sexual relations, 117–118
Democracy

does not achieve dealienation, 102
essence of all political constitutions, 101, 102
resulted from Christianity? 100–101
Demystification, 162
Descartes, René, 9, 246
Despot, 100
Despotism, 315
Dialectic
concept of, not worked out by Marx, 208
Marx's critique of, 204–205
triadic form in Marx, 64 n. 18, 316–317
Dictatorship of the proletariat, 106, 259–263, 290–291
Diderot, Denis, 175, 178
Division of labor, 56–59
abolition of, unclear, 255–258
effect upon the family, 115–117
evil? 58
identical with private property, 59 n. 12, 309
inseparable from private property, 59, 67
natural eventuality, 59
promoted advance of mankind, 59
single cause of alienation, 58
Drives, 133, 133 n. 16, 219, 289–290
Dualism
"abstract," 101, 328
as result of alienation, 125
Axelos's basic criticism of, xxiii–xxiv
Christian, 328
exploiter vs. exploited, 60, 65
flaw in Marxian program? 328
natural need and technique, 78
not transcended in Marx, 50, 146, 148, 151, 153–154, 156, 162, 171, 226, 263, 275, 284, 327–328

Earth, relation to labor, 218, 328
Economic, the
moving agency in development, 80
primacy of, 63, 92, 304, 306–307
Economic activity, essentially social, 56
Economic and Philosophic Manuscripts of 1884, principal source, 25, 46
Economic determination, ultimately decisive, 157 n. 21, 175–176
Economic life, to be abolished, 238
Economic movement, narrow vs. broad conception, 80
Economic powers, controlled in communism, 254–255
Economic structure, the real base, 166, 290
Economism, crude, 304
Economists, represent bourgeois class, 236
Education of children, communal, 119
Egocentricity, to be transcended, 222, 255
Ellul, Jacques, xv n. 11, xvi, xxii
Emancipation
human, 222, 232, 233, 264
political, 232
Empiricists, 198
Encyclopedists, French, 9
Energeia (ἐνέργεια—"act"), 246
Engels, Friedrich, 64 n. 18, 114 n. 6, 157, 262–263, 291
Enlightenment, 9
Entelechy, 246
Epicureanism (Greek), 180
Epicureans, 8
Epochs of human history (Marx), 62, 65
Epos ("epic"), 181
Equality, abstract, 260
Ergon (ἔργον—"action, deed, work"), 190, 192
Errance, 12, 22, 50, 154, 283, 292, 293, 323

sense of, explained, xxvii–xxviii, 12 n. 2
two senses in Axelos, xxviii n. 72
Estates, transformation into classes, 69
Ethos, 141
Exchange-value, 59 n. 14, 291
reified, commercialized quantity, 184
Expropriation, 127
of the expropriators, 248, 278, 307
Externalization
alienated, 219–220
defined, xxiii
in Hegel, 36

Facts
for empiricists, 198
for Marx, 195, 199
Family
abolition of, 115–116, 119
dissolution of, 115–116
slave relationship within, 114, 115
supposed foundation for the State, 115
third condition for history, 113
Fatalism, 267
Feelings, ontological import, 277
Fetishism, 129
Feudal society, technical contrast with capitalist society, 293
Feuerbach, Ludwig, 39–41, 161, 197
theses on:
I, 136, 137
II, 149
IV, 162
V, 283
VIII, 50, 162, 283
IX, 225
X, 225
XI, 214
First historical act, 77, 113–114
Forces, the elementary (Axelos), xxix n. 79
Foundation
not a primary concern, 55, 78
not reached, 155, 275, 277, 281, 327–328
quest for, xxviii
Franklin, Benjamin, 77
Freedom
achieved beyond production, 333
and necessity, 243, 309–310
need for, 329
of the press, 186–187
Freedom, Equality, Brotherhood! 260
Free will, law not based on, 94
Freud, Sigmund, compared with Marx, 19 n. 3, 133 n. 16
Future, not clearly delineated, 99, 102, 107, 119, 121–122, 308. *See also* Marx, Karl, future-oriented analysis

Garden of Eden. *See* Paradise
Generatio aequivoca, 218, 296
German Ideology, The
history of ownership, 68–70
Marx's theory of ideology, 150–152, 154
principal source, 25
Germany, emancipation of, 247
Givens, natural, 299
God
death of, 165, 234
hidden in Hegelian thinking, 203
Gods, Greek, 180, 189
Goethe, Johann Wolfgang von, 175, 178, 329
on money, 74
Golden Age, 64 n. 18, 240
Greek thought, characterized, 7, 9, 328
Ground, obsessive phantom of, 27. *See also* Foundation

Having, 68, chap. 7 *passim*, 290
Head and feet, in Marx's dialectic, 321–322
Hegel, Georg Wilhelm Friedrich, 9, 29–50 *passim*, 161, 170, 188,

197, 246, 319, 320
  chief mistake, 209
  consummates Western intellectual tradition, 29, 138, 156, 202
  errors of, according to Marx, 202–204, 208, 214
  justifies existing state of affairs, 95, 202
  last philosopher, 33 n. 4
  maintains double alienation, 203–204
  Marx's critique of Hegel's dialectic, 96, 204–209
  Marx's critique of Hegel's philosophy as philosophy, 201–209, 213–214
  negated by Marx, 42. *See also* pp. 202–207
  on art, 190–192
  outstanding achievement of *Phenomenology*, 55–56
  philosophy of history, 201–202
  political philosophy
    criticized, 95–97
    realistic, 98
    reversal by Marx, 135–136, 137–139, 206
Hegelian dialectic, revolutionary importance, 204, 207
Hegelians, Left, 136, 170
  overall criticism of, 39–40, 161–162
Heidegger, Martin, xxi n. 42, xxvii, 12 n. 2, 27 n. 1, 321 n. 65
  on Marx and Nietzsche, 33 n. 4
Heraclitus, 7, 273
  on money, 74–75
  on ultimate play, 286
  on unity, 156
Herd-consciousness. *See* Consciousness, herd-
Historical point of view, 11–12
Historicism, 281
Historiography, 227, 311
History
  as history of economy (Marx), 38
  as history of Spirit (Hegel), 38
  as history of technique, 27
  as history of technique and alienation, 224
  as nature-opposing, 120
  beginning of, xi, 54, 218
  constitutive cofactors for, 113–114, 145–146
  for empiricists, 198
  for idealists, 198
  is history of alienation, 148, 217, 220, 227, 239, 241
  more than just evolution, 114
  natural, human, social life, 299
  no end to, 241
  no superstructural history, 94, 169, 175, 212
  of nature and of man, 282
  oriental and Asian, 180
  originating act of the World, 219, 226–227
  philosophy of. *See* Philosophy, of history
  production of, 299
  progressive universalization through technique, 196
  realization *and* alienation, 126
  riddle of, 221
    real productive forces, i.e., technique, 203
    solution, 17, 64 n. 18, 223, 230, 239, 266
    solution must cover all spheres, 230: economic, 230–231; ideological, 236; political, 232–233; religious, 234
  solution total reconciliation, 238
  unpenetrated, 108
  theoretical awareness of, 172
  to be itself transcended, 226
  transformation of human nature, 113, 217

384 Index

transformed into world, 325
universal
   absorbs national States, 313
   and the proletariat, 247–248, 251
   in Hegel, 205
   "so-called," 227
   yet to come, 70
   Western, three major phases, 70, 179
world
   beyond local histories, 287
   has never yet existed, 300
   produced by big industry, 81–82
   properly speaking, 247
Holy, the, 165
Home and housing, 141
Homer, 178
Homo oeconomicus, transcended, 255
Human being
   defined in terms of relationship/activity, x
   manifested through making, 123
   producer and product, 120
   protagonists of becoming, 111
   thrust toward the universal, xii, xiii–xiv
Human body, condition for labor, 54
Human essence
   explained, 41, 44, 53
   nature-opposing, 120, 218
   never yet fully manifested, 223
   See also Man, essence of
Humanism
   in the perspective of socialism/naturalism, 324–325
   meaning of true, full, 230
   positive, 235, 237

Idea, in Hegel, 156, 190
Ideals, 170
Ideas
   after conquest of alienation, 193, 194
   conductor agencies, 149
   historical, not eternal, 169
   intellectual instruments, 150
   Marx's criticism of, 168–173
   take on form of universality, 196
   term synonymous with *thoughts*, 168 n. 16
Ideology
   camera obscura, 152
   complement to real world, 193, 194
   ideological superstructure, 62, 166–167, 290
   Marx's theory of, 150–152, 154
      ambiguous, 155
      reduced to the economic, 157
   reflects practice, 166, 167, 170
Illusion of the epoch, 303, 311
Individual, and the mass, 321 n. 63
Individualism
   consummated in collectivism, 267
   generalized in crude communism, 231
   transcended, 225
Industrialism, 80, 81
Industry
   big, 81, 250
   complete form of labor, 81
   consummates alienation, 212
   history of, 81, 186
   makes history universal, 82
   manifests basis of science, 279
   manifests relationship of nature to man, 211, 279
   medium for transformation of human life, 212
   most highly developed form of technique, 212
   ontological import of, 278
   prepares dealienation of man, 212
   prototype of human activity, 187

reveal's man's essential powers, 81, 86, 224, 279
Instinct for production, 290, 298–300, 324
Instincts, 133. *See also* Drives
Intellectualism, not avoided, 269
Intellectuals, bourgeois, 171–173

Jonas, Hans, xvii n. 28, xxii–xxiii
Judaism, Marx's criticism of, 162–164
Judeo-Christian tradition, 8, 9–10, 246, 274

Kant, Immanuel, 9, 246, 319, 320
Kepler, Johannes, 175
Knowledge
  absolute, in Hegel, 29–39, 205, 213, 246, 272
  interchangeable with *consciousness* and *thought*, 145
Kugelmann, Ludwig, 259, 263

Labor
  abolished in its generalization, 266
  abolition of, 255
  alienation of, answer to riddle of history, 17
  converts nature into history, 204, 218, 290
  eliminated, not redistributed, 251
  essentially practical and objective, 136–137
  evil? 58
  form correlative to mode of exploitation, 68
  ground of being, 287
  in private property basis of all alienation, 139
  inseparable from division of labor and private property, 59
  intellectual
    distinguished from practical, 150
    itself divided, 169–170
  machine labor most suitable for development, 79
  not all alienating, 56
  productive and nonproductive, 190 n. 35
  social in essence, 53
  subjugation of Nature, 55
  technicist, 84
  the essence of man (Hegel), 56
  theoretical, derives from practical, 137
  universalized in capitalist society, 84–85
Labor class, not properly a class, 90
Labor force, 59 n. 14, 60
Landed property, first major form of private ownership, 67
Language, 145–147, 150, 155 n. 17
  and the development of consciousness, 147
  and thought arise from practical need, 145
  distinguishes man from animal, 145, 150
  material in nature, 145
Lassalle, Ferdinand, 75 n. 11, 185
Law
  inequality of, 260
  only illusorily based on free will, 94
Law, juridic: mystified expression of economic laws, 94
Laws of world trade, 46
Leaders, role of, 263–264
Leibniz, Gottfried Wilhelm von, 9
Lenin, Vladimir Ilich, 107, 108 n. 27, 159, 172–173
  on imperialism, 248 n. 51
  on proletarian revolution, 263 n. 86
Liberty, only for dominating class, 90
Lobkowicz, Nicholas, xxxi
Logic, 197, 198–199, 205
Logic of fact vs. fact of logic, 323
Logos (λόγος—"word, reason"), 7, 34, 156, 192, 199, 202, 207, 213, 218, 246, 286, 331
  dissociation from *physis*, 274
  distinction from *praxis*, 331
  narrowed in Marx's theory, 157

386 Index

not original in man's nature, 54
not reconciled with *praxis*, 244, 320
of being ignored, 277
of *technē* (techno-logy), xv, xviii
Love, 114–122
    as a living expression of self, 116
    as passion, 118–119
    resists theoretical construction, 118–119
Lukács, Georg, on materialism as historical, 291 n. 1

Machine
    Machine Age completes alienation, 78
    Marx's view of, 293–294
    positive force yet alienating, 79
    properly, only recent, 293
    synthesis of all instruments, 79
Making, in mechanized form always alienating? 130
Making and being, separated and crushed by having, 290
Making and having, 123, 126, 140
Man
    alienated from the beginning, xix, 131, 135, 136, 141, 176, 217, 219
    a natural-human-social being, 132, 246
    and nature, primitive identity, 147
    as slave to machine, 79
    being coterminous with technique, 330. *See also* Human being
    center of Marx's thought, 44, 217, 222–223
    distinguished from other animals, 77, 145, 150, 153
    driven by an objective power, 133, 217, 219
    essence of

        metaphysical premise for reconciliation, 241
        premise for transcending alienation, 220–221
    essential dependence in nature, 203, 217–218, 240
    foundation for, 176, 229, 235, 315, 324–325, 329
    indissolubly natural and social, 217–218
    integration with nature and society, 241, 246
    less subject or object than *activity*, 126
    natural and human being, 124, 197
    product and producer, 245, 290
    relationship to nature, 217–218, 240, 279
    self-creation, 203, 316
        era of, 328–329
    the essentially practical being, 239
    the social animal of reason, 132
    to become total in total society, 105, 222
    total, 317
        and technique, 318
        and Totality, 281
        practical, communist, materialist, 318
    zoo-technological definition, 77, 219
Manichean dualism, in Marx, 65
Manifestation, becomes alienation, 126
Mao Tse-tung, 157
Marriage, 115–117
Marx, Karl
    abolition of metaphysics, 326–327
    a bourgeois intellectual, 171
    antinaturalistic naturalism, 54
    as against Marxism, 98, 291 293, 314
    aspirations of unification unfulfilled, 156

consummates and negates Western tradition, 105
continues modern radicalism, 121
criticism, general intent, 161
critique irresistible, 292
critique of bureaucracy, 97–98, 102–104
critique of Hegel's dialectic, 204–209
critique of Hegel's political philosophy, 95–97
critique of religion, 159–166
dialectic characterized, 206–208
doctoral thesis, 179, 271
does not depreciate Hegel, 95–96
early Marx vs. later Marx, 23, 45
early poems, 178
economic expectations unfulfilled, 60–61
economic thought ambiguous and polyvalent, 307–310
enemy of utopia, 334
fulfills modern metaphysics, 327
fundamental idea, 120
future-oriented analysis of present, 38, 59, 66, 102, 105
generalizes what is true of one epoch, 91–93
hope in future technique, 294
humanistic radicalism based on naturalism, 53–54
idealism of, 98
interest in history, 46
interpretation of Hegel creative and distorting, 138
not an orthodox Marxist, 302
one-sided by conviction, 294 n. 8
on his original contributions, 259 n. 79
on love, 116–122
on Russia, 85
on the United States, 84
optimism of, 317
passion for nothingness, 292, 326
political philosophy holds open questions, 107
position on art ambiguous, 176–178, 179, 186–187
practical character of his politics, 107 n. 25
present historical world his base, 42, 43, 51, 55, 56, 63, 64, 66, 72, 78, 92, 93, 106
qualified empiricism, 123
recasting of Hegel's dialectic, 33–35, 206–207
restricts focus to West, 165
romanticism of
  contrast with realism, 125
  contrast with positivism, 79, 82
starting point
  historical man, 64 n. 18
  man, 289, 297
  metaphysical idea of man, 132, 225
  to be free of metaphysics, 78
stronger in negative analysis than positive prescription, 120, 124, 132, 135 141
technicism of, 297
the "last philosopher," 331
thought of
  absence of ontology, 289
  affected by what it negates, 275
  centered in man, 44, 217, 222–223
  continues bourgeois thought, 275
  continues metaphysics of subjectivity, 153, 226, 246, 277
  continues Western metaphysics, 7, 105, 284
  culminates Western meta-

physics, x, xxiv, 226, 246
essentially historical, 78
extends to all history, 301–303
extremely problematic, 107
follows upon traditional dissociations and concepts, 274–275. *See also* Dualism, not transcended in Marx
foundation, not ontological but historical, 55
greatness of, 295–296
ideological lapse in, 257
metaphysical, 132, 135, 141, 152–153, 156, 224, 225–226
metaphysical despite disclaimer, 135, 136, 146, 167–168, 289, 323
not itself science, 212–213
not ontologically grounded, 64
not self-critical, 228–229
unaware of its presuppositions, 274, 278–279
to be transcended in thought and practice, xxiv
uncertain about status of thought, 236, 245, 247, 252–253, 263, 275, 284, 321, 322, 327
view of religion shortsighted, 165
warning against false communism, 251
will for transcendence, 292, 326
Marxism
crude economics of Marxists, 293
ideology still important for Marxists, 157
Marxist (adj.) vs. Marxian, 22–23
Marx's thought as, 7, 22–23, 107
metaphysical foundation of, 275
Masses, role in revolution, 248
Materialism
dialectical, attributable to Engels, 291
historical (Lukács), 291 n. 1
Marx's, 226
heir to positivism, 306
historical, 61, 63, 207
matter means human labor, 291
matter means *praxis*, 136
metaphysical character of, 141, 153
not dialectical, 63
not ontological, 207
practical, 200
practical, not ontological, 244 n. 39, 323
practical materialist = communist, 238, 276
materialist metaphysics criticized, 136
vanishes together with spiritualism, 276
Materiality, sense for human being, 127–129
Meaning
and cultural objects, xi n. 6
and the sensuous
distinguished, 50, 125, 223, 276, 284
reconciled, 226, 243, 245, 287
reduction, 46, 245–246, 276
Mechanization, role in alienation, 78–85, 221–222
Mediation
as alienation, xix n. 34
opposed by Marx, 208
transcending, 235, 237
Mediations, in ultimate economic determination, 176
Mészáros, István, xii n. 7, xix n. 34, xix n. 35, xix n. 37, xx n. 38, xx n. 39, xxxii n. 83
Meta-Marxism, 108
Metaphysics

based on sensuous/spiritual distinction, 326
deficiencies of, 11
dualism of the metaphysical with the physical inverted, 152–156, 327. *See also* Dualism, not transcended in Marx
major historical stages, 7–25, 246, 274
Marx's program for, 136, 271–288
Marx's rejection of, 199
misses the essential contradiction, 209
realized in technique, 289
sense of the metaphysical, 225
social physics, 43, 226
superseded by historical physics, 55
to become historical physics, 274
transcendence in man denied, 54, 121
Middle Ages, horse and sword in, 311
Modern age, alienation pushed to its limits, 103
Modern radicalism, continued in Marx, 121
Modern thought, characterized, 10–11
Money, 72–75
essence of, 72–73
res par excellence, 72
secularized Jewish god, 164
the universal prostitute, 116
Moral conscience, 139
Morality, 139–140
Mortality, in terms of species being, 241–242
Motor of appropriation, majority of mankind, 248
Motor of becoming, technique, 324, 331
Motor of converting nature into history: natural needs, 277
Motor of development
productive forces, 80, 225
technique, 167, 208, 212
Motor of historical development
action of man modifying the natural, 198, 212, 219
technique, 212, 289, 325
Motor of history
actual revolution, 236
negativity, 294 n. 8
productive forces, 80, 114, 203, 249
technique, 203, 289, 295, 296, 304, 325
technological development, 294
Motor of science, technique, 19, 209 n. 33
Mystery, 160
Mystical feeling, 195–196
Mystification in Hegel, 95
Mythology
Greek, presupposed by Greek art, 182
relation to technical underdevelopment, 182

Nations, conflict of, 107–108
Naturalism
alone can understand history, 219
fully achieved = technicism, 256
in plan of reconciliation, 324
Marx's
content of, 277
denaturalization of nature, 198
Naturalism-humanism-communism, summation of abolishing action, 326
Nature
abstract sense = in itself, 197
always social for man, 219
as externalization of Spirit (Hegel), 29, 35
becomes human, xxi, 279
becomes man, 219, 227, 240, 282
becomes objective by human labor, 194
break with, in division of la-

bor, 58
denaturalization of, 81–82, 193, 198, 256, 267, 326, 335
exploitation of, and exploitation of men, 58, 68, 79, 305
human essence of, 240
immediate object for science, 280
in (Christian) Redemption, 274
industrialization and socialization of, 256
integration with society and man, 241, 246
mastery of, 10, 18, 19, 194
never outside history for man, 211, 218
no autonomous becoming, 279
"nothing" without man, 240
not Marx's starting point, 289
of interest only as in grip of technique, 198
opposition to
  achieved in industry, 81
  constitutive of man, xviii–xix
  dissolved in universal reconciliation, 221
  established by the tool, 77, 166
  natural to man, 54, 218
  resolved in communism, 239
  transformative situation, 113
preexistence of
  meaningless, 197
  not of interest, 116, 211, 218, 277
resurrection of, 240, 243, 245
struggle with, Marx's point of departure, 78
technification of, 245
the being of the totality in its becoming, 218

the first object, 127
transformation (conversion) into history, 200, 204, 211, 218, 245, 277, 316
  by labor, 204, 211, 290
  by technique, 153, 325
  culminates in communism, 325
  Marx's interest, 198
transformation of, through technique, 192, 193, 298
transformed into material for social labor, 212
true for man as developed through industry, 211
vanquished in industrialization, 82

Need
  active passion, 129, 133–134
  dialectic of new needs, 113, 289–290
  distribution of goods according to, 265, 266, 308
  for rest, 134
  satisfied in communist society, 265, 266
  spiritual, 149
  unfulfilled basis of religion, 165

Needs
  elementary *and other*, 329
  false metaphysical, 297
  filling by elemental technical performance, xi
  in reciprocal relation with productive techniques, 77–78, 113–114
  natural: motor agency in conversion of nature into history, 277
  production of new, xi, 77–78, 113, 294, 295, 298, 324
  radical, 329
  real, contrasted to artificial, 83–84
  satisfaction of, never ended, 54

Negation
  affected by the negated, 122,

260, 305–306, 334
  itself negated, 227, 234, 237, 241, 268
  the motor of history, 294 n. 8
Negativity in Marx's critique, 41–42
Neo-Hegelians, 197
Nietzsche, Friedrich W., 133 n. 16, 168
Nihilism in Marx, 292
Noneconomic, the: also constitutive and organic, 92, 93

Objectification
  alienation of, xix n. 37
  defined, xxxi
  in Marx, 128
  necessity of, xx n. 39
  of the human essence, 121, 124, 125–129, 193–194, 245, 325
Objectivity
  bad, 128
  implies *praxis* for Marx, 136
  simple, to be transcended, 222
Objects
  alienation of, 219–220
  cease to be objects, 330
  cultural, xi n. 6
  Gegenstand, 124 n. 2
  needed by human subject, 127
  nonreified, 297
  retain supremacy, 277–288
Ontology, deficiencies of, 11
Onto-theo-logy, 27
Operative concepts, not clarified in Marx, 135, 168
Oppositions, philosophical: origin in alienated abstraction, 195
Origin, the involvement of nature and man as, 219
Original sin, 17–19
Origins, inquiry into prospective, not retrospective, 224–225
Owl of Minerva, 214
Ownership
  brief history, 68–70

  collective, in communism, 308–309
  communal, 68–69
  expropriation, 68
  tribal, 68
  See also Property

Paradise, no return to, 240–241, 308. *See also* 64 n. 18
Parmenides, on unity, 156, 205
Passion
  active, 133–134, 243
  ontological being of, 278
Persona ("person"), confused with *res*, 315
Phenomenology of Spirit (Hegel), 204
Philistines (aesthetic), 179
Philosopher, the, 196–197
Philosophia (φιλοσοφία—"love of Wisdom"), 202
Philosophic mind, 195
Philosophy
  abolition/transcendence of, 201, 230, 237–238, 271–288, 323
  always arrives at the end, 213–214
  and abolition of proletariat, 272
  and the proletariat, 230
  as a profession, 201 n. 16
  as instrument of revolution, 230
  as inversion of naïve consciousness, 31
  becoming world, 202, 209, 272, 323
  communist, nonsense concept, 285
  conversion into science, 279–285
  counterideological role, 230
  crowns ideological alienation, 195
  expresses the ideas of the dominant class, 196, 200
  final task is to unmask alienation, 200–201
  German, Marx's critique of, 201–207

guide to practice, 273
Hegel's idea of, 190
highest science in Hegel, 32
history of, not possible, 200
intellectual onanism, 195, 273
light for revolution, 273
Marx's critique of, 195–209, 213–214
of history, 197, 198, 205
of nature, 197, 198, 205
of philosophy, Marx's, 273
to be critical and therapeutic, 274
ultimate role for, 284–285
Physical, the
dualism with the metaphysical, 152–156, 327
privileged in Marx, 146, 327
Physis (φύσις—"nature"), 7, 9, 190, 192, 198, 240, 246, 296
displaced by Christian creation, 328, 330
dissociation from *logos*, 274
Planning, in communist society, 253, 265, 309
Plato, 7, 190, 246
on truth being practical, 108–109
origin of Western European metaphysics, 155
Play
art as, 194
as future human activity, 258
Axelos's concept of, xviii n. 30, xxvi–xxix
of the World, 165 n. 12
supreme, 286
transcending politics, 313–314
Plotinus, 8
Poetry, 186–187, 188, 190
Poiesis (ποίησις—"making, production"), 190, 192, 194
Political economy, 209
ahistorical, 304
and production/consumption, 299
bourgeois form characterized, 67, 75
develops its own dialectic, 108
science par excellence when historical, 67
Politics
form organizing economy, 87, 92, 93, 167
political, the: meaning of, narrowed by Marx, 93
political forms
also determinative and constitutive, 93
return effect upon economic basis, 89
political power repressive, 261
Polytheism, 165
Poverty, 134
human significance of, 129, 243
Power relationship, for Axelos as compared to Marx, xviii n. 30
Powers, the great (Axelos), xxix n. 79, 165 n. 12
Practical, the: privileged position retained, 276
Practical activity, distinguishes man from other animals, 77
Practical energy, 244, 247, 271, 272
Practical reason, 63
Practice
basis for religion, 159–160, 164–165
sole source of truth and reality, 165
transcended by thought. *See* Consciousness, transcends real action
true, the criterion of truth, 149
will prevail over the theory, 243
Praxis (πρᾶξις—"practice"), 136, 192, 244
and the understanding of *praxis*, 283–284
basis for ideas, 168
distinction from *logos*, 331
dualism with *logos*, 320
not explicated by Marx, 168

practice as a conquering
    force, 202
reconciled with its destiny,
    272
technical, making all be
    through labor, 246–247
total, 273, 326
total liberation of, Marx's
    aim, 205–206
transformative = Technique,
    325
true, 137, 149
Predicate, converted into subject,
    323–324
Premises, Marx's, 225
Pre-Socratics, 7, 156, 246
Primitive man, of little interest to
    Marx, 78
Private life, split from public life,
    100
Private ownership, cause of all
    alienation, 176, 194, 223. See
    also Private property
Private property, 67–75
    abolition of, means appro-
        priation of human life,
        117
    alienation explained, 67
    effect upon the family, 115–
        117
    elimination of, not generali-
        zation, 117, 230–232,
        243
    elimination of, *the* premise
        for humanism, 235
    first opponent to fight, 230
    identical with division of
        labor, 59 n. 12, 309
    need for suppression of,
        222
    particularizes the universal,
        67
    positive meaning of, 278
    produces man capable of
        liberation, 316
    promoted advance of man-
        kind, 59
    real premises for abolishing,
        247–248
    root of all alienation, 243

    what its abolition means, 255
Problems, mankind's setting of, 62,
    286
Procreation, 114. *See also* Repro-
    duction
Production
    absolutely thetic power, 316
    appropriation of nature, 307
    as process privileged over
        product, 246, 291, 330
    common source of the ma-
        terial and the intellectu-
        al, 151
    creator of subject and object,
        298
    essence hidden, 297
    first human act, xi
    material
        as well as intellectual, 264–
            265, 287, 327, 328,
            333–334
        essence of being and his-
            tory, 310
        to be transcended? 310,
            324
    modern, object of Marx's
        study, 300
    motor force of negativity,
        316
    of ideas, 151, 167
    produces history, 299
    produces object for subject
        and subject for object,
        188
    product distinguished from
        other objects, 296–297
    products and form of pro-
        duction, 295
    religion a mode of, 159
    spiritual, 167, 287
    superstructural forms
        modes of, 176
Production/consumption
    cycle of, 293–300
    subject-object dialectic of,
        295, 296–297, 299
Productive activity, ground of be-
    ing, 287
Productive forces
    appear independent, 90

motor of historical development, 80
opposition to nature, 166
Productive labor, central to Marx's thought, 64
Productive process, not univocally determinative, 65–66
Productive techniques, in reciprocal relation to needs, 77–78
Productivity
  absolute, 295–296, 325
  basic role for human development, 290
  makes all be. 246
  the conquering power, 330
Progress, concept of, 181
Proletarians, heroes of human emancipation, 230
Proletariat
  and revolutionary *ideas*, 171–173
  character and role, 249–250
  emancipation of, by political action, 107
  nonclass, 260
  role in realizing technique, 297
  universal class, 60
Property
  feudal, 69
  first instance slavery in the family, 115
  in communist society, 278, 307–309
  modern, 69
  primitive communal, 64 n. 18
Prostitution, 116, 119, 233, 234
Proudhon, Pierre Joseph, 294
Psychoanalysis of human history, 274
Purposes, highest, 329

Quantity, reign of, 83
Questions, the great: uneliminable, 286–288

Radical, meaning of, 229
Ratio ("reason"), 8, 12, 105, 275
Reality
  meaning in Marx, 128, 136, 166
  not fully defined in Marx, 148, 275
  total, 204
  true: objective but not reified, 129
Real (*wirklich*) means "active" (*wirkend*), 136, 148, 166
Reconciliation
  and science, 280
  consummate technicism, 325–326
  difficulties of, 324–332
  in Hegel, 36–37, 208–209
  second phase of socialism, 265
  to abolish all antagonisms, 239
  truth and passion, 272
  universal, 188–189, 201, 212, 213, 221, 223, 224, 227, 244, 247
    a metaphysical concept, 225–226
    communism, 239, 265
    meaning of, 215, 200
Redemption, 8, 274
Reduction in Marx, 44, 167, 243, 278, 284
  Marx's passion for, 305
  of being to human *praxis*, 244–245, 245–246, 283
  of being to production, 291, 333
  of intellectual production to material, 152–155
  of meaning to the sensory, 46, 245–246, 276
  of philosophical problem to empirical fact, 195, 199, 226
  of religion to practice, 164
  of supersensuous to sensuous, 326
  of the ideological to the economic, 157
  of thought to practice, 322
  of world to technical production, 331

395 Index

Reflection
   grasp of alienation necessary, 220
   ideas reflect dominating class interests, 168–170
   precondition for communist movement, 236
Reification, 55, 100, 116, 120, 128–129, 134, 184, 219, 287
   conversion into commodity, 308
   defined, xxxii
   inevitable in Marx, 128
   necessity of, xx n. 38
   of men, 315
   transcending of, 222, 309
Relations of production, 61, 89, 166
Religion
   Asian, 165
   Hegel's idea of, 190
   "natural," 159, 164, 165
   opium of the people, 160
   origin, 159–160, 164–165, 170
   rests on defect, 159–160, 162
Religiosity, 162, 165
Reproduction
   linked to production, 18, 114, 114 n. 6, 145, 217, 329
   organs of, 114
Res ("thing"), 8, 72, 307, 315
Res cogitans, 8, 12, 246, 275
Res extensa, 8, 12, 246, 267, 275
Return and reintegration, 218, 239, 240, 316, 330
   meaning of, 220–221
Revolution
   motor force of history, 236
   necessary, 252, 259, 265
   permanent, 137, 233, 297
   previous occurrences, 251
   proletarian, 290
      economic dimension, 253–258
      political dimension, 258–267
   revolutionizing existing world, 238
   true communist: human and social, not political, 233
   will dealienate labor, 221
Revolutionary practice, 140, 141, 153
Ricardo, David, 57
Rights
   generalization of, 233
   of man, 233
Rimbaud, Arthur, 193 n. 40
Romanticism, Marx against, 184, 319
Ruge, Arnold: letters to, 274, 305, 314, 329
Russia and technicist labor, 84–85

St. Augustine, 8
Salvation beyond alienation, 106–107
Saving, 130–131
Scepticism, Greek, 180
Science
   absolute, 210
   alienation of, 209–213
   and human nature, 279–280
   and nature, 279–280
   and technique, 212
   as alienated is unnatural, antihistorical, and inhuman, 211
   as revolutionary, 237
   a theoretical technique, 283
   basis of relationship of man to nature, 279
   communistic: cannot absorb consciousness, 282, 285
   compartmentalization of, 210–212, 213
   dealienated: meaning of, 279–285
   extreme point of alienation, 213
   has no proper history, 175, 209, 212
   historical, 209, 211, 280
   knowing that rests on *doing*, 212
   not to be abolished for Marx, 212
   product of industry, 209 n. 33

396 Index

retained beyond supersession of alienation, xxiv n. 53
sense in Hegel, 32, 38. *See also Wissen* and *Wissenschaft*
the *one* science, 279–281
  essentially historical, 282–283
  theory of, not to be found in Marx, 212
  true only as proceeding from sensuous consciousness, 211
Sciences
  and division of labor, 209–210
  natural, 209, 211, 212, 280
  of man, 280
  separation of, not valid, 211
  specialization of, 210–212, 213
Self-consciousness
  the achievement of, in history (Marx), 205
  the illusions of, 139
Self-criticism
  and historical perspective, 303–304
  leads to negation criticism, 304
Sense consciousness, 147, 154
Senses
  the human, material *and* spiritual, 124–125
  will prevail over thought, 243
  See also Meaning, and the sensuous
Sensibility (*Sinnlichkeit*), meaning of sensuousness, xxxi
Sensuous, the: privileged position of, retained, 276, 327
Sensuous activity
  basic, 197, 243–246
  ontological import, 279
  privileged, 279
Sensuousness, implies *praxis* for Marx, 136
Sexes, relationship between, 114, 116, 117–118, 120

Sexual act and division of labor, 114–115, 120
Sexual activity, 117
Sexual forces and productive forces, 114
Sexuality and love
  not dissociated, 118
  referred to technique and production, 120–121
Shakespeare, William, 175, 178, 181
  on money, 73–74
Skeptics, 8
Slavery, 315
Smith, Adam, 57
Social Contract (Rousseau), 262
Socialism
  and atheism, 234
  in strict sense = communism, 325
  two phases, 264–266
Socialists, theoreticians of proletariat, 236
Social mind, 148
Society
  basis for the State, 259
  bourgeois, rests upon egoistic commerce-minded interest, 164
  classless, 253, 259
  communist, "idyllic" activities in, 257–258, 262
  integration with man and nature, 240, 246
  natural, technically underdeveloped, 257
  not well explicated by Marx as *totality*, 99
  universal, 165
Sociologism, 281, 325
Space, and time, 283
Spartacus, 175
Species being (*Gattungswesen*), 111, 113, 117–118, 124, 132, 186, 312, 314
  alienated in speculative thought, 195
  a metaphysical concept, 225
  premise for universal reconciliation, 221, 225

Species essence, 123
Species life, 136
Species man, 56
    defined, 100
    has never yet existed, 315
    highly metaphysical idea, 135
    ideological concept? 317
    not well explicated by Marx, 99, 317
Spirit, Absolute (Hegel), 37–38
Spiritual, the
    after alienation superseded, 153–158
    in Marx's metaphysics, 141, 153–158. See also Production, spiritual
Spiritualism, 276
Spirituality, 226
Stalin, Joseph, 157
State
    and its enemies, 91, 97
    annihilation of, 258
    armature of civil society, 97
    as all-powerful absolute subject, 105
    autonomization of, 90–91, 93, 94, 100
    bourgeois, focus of Marx's attack, 93
    exists for the sake of private property, 94
    free, Engels's description of, 262–263
    future, no prediction of, 262
    illusion of, 102
    instrument and weapon of dominant class, 90
    intermediary role of, 232
    not a universal reality, 232
    résumé of practical struggles, 311
    superstructural, 91
    to wither away? 262–264
Stirner, Max, 39, 161, 176 n. 3
Stoics, 8, 180
Strauss, David Friedrich, 39, 161
Subject
    absolute (Hegel), 30, 34
    absolutely productive, 275
    and object
        dialectic in production/consumption, 295, 296–297, 299
        joint transcendence of, 330
        meaning of, in Marx, 125–129
    nonsubjectivist sense, 297
    productive, "metaphysical" ground for technique, 330
Subjectivity
    generalized and socialized by Marx, 246
    human, 10, 12, 15
    philosophy of, not transcended, 127, 133. See also Marx, thought of, continues metaphysics of subjectivity
    to be transcended, 222
Sublimation
    ideas a sublimated expression, 168, 290
    metaphysical needs, 297
    of practice in religion, 165
    superstructure a sublimation of impotence, 298
Suffering, 133–134, 243
Superman, 235
Supersession, 22, 204, 323
Superstructure
    general account, 166–173
    interaction with base structure, 276
    not finally abolished, 156
Supply and demand, 254
Surplus labor, 59 n. 14
Surplus value, 59 n. 14
Systematic point of view, 11

Technē (τέχνη—"art, technique"), 9, 190, 192, 331
    logos of, xv, xviii
    meaning of, 175
    not clarified by traditional dualisms, 274
    privileged component in Western tradition, 63

Technicism
    and bureaucratism, 103
    consummate in reconciliation, 325–326
    = fully achieved naturalism, 256
    immense conquest for mankind, 81
    Marx's, 297
    transmuted into communism, 105
    See also Civilization, technicist
Technification of the natural, 245
Technique
    alienated, xxi–xxiv, 85
    and art, 181–183, 185, 192
    and definition of totality, 330
    and religion, 162
    and speculative thought, 319
    and the political, 312–313
    and total planning, 309, 312, 330, 334
    as fulfillment of modern philosophy, 286
    as play, xxviii–xxix
    associated with science, 212, 284
    atomic, 318
    Axelos's concept of, compared to Marx's, xv n. 14, xviii n. 29
    basis for science, 279, 283
    commands drives, 296
    determines art, 175
    development of: positive and negative features of, 309, 315–316
    explained in terms of *technē*, 192
    foundational to communism, 105
    freed, 84, 333, 334
    ground of being, 331
    has always been alienated and alienating, 220
    high development of, and division of labor, 252, 254, 257–258
    in contradiction with superstructural forms, 93
    in science as compartmentalizing, 11, 14
    linked with science in conquest of planet, 226
    makes dealienating revolution possible, 221
    modern
        and class conflict, 300
        and unfolding of true ideas, 171
        held back, 297
        quasi-creative production, 295
        role in revolution, 249
        transformation of the elementary tool, 82
    motor of becoming, 324, 331
    motor of development, 167, 208, 289
    motor of history, 203, 212, 289, 295, 296, 304, 325
    motor of productive forces, 293
    motor of science, 19, 209 n. 33
    nondevelopment of, 309
    no purposefulness beyond, 330
    not derivative, 294
    not freedom from technique but freed technique, 84 n. 10
    not just machine and industry, 325
    only a modern thing, 293
    opposition to, and conquest of nature, xi–xii, xxiii, 55, 120, 132, 182, 194
    planetary, 200, 215, 247, 256, 257, 295, 312
        conquering force, 332
        epoch of, 296
        following total plan, 330, 334
        lever to move the world, 292
        metaphysical foundation of, 275

mobilization of all productive forces, 267
prodigious development in bourgeois capitalism, 297
productive, to solve all questions, 296
progress of, coincides with progress of alienation, 141
proliferation of techniques, 193
properly speaking, 293
removes man from his own nature, 120
riddle of history, 64 n. 18, 203
scientific, 212
secret of history of love, 120
secret of the history of man's nature, 120
source in metaphysics? 63
spring of the economic, 307
summation of role, 325–326
to be dealienated, 208
to be transcended? 324
transforms Nature into History, 153, 325
underdevelopment of, 257, 290, 325
  and art, 192
  and mythology, 182
  cause of alienation, 298
universalizes history, 196, 300, 302
utterly transforms objects, 330
Technology
  and the "break" of alienation, xix–xxi
  consummately errant, xviii
  culmination both positive and negative, xiii
  development of
    base for superstructure, 143
    motor of history, 294
  distinction from technique, xv–xviii
  = engineering, xvii n. 26
  English usage of the word, xvi n. 23
  full meaning of, 331
  fusion of science and technique, xvii–xviii, xxiii
  power stance, xxiii, xxiv
  priority over ideology, 143
  proper sense, xv–xvi, xvii–xviii, xxii
  source of double alienation, 331
  strict meaning of, xv
  thrust toward the universal, xii, xiv, xxii
  to absorb ideology, 331
  transformation unlimited, xviii, xxi
Theology, 166
Theoria (Θεωρία—"theory"), 192
  and practice
    coinciding not clear, 162
    distinction based on concept of technique, 331
    distinction implies alienation, 148
    separated in alienated science, 213
  and revolutionary action, 173, 236, 249
  as fused with social revolution, 237
  derivative but creative, 236
  practical instrument for Marx, 229
  role in revolution, 171–173
  theoretical awareness of history, 172
"Thing-ification" (*chosification*), 128, 129, 276, 308
Thingism (*chosisme*), 128
Thingness (*Dingheit*), 137
Things, as human products, 243
Thought
  abstract, expresses dominant class, 196
  and reality
    dualism of, basic ideological contradiction, 319
    dualism of, not transcended, 171–173,

281–282
  the dualism of, 186, 220, 319–324
  as play, xxvii–xxviii
  as superstructure, influences base, 157
  guides action in communism, 236
  has existed only in alienation, 149, 152, 157
  has part in historical becoming, 145
  revolutionary
    aims of, 274
    guide to action, 253
    ideological, 236
  speculative, to be surmounted, 108
  term interchangeable with *consciousness* and *knowledge*, 145
  transcends nature in Western metaphysics, 155
  unity of thinking (νοεῖν) with being (εἶναι), 321
  See also Consciousness
Thoughts, term synonymous with *ideas*, 168 n. 16
Thrift, 130–131
Time, and space, 283
Time in Hegel, 204
Tongue a tool, 146
Tool, 77
Totalitarian, contrast with totality-minded, 315
Town and country, separation of, 57, 81
  abolished, 256
  dawn of history in emancipation from nature, 70–71
  led to administrative and political setup, 90
Tragedy
  and comedy in historical movement, 188–189, 244, 313–314
  end of, 230
  in history, 42
  political, 103

Truth
  and error in errance, xxvii–xxviii
  criterion and source of: true practice, 149, 165
  Hegel's concept of, 30, 36, 156
  not defined fully by Marx, 148

Underdeveloped countries, 85
Unification as enterprise, 242
United States, 84
Unity
  nonempirical principle, 316
  proclaimed but unexamined, 320–321
Use-value, 60 n. 14, 184, 291
Utilitarianism transcended, 222, 255
Utopia, 312

Vernunft ("reason"), 190
Violence
  and war, 313
  in the revolution, 230, 233
Voluntarism, 267

Wage labor, 308
Westphalen, Jenny von, 118 n. 17, 178
Will
  as *causa prima*, 105
  power-filled, 273
  to become social and collective power, 105
  to total activity, 49
Will to power, 9, 13, 275
  unrecognized by Marx, 133 n. 16, 305, 313
  with technique gains control of economy, 108
Wissen, *-schaft* ("know, -ledge"), 32, 190
Woman
  as instrument for (re)production, 119
  as prey or spoil, 116, 117
  as serving pleasure, 116, 119
  communalization of, re-

jected, 117, 119, 233
  figure of human alienation, 233
  to be no longer dependent upon man, 119
Workday, shortening of, 310
Worker
  figure of human alienation, 233
  terminal point for development, 71
Workers
  exploited class, 60
  gravediggers of capitalism, 71
  negators of negation, 64
World
  as play, xxvii–xxviii
  becomes human history, 218
  becomes philosophy, 202
  becomes *world* with the Greeks, 180
  characterized, 15
  Christian view of, 328
  conquest of
    by man, 296
    by technique, 287
    in universal reconciliation, 215, 223
  dualistic, for Marx, 42–43
  errant star, 292
  for Marx, 27
  Greek view of, 328
  has never yet been world, 300
  historical product, 197
  indissolubly natural, human, and social, 239, 246
  material of human labor, 200
  modern
    culmination of alienation and technique, xiii
    favorable to technique, 302
    not a world, 103
    Western and universal, 302
  natural, replaced by technified world, 84
  nature's conversion into history, 200
  revealed through human activity, 42, 215, 218, 223, 227, 239, 275
  the being of totality in its becoming, 218, 223, 227, 239
  upside-down and perverted, 94–95
World literature, 186
World market, 185, 303
World spirit (Marx), 205

Young Hegelians, 161. *See also* Hegelians, Left